Environmental Law Before the Courts

Giovanni Antonelli • Michael Gerrard •
Sara Colangelo • Giancarlo Montedoro •
Maurizio Santise • Luc Lavrysen •
Maria Vittoria Ferroni
Editors

Environmental Law Before the Courts

A US-EU Narrative

Springer

Editors
Giovanni Antonelli
Italian Ministry of Culture
Rome, Italy

Sara Colangelo
Georgetown University
Washington, DC, USA

Maurizio Santise
Tar Campania
Naples, Italy

Maria Vittoria Ferroni
Department of Political Sciences
Sapienza University of Rome
Rome, Italy

Michael Gerrard
Columbia University
New York, NY, USA

Giancarlo Montedoro
Consiglio die Stato
Rome, Italy

Luc Lavrysen
Constitutional Court of Belgium
Ghent, Belgium

ISBN 978-3-031-41526-5 ISBN 978-3-031-41527-2 (eBook)
https://doi.org/10.1007/978-3-031-41527-2

This Springer imprint is published by the registered company Springer Nature Switzerland AG
The registered company address is: Gewerbestrasse 11, 6330 Cham, Switzerland

Paper in this product is recyclable.

Foreword

Environmental law comprehends an enormous swath of the law—and life—incorporating distinct elements of the environment such as water and air quality, biodiversity, light and noise pollution, and much more. It implicates substantive human rights such as the right to food and the right to culture as well as procedural rights including rights of participation and freedom of speech and association. Increasingly, environmental law includes the rights of nature itself, whether as a particular element (such as the rights of an animal or animal species) or the rights of entire ecosystems like rivers and forests. Environmental law has wide-ranging economic and political implications, as well as profound cultural, personal, and spiritual aspects. And, unlike other areas of law, environmental law has temporal dimensions as we become increasingly cognizant that environmental resources are finite and therefore that future generations, who have the same moral entitlement to environmental benefits as present generations do, are likely to have far less access and far less enjoyment of them. As the Federal Constitutional Court of Germany has expressed it this way: "One generation must not be allowed to consume large portions of the CO2 budget while bearing a relatively minor share of the reduction effort, if this would involve leaving subsequent generations with a drastic reduction burden and expose their lives to serious losses of freedom." Environmental law is how we negotiate the distribution of environmental benefits and burdens.

All of this is exacerbated by the dire consequences of climate change, both immediate and impending. Indeed, as Maria Vittoria Ferroni and Tiziana Bandini remind us in their chapter on biodiversity law, "human societies might have already exceeded some of the planetary limits by placing themselves in a danger zone, [that] could lead, if not quickly reversed, to the sixth largest extinction event in the history of life on earth, caused by the impacts of human activities on the planet." If we are truly going to preserve the dignity of all living things on earth, we must use all the powers of the law to ensure equal and equitable opportunities for them to thrive now and for all time.

But throughout the conversation about environmental values, the problems of environmental justice and injustice are pervasive and pronounced: neither the human nor the non-human elements of our planetary environment experience the burdens of environmental abuse equally, equitably, or according to responsibility: it is invariably the most fragile beings among us who are the most vulnerable to environmental harms and who have the fewest resources to resist them. But those most vulnerable are, by definition, the least likely to be protected by law or politics. Perhaps because they are too young to vote (as in the children-led climate litigation cases), or because they are too poor or socially marginalized to be effectively represented in a political world that values money and power over the lives and wellbeing of human beings. Or because they are non-human beings who have no voice at all that we can hear. Judicial engagement is therefore nowhere more necessary than it is here, to protect those who cannot protect themselves through the political process.

And yet, judicial engagement presents its own set of complexities. All of the challenges that pertain to litigation generally are particularly prominent in environmental cases. Environmental cases tend to be unusually complicated, involving the interplay of multiple statutory constructs, administrative law rules and regulations, and constitutional principles, often at different levels of government, as Cale Jaffe and Aspen Ono show in their chapter on the Clean Water Act in the United States. As they explain, "the interconnectivity of the Clean Water Act and other federal statutes reflects the interconnectedness of the things they regulate"—an observation that is equally apt throughout environmental jurisprudence. Just on the specific issue of protected areas within the EU, for instance, there is an "almost infinite array of approaches" to protected sites, according to Colin Reid in his chapter on "The Court of Justice and Protected Areas in the European Union." At its best, the "holistic review" that environmental law demands is necessary to ensure the protection of environmental and non-environmental resources, over time, and across the human populations that interact with them. But the necessity for complex analysis leaves us vulnerable: courts are not typically experts in complex thinking, in the resolution of "wicked problems," as they are sometimes called, and if a court limits its review or fails to address interconnected interests, the whole "daisy chain" falls apart, as Jaffe and Ono demonstrate.

Even before the litigation begins, the challenges of adjudication are evident. As John Dernbach notes about environmental rights provisions in state constitutions in the United States, plaintiffs first have to persuade a court that the provisions are self-executing—that is, that they can be enforced directly without the need for implementing legislation—and unfortunately, courts in the United States have not been enthusiastic about enforcing such provisions. Even if there is a constitutional right to anchor a claim to, it is supremely difficult for plaintiffs to establish standing in courts in the United States and the costs (mostly in sustaining the evidentiary burden even at the liminal stage) can be inordinate. And it can be challenging to bring the appropriate defendant into court: government defendants often have immunities that protect them from litigation and corporate defendants can invariably outspend and outlitigate environmental, public interest plaintiffs. Once in court, it can be nearly impossible to prove that a particular defendant caused the particular actionable injuries of the plaintiff or her class. And in those rare instances where a plaintiff does prevail on the merits, remedies are especially difficult to enforce against political intransigence and economic incentives to evade environmental obligations.

Courts simply have very few levers to hold private and public entities accountable for the harms that ensue from environmental degradation and changes in the earth's climate.

It does not matter that environmental matters are adjudicated in a variety of different judicial structures, including both state and federal courts in the United States and national and regional bodies in Europe, including at the highest levels at both the Court of Justice of the European Union and, increasingly, the European Court of Human Rights, as Luc Lavrysen, Jan M. Passer and Věra Pazderová, and Natalia Kobylarz show, respectively. Moreover, environmental law has a home not only in courts of general jurisdiction but also in specialized tribunals (as Merideth Wright shows in the USA), and these may be judicial or may be administrative or constitutional. They may be national or transnational, like the Investor-State Dispute Settlement Tribunals that Matteo Fermeglia describes in his chapter. And like those panels, most tribunals raise serious concerns about institutional legitimacy and capacity to dispense justice as well as to use Fermeglia's blunt phrase, "departure from basic elements of the Rule of Law." And, as several of the chapters included here ably demonstrate, the problem of time persists: courts are used to remedying wrongs that happen in the here-and-now, but environmental harms evolve over time and remedying environmental harms can take years or decades—far beyond the temporal horizons of most court cases, or the patience of must judges.

To examine environmental law in any one social or political context is difficult enough; to examine it in two is more than twice as difficult. Indeed, comparative work, particularly in the context of environmental law, requires both a broad lens to recognize common themes and shared challenges that span cultures and geographies and a tight focus on the particular concern, the specific locality, and the unique conditions that become the subject of litigation and contestation.

It is therefore difficult for any one voice or perspective to provide insight from all of these vantage points. The transnational analysis of environmental adjudication is therefore well suited to a collection of narratives that, together, provide insights into the challenges and opportunities of contemporary environmental law as it is adjudicated in courts.

The essays collected here provide such a narrative. They describe how courts in the United States and Europe are addressing matters concerning the environment. They highlight some of the important gains that have been made and explain how some of these environmental victories can be models for further judicial development. Colin Reid, for example, shows the significance that the CJEU has "consistently placed on ensuring that the environmental commitments made by the Member States are satisfied." Judge Merideth Wright describes the experiences of specialized "green" tribunals in the United States in the states of Vermont and Hawai'i.

Some of the contributions here make specific normative arguments, as Luc Lavrysen does in advocating for a specialized corps of judges with environmental expertise or as Sara A. Colangelo does in demonstrating "the need for an expansion of existing law or new approaches to address the entrenched harms that characterize environmental racism and injustice." Indeed, much of the implicit and explicit advocacy in these chapters grows out of the failure of the law as it has been to meet the challenges of the planet as it is and will be. The law has not envisioned "species extinction [that] is already at least tens to hundreds of times higher than the average rate over the past 10 million years and is accelerating," as Ferroni and Banderi explain. It was not designed to remedy environmental injustices that affect not only black and brown communities within a single country but also transnational and historic human and environmental rights abuses. And it was certainly not designed to protect people and the environment from the "environmental apocalypse," that is being hastened as we refuse to engage (as the United States Court of Appeals for the Ninth Circuit said in Juliana v United States (in refusing to engage).[1]

Other chapters remain in the descriptive realm, allowing readers to draw normative conclusions for themselves. Matthias Keller's brilliant chapter examining "How German judges decide environmental cases under public law" is an insider's guided tour through the German judicial system. It provides insights useful even beyond the environmental contexts, although the examples of specific applications relate to environmental law, including the recent "epoch-making" Climate Decision from the German Federal Constitutional Court. Natalia Kobylarz's chapter shows how a new right to a healthy environment that could be adopted as a new protocol to the European Convention on Human Rights would likely affect the existing normative framework concerning admissibility requirements and, in a complementary mode, Jan M. Passer and Věra Pazderová provide an overview of the protection of the environment in the Court of Justice of the European Union, particularly in the context of air pollution, thus providing us with an overview of how both principal tribunals in Europe are responding to the problems of environmental degradation, from a human rights and an environmental law perspective.

Each chapter in this collection shines a spotlight on one section of the tapestry of environmental protection in the courts of Europe and the United States. Individually, the chapters provide moments of profound insight and lucid analysis. Together, the threads of these narratives weave a cohesive story, though not always a joyous one. In the regions of the world most responsible for the environmental apocalypse and with the most power and resources to change the course of global environmental governance, there is little appetite to address the situation with the commitment it demands. And whether the fragmentation of the law is its strength or its weakness, it is clear that the law has not—*so far*—proven itself to be up to the task. The essays included in this collection constitute an indispensable tool for practitioners and scholars alike: they provide a detailed and expert roadmap for environmental adjudication in the United States and Europe, and, collectively, they show a clear path forward on issues as diverse, interconnected and complex as the environment itself.

Widener University Delaware Law School
Wilmington, DE, USA
9 July 2023

Erin Daly

[1] Juliana v. United States, 947 F.3d 1159, 1164 (9th Cir. 2020).

Contents

How the Courts Have Interpreted the Relevance of the U.S. Constitution to Environmental Law

Michael Gerrard, Jonah Baskin, Jennifer Kim, and William Kovach

Abstract The U.S. Constitution provides much more assistance to industrial interests than to environmental interests. The U.S. courts have not found the Constitution to create any rights to a healthful environment for human beings, or to create any rights of nature for its own sake. As interpreted by the Supreme Court, several clauses inhibit the abilities of governments and citizens to act on climate change by, for example, holding that certain state actions are preempted by federal law; requiring the government to compensate those whose property it takes; and restricting who has access to the courts.

1 Introduction

The U.S. Constitution contains no explicit reference to the environment or environmental rights. Nor have the federal courts found environmental rights to be implicit in the Constitution, despite several attempts by litigants to secure such findings.[1] However, the courts have found several clauses in the constitution to be relevant to

[1] *See, e.g., Tanner v. Armco Steel Corp.*, 340 F. Supp. 532, 536–37 (S.D. Tex. 1972) (holding that "no legally enforceable right to a healthful environment, giving rise to an action for damages, is guaranteed by the Fourteenth Amendment or any other provision of the Federal Constitution."); see also *Gasper v. La. Stadium & Exposition Dist.*, 418 F. Supp. 716, 720–21 (E.D. La. 1976); *Pinkney v. Ohio EPA*, 375 F. Supp. 305, 310 (N.D. Ohio 1974); *Ely v. Velde*, 451 F.2d 1130, 1139 (4th Cir. 1971).

M. Gerrard (✉)
Columbia University, New York, NY, USA
e-mail: Mgerra@law.columbia.edu

J. Baskin · J. Kim · W. Kovach
Columbia Law School, New York, NY, USA
e-mail: jonah.baskin@law.columbia.edu; jennifer.kim@law.columbia.edu; william.kovach@law.columbia.edu

G. Antonelli et al. (eds.), *Environmental Law Before the Courts*,
https://doi.org/10.1007/978-3-031-41527-2_1

environmental protection. Frequently these interpretations inhibit rather than advance such protection.

This chapter discusses how the courts have interpreted elements of the U.S. constitution to be relevant to environmental protection.

2 Commerce Clause

Article I, Section 8, Clause 3 of the Constitution provides that Congress has the power "To regulate Commerce with foreign Nations, and among the several States, and with the Indian Tribes." The courts have clarified over time that this grant, in its "dormant" form, [2] also limits state and local power.[3]

The "dormant" commerce clause's touchstone is discrimination: "Whatever … [a State's] ultimate purpose, it may not be accomplished by discriminating against articles of commerce coming from outside the State unless there is some reason, apart from their origin, to treat them differently."[4] "Where simple economic protectionism is effected by state legislation, a virtually per se rule of invalidity has been erected."[5] But "[w]here the statute regulates evenhandedly to effectuate a legitimate local public interest, and its effects on interstate commerce are only incidental, it will be upheld unless the burden imposed on such commerce is clearly excessive in relation to the putative local benefits."[6]

In one line of cases, the Supreme Court has struck down certain regulations restricting the import or processing of solid waste from out of state. For instance, in *Philadelphia v. New Jersey*[7] the Supreme Court invalidated a New Jersey statute that "prohibit[ed] the importation of most 'solid or liquid waste which originated or was collected outside the territorial limits of the State …'" as violating the Commerce Clause.[8] The ban was in effect "until the commissioner [of the State Department of Environmental Protection] … determin[ed] that such action can be permitted without endangering the public health, safety and welfare and has promulgated regulations permitting and regulating the treatment and disposal of such waste in this State."[9] The Commissioner adopted such regulations, but apart from four "narrow exceptions," New Jersey's borders remained closed "to all waste from other States."[10] Private New Jersey landfills and cities in neighboring states that had contracted with

[2] *Energy & Env't Legal Inst. v. Epel*, 793 F.3d 1169, 1170 (10th Cir. 2015).

[3] *See, e.g., Philadelphia v. New Jersey*, 437 U.S. 617, 623 (1978).

[4] *Id.* at 626–627; *see also Epel*, 793 F.3d at 1173.

[5] *Philadelphia v. New Jersey*, 437 U.S. at 623.

[6] *Id.* at 624; *Pike v. Bruce Church, Inc.*, 397 U.S. 137, 142 (1970).

[7] 437 U.S. 617 (1978).

[8] *Id.* at 618; 629.

[9] *Id.* at 618–19.

[10] *Id.* at 619.

landfills sued, alleging several state and federal violations.[11] The federal Supreme Court based its opinion on the Commerce Clause.[12]

The Court first noted that "commerce" in the dormant Commerce Clause context has the same, very broad, definition as in the clause's affirmative grant of congressional power.[13] So, "commerce" encompasses the waste at issue in the case.[14] However, the Court noted that if the relevant articles' "worth in interstate commerce [i]s far outweighed by the dangers inhering in their very movement, States could prohibit their transportation across state lines."[15]

Because "[t]he New Jersey law block[ed] the importation of waste in an obvious effort to saddle those outside the State with the entire burden of slowing the flow of refuse into New Jersey's remaining landfill sites," the Court ruled that it violated the Commerce Clause.[16] It did "not matter that the State ha[d] shut the article of commerce," here landfill space, "inside the State" rather than outside of it.[17]

In *C & A Carbone v. Town of Clarkstown*,[18] the Supreme Court held that a "flow control ordinance, which requir[ed] all solid waste to be processed at a designated transfer station before leaving the municipality" at issue, violated the Commerce Clause.[19] The city had contracted for a local private contractor to construct the transfer station and manage it for five years.[20] In exchange, the city "guaranteed a minimum waste flow of 120,000 tons per year" at a rate that exceeded "the disposal cost of unsorted solid waste on the private market"; if the facility did not meet that threshold, the city agreed to make up the fee deficit.[21] The city enacted the challenged ordinance to ensure it could provide the waste flow it guaranteed to the facility.[22]

The Court struck down the ordinance as "discriminat[ing] against interstate commerce"[23] for two reasons. The ordinance impacted interstate commerce because the waste processing facilities it restricted "receiv[ed] and process[ed] waste from places other than Clarkstown, including from out of State."[24] The requirement that such facilities send their waste to the new facility imposed an additional cost,

---

[11] *Id.*

[12] *Id.* At 620–21.

[13] *Id.* At 622.

[14] *Id.*

[15] *Id.*

[16] *Id.* at 629.

[17] *Id.* 628.

[18] 511 U.S. 383 (1994).

[19] *Id.* 385.

[20] *Id.* at 387.

[21] *Id.*

[22] *Id.*

[23] *Id.* at 390.

[24] *Id.* at 389.

"driv[ing] up the cost for out-of-state interests to dispose of their solid waste."[25] The ordinance also "depriv[ed] out-of-state businesses of access to a local market" by "prevent[ing] everyone except the favored local operator from performing the initial processing step."[26] To sum up its holding, the Court said, "State and local governments may not use their regulatory power to favor local enterprise by prohibiting patronage of out-of-state competitors or their facilities."[27]

Some lower federal courts have upheld regulations restricting the use of electricity in a state generated through the combustion of fossil fuels, while acknowledging the out-of-state effects of such regulations. *Energy & Env't Legal Inst. v. Epel*[28] was a Tenth Circuit case considering a Colorado state statute "requir[ing] electricity generators to ensure that 20% of the electricity they sell to Colorado consumers comes from renewable sources."[29] "Because electricity can go anywhere on the grid and come from anywhere on the grid, and because Colorado is a net importer of electricity,"[30] the statute affected interstate markets, including by harming out-of-state fossil fuel produces.[31] The court considered the statute under a line of dormant commerce clause cases that rendered "certain price control and price affirmation laws that control … conduct outside the state's borders … almost per se invalid."[32] It determined that the instant statute lacked the "three essential characteristics" to strike down a state statute under those cases: "(1) a price control or price affirmation regulation, (2) linking in-state prices to those charged elsewhere, with (3) the effect of raising costs for out-of-state consumers or rival businesses."[33] A central point was that the Colorado law treated in-state and out-of-state electricity generators the same; there was no discrimination. It is of note that the author of this decision, Neil Gorsuch, now sits on the Supreme Court.

In *Rocky Mt. Farmers Union v. Corey*,[34] the Ninth Circuit considered a California Low Carbon Fuel Standard regulation that imposed reporting requirements and, beginning in 2011, "established a declining annual cap on the average carbon intensity of California's transportation-fuel market."[35] The regulation treated fuel differently based on its carbon intensity,[36] which "is the amount of lifecycle greenhouse gas emissions caused by production and transportation of the fuel, per unit of energy

[25] *Id.*

[26] *Id.*

[27] *Id.* at 394.

[28] 793 F.3d 1169 (10th Cir. 2015).

[29] *Id.* at 1170.

[30] *Id.* at 1171.

[31] *Id.* at 1173–74.

[32] *Id.* at 1172.

[33] *Id.* at 1173.

[34] 730 F.3d 1070 (9th Cir. 2013).

[35] *Id.* at 1080.

[36] *Id.* at 1089.

of fuel delivered."[37] "[T]he Fuel Standard consider[ed] location, but only to the extent that location affect[ed] the actual GHG emissions attributable to a [fuel's] default pathway."[38] The principal effect of location was that the further the fuel had to travel from where it was generated to where it was used, the greater were the GHG emissions in transporting it.

The court determined that the policy did not unduly discriminate; the Commerce Clause did "not require California to ignore the real differences in carbon intensity among out-of-state ethanol pathways, giving preferential treatment to those with a higher carbon intensity." [39] The court also rejected a claim that the standard impermissibly regulated out-of-state-conduct, noting that "California may regulate with reference to local harms, structuring its internal markets to set incentives for firms to produce less harmful products for sale in California."[40] The regulation required approval only where sales occurred in California.[41] And "[s]o long as California regulat[ed] only fuel consumed in California, the Fuel Standard" did "not present the risk of conflict with similar statutes" in other states.[42]

Nonetheless, the Ninth Circuit "remand[ed] the case for the district court to determine whether" certain of the regulation's "provisions discriminat[ed] in purpose or effect and, if not, to apply the ... balancing test" from *Pike v. Bruce Church, Inc.*[43] On remand, the district court dismissed the claims, and the Ninth Circuit affirmed the dismissal.[44]

The cases discussed above illustrate how the Commerce Clause functions to restrict state environmental action that discriminates against other states. However, the clause also limits the federal government's prerogative to address certain environmental concerns. In *Solid Waste Agency v. United States Army Corps of Eng'rs*,[45] the Corps of Engineers had interpreted Section 404(a) of the Clean Water Act[46] "to confer federal authority over an abandoned sand and gravel pit in northern Illinois which provid[ed] habitat for migratory birds."[47] The Court decided in 2001 that the statute did not grant such authority, and so it avoided a ruling directly based on the Commerce Clause.[48]

---

[37] *Id.* at 1080 n.1.

[38] *Id.* at 1089.

[39] *Id.* at 1093.

[40] *Id.* at 1104.

[41] *Id.*

[42] *Id.* at 1105.

[43] 397 U.S. 137, 142 (1970); *Corey*, 730 F.3d at 1107.

[44] *Rocky Mountain Farmers Union v. Corey*, 913 F.3d 940, 956–58 (9th Cir. 2019).

[45] 531 U.S. 159 (2001).

[46] 33 U.S.C. § 1344(a).

[47] *Solid Waste Agency*, 531 U.S. at 612.

[48] *Id.*

Nonetheless, the Commerce Clause played a role in the decision. The Court noted that the Corps' interpretation unambiguously exceeded the statutory grant.[49] But it also stated that "where an otherwise acceptable construction of a statute would raise serious constitutional problems, the Court will construe the statute to avoid such problems unless such construction is plainly contrary to the intent of Congress."[50] And an administrative interpretation does not change that.[51] The Court found no clear congressional statement that § 404(a) "reached an abandoned sand and gravel pit" like that at issue.[52] The clear implication was that if the Clean Water Act had authorized the Corps to regulate isolated waters whose only contact with interstate commerce was that migratory birds occasionally landed there, the Clean Water Act could have exceeded the power of Congress under the Commerce Clause. In 2023 the Supreme Court issued another decision limiting the scope of Section 404(a), *Sackett v. Environmental Protection Agency*.[53] The Court's decision relied on a reading of the statute, but the concurring opinion by Justice Clarence Thomas, joined by Justice Neil Gorsuch, explicitly relied on the Commerce Clause.

3 Case and Controversy Clause and Standing Doctrine

The Constitution's Case and Controversy Clause enumerates the situations to which "[t]he judicial Power" of the United States extends.[54] Most relevant here, it limits the "judicial power" to "Cases" or "Controversies."[55] That limitation underlies standing doctrine.[56]

A litigant must demonstrate standing to bring a claim in federal court.[57] As interpreted by the Supreme Court, Constitutional standing has three elements: "injury in fact," "a causal connection between the injury and the conduct complained of," and redressability.[58] To satisfy the injury requirement, a litigant must demonstrate that they have experienced an "actual or imminent" injury that is "concrete and particularized."[59] The causation requirement demands that the relevant injury be "fairly … trace[able] to the challenged action of the defendant, and not … the result [of] the independent action of some third party not before the court."[60] Finally,

---

[49] *Id.* at 172.

[50] *Id.* at 173.

[51] *Id.* at 172.

[52] *Id.* at 174.

[53] 143 S.Ct. 1322 (May 25, 2023).

[54] U.S. Const. art. III, § 2, cl. 1.

[55] *Id.*

[56] *Lujan v. Defs. of Wildlife*, 504 U.S. 555, 560 (1992).

[57] *Id.* at 561.

[58] *Id.* 560–61.

[59] *Id.* at 560.

[60] *Id.* (internal quotation marks omitted).

standing requires that it "be 'likely,' as opposed to merely 'speculative,' that the [litigant's] injury will be 'redressed by a favorable decision.'"[61]

Over time, the doctrine's evolution, and particularly the "injury in fact," requirement, have limited the ability of individual plaintiffs to bring lawsuits to alleviate environmental harms.[62] As such, standing can often present a formidable, though not necessarily insuperable, barrier to bringing environmental litigation.

In *Sierra Club v. Morton*,[63] the Supreme Court rejected a theory of standing based on the suit being a "public action" and the plaintiff, Sierra Club, possessing a "longstanding concern with and expertise in" environmental issues, thereby being able to sue.[64] The U.S. Court of Appeals for the Second Circuit had adopted such a "public interest" theory in *Citizens Committee for the Hudson Valley v. Volpe*.[65] In *Morton*, Sierra Club challenged the National Forest Service's permitting of a "ski resort that would also serve as a summer recreation area" in the Mineral King Valley.[66] The valley was a part of the Sequoia National Park and was designated as a national game refuge.[67] Sierra Club's specific legal challenges included an allegation that the construction permit exceeded the maximum-acreage limitation, and a challenge to a proposed permit for highway and electric lines through the national park to support the resort.[68]

The Court, however, found that Sierra Club lacked standing to bring the suit.[69] Notably, the opinion is clear that "[a]esthetic and environmental well-being, like economic well-being" are cognizable.[70] But a cognizable interest is not enough for standing; rather, "the party seeking review" must be "among the injured."[71] The Court noted that the alleged injury in this case would "only" fall directly upon "those who use Mineral King and Sequoia National Park."[72] Sierra Club failed to demonstrate that it was within that group because it did not allege "that it or its members would be affected in any of their activities or pastimes by the" development.[73] The Court expressed fear that if it recognized Sierra Club's right to sue based on a "special interest" in environmental protection, it would be unable to prevent anyone with some "bona fide special interest" from bringing suits.[74] But allowing people to sue to "vindicate their own value preferences," the majority asserted, would risk undermining the standing doctrine's goal of ensuring that

[61] *Id.* at 561.

[62] Belitz (2017), p. 166 n.176.

[63] 405 U.S. 727 (1972).

[64] *Morton*, 405 U.S. at 736.

[65] 425 F.2d 97 (2d Cir. 1970). *Morton*, 405 U.S. at 736 n.9.

[66] *Id.* at 729.

[67] *Id.* at 728.

[68] *Id.* at 730 n.2.

[69] *Id.* at 741.

[70] *Id.* 734–35.

[71] *Id.*

[72] *Id.* at 735.

[73] *Id.*

[74] *Id.* at 739–40.

litigants are "those who have a direct stake in the outcome" of the instant case.[75] Had the Sierra Club identified even a single member who actually utilized the area and could claim to be injured by the proposed action, the lawsuit would probably have been allowed to proceed.

Morton made clear that in order to bring environmental lawsuits, a plaintiff must demonstrate direct, personal injury. But the question remains: how might a litigant demonstrate such injury? One often-cited case that helps answer that question *is* *Lujan v. Defenders of Wildlife*,[76] Under Section 7 of the Endangered Species Act of 1973, federal agencies must, "in consultation with and with the assistance of the Secretary" of the Interior "insure" its actions are "not likely to jeopardize the continued existence of any endangered species or threatened species" or harm the habitats of such species.[77] A 1979 regulation interpreted Section 7 to apply to actions taken in foreign nations. But in 1983 the Department of the Interior issued a regulation reinterpreting Section 7's consultation requirements to extend only to actions within "the United States or on the high seas."[78] "[O]rganizations dedicated to wildlife conservation and other environmental causes" sued the Secretary of Interior, challenging the reinterpretation.[79]

Speaking for the Court in relevant part, Justice Scalia's opinion reaffirmed that to have standing the plaintiff organizations had to demonstrate "that one or more of" their "members would … be directly affected" by the new interpretation, "apart from their special interest in the subject."[80] The Court found that the plaintiffs had failed to make such a showing. Two members of Defenders of Wildlife had traveled to foreign nations and observed endangered species and their habitats; they said they intended to return to observe the habitats again but gave no specific plans. But neither "[p]ast exposure to illegal conduct" nor the affiants' non-specific intentions were enough to establish the required "actual or imminent" injury.[81]

The Court also rejected three other theories of standing. First, the plaintiffs argued "that any person who uses any part of a 'contiguous ecosystem' adversely affected by a funded activity has standing even if the activity is located a great distance away."[82] Such a theory, the Court determined, was inconsistent with *Lujan v. Nat'l Wildlife Fed'n*,[83] "which held that a plaintiff claiming injury from environmental damage must use the area affected by the challenged activity and not an area roughly 'in the vicinity' of it."[84] The other two theories were based respectively on

---

[75] *Id.* at 740.

[76] 504 U.S. 555 (1992).

[77] *Id.* at 558 (quoting 16 U.S.C. § 1536(a)(2)).

[78] *Id.* at 558–59.

[79] *Id.* at 559.

[80] *Id.* at 563 (internal quotation marks omitted).

[81] *Id.* at 564.

[82] *Id.* at 565.

[83] 497 U.S. 871 (1990).

[84] *Defs. of Wildlife*, 504 U.S. at 565–66.

general interest in observing endangered animals and on professional interest in such animals.[85] The Court dismissed those theories, saying that it went "beyond the limit," "and into pure speculation and fantasy, to say that anyone who observes or works with an endangered species, anywhere in the world, is appreciably harmed by a single project affecting some portion of that species with which he has no more specific connection."[86]

Though the plaintiffs in *Morton* and *Defenders of Wildlife* failed to establish standing, later cases suggest certain situations in which standing is possible for those litigating environmental harms. In *Friends of the Earth v. Laidlaw*,[87] for instance, the Court found standing for such a claim.[88] There, the defendant, Laidlaw, purchased "a hazardous waste incinerator facility … that included a wastewater treatment plant."[89] The South Carolina Department of Health and Environmental Control (DHEC) granted Laidlaw a National Pollutant Discharge Elimination System (NPDES) permit under the Clean Water Act (CWA), authorizing the discharge of certain pollutants into the North Tyger River.[90] Laidlaw exceeded discharge limits set by the permit.[91] Several environmental organizations sued under Section 505(a) of the CWA, which says that "any 'citizen,' defined as 'a person or persons having an interest which is or may be adversely affected'" may sue to enforce "any limitation" in an NPDES permit.[92]

To demonstrate standing, particularly the injury prong, the plaintiff organizations provided affidavits from members stating that they would like to utilize the river and surrounding area in various ways, including fishing, camping, walking, birdwatching, and purchasing a home nearby.[93] They averred that they did not use the area in such ways because of Laidlaw's discharging pollutants into the stream.[94] The Court determined, with limited discussion, that such "conditional statements"—asserting that one would use the space if the discharge were to cease—were distinguishable from the "speculative" intentions found inadequate in *Defenders of Wildlife*.[95]

In addition to an injunction the Laidlaw plaintiffs sought civil penalties to be paid to the government.[96] "[A] plaintiff must demonstrate standing separately for each form of relief sought,"[97] so the Court separately considered whether civil

[85] *Id.* at 567.

[86] *Id.*

[87] 528 U.S. 167 (2000).

[88] *Id.* at 189.

[89] *Id.* at 175.

[90] *Id.* at 175–76.

[91] *Id.* at 176.

[92] *Id.* at 174; 177.

[93] *Id.* at 181–83.

[94] *Id.*

[95] *Id.* at 183–84.

[96] *Id.* at 185.

[97] *Id.*

penalties satisfy the redressability prong of standing. It stated that civil penalties can satisfy redressability by virtue of their "deterrent effect,"[98] where violations are "ongoing" and "could continue into the future if undeterred,"[99] The determination was based in part on the fact that Congress had "found that civil penalties in Clean Water cases … deter future violations."[100]

Morton, *Defenders of Wildlife*, and *Laidlaw* thus make clear that to bring an environmental claim an individual or organizational plaintiff must demonstrate a direct and individual injury; if the injury is ongoing or at all speculative, the plaintiff must also demonstrate that the injury is "imminent" by proving something more than a "some day" intention.[101] But a later case, *Massachusetts v. EPA* suggested that the burden may not be as high for plaintiffs that are States.[102]

In *Massachusetts v. EPA*, "a group of States, local governments, and private organizations alleged … that the Environmental Protection Agency (EPA)" had "abdicated its responsibility under the Clean Air Act to regulate the emissions of four greenhouse gases, including carbon dioxide."[103] The specific questions in the case were: (1) whether Section 202(a)(1) of the Clean Air Act granted EPA "the statutory authority to regulate greenhouse gas emissions from new motor vehicles; and if so," (2) "whether its stated reasons for refusing to do so are consistent with the statute."[104]

Because "[o]nly one of the petitioners needs to have standing to permit" the Court "to consider" a "petition for review,"[105] the standing portion of the opinion emphasized Massachusetts' standing. [106] In particular, the opinion stated that "Massachusetts' well-founded desire to preserve its sovereign territory" "supported federal jurisdiction."[107] Upon entering the union, a State "surrenders certain sovereign prerogatives," including its ability to declare war and the supremacy of its laws relative to federal ones.[108] Massachusetts's "quasi-sovereign interests" and Congress's having "recognized a … procedural right to challenge the rejection of" EPA's "rulemaking petition as arbitrary and capricious," the Court asserted, warrant giving a state "special solicitude in … standing analysis."[109]

But the Court also explained "why Massachusetts met every element of conventional standing doctrine."[110] The portion of the opinion that focused on injury emphasized that "climate-change risks" being "'widely shared'" did "not minimize

---

98 *Id.*

99 *Id.* at 188.

100 *Id.* at 185.

101 *Defenders of Wildlife*, 504 U.S. at 564.

102 549 U.S. 497, 518; 520 (2007).

103 *Id.* at 505.

104 *Id.*

105 *Id.* at 518.

106 *Id.* at 518–21.

107 *Id.* at 519.

108 *Id.*

109 *Id.* at 520.

110 Green (2012), p. 36.

Massachusetts' interest in the outcome of" the case.[111] It then explained that rising sea levels resulting from climate change would continue "swallow[ing]" coastal land owned by Massachusetts. [112] The opinion went on to make clear that, even if regulating the auto industry's emissions had only an "incremental" effect in limiting climate change, plaintiffs could attack "incremental" contributors to their injury.[113] Given those considerations, the Court found that Massachusetts had standing to bring the suit.[114]

In sum, while the Case and Controversy Clause presents a barrier to climate change litigation, *Laidlaw* and *Massachusetts v. EPA* provide examples of plaintiffs who overcame it. Nonetheless, *Morton* and *Defenders of Wildlife* in particular impose serious barriers and costs to bringing such litigation. There may be good reason to support the ability of organizations dedicated to sustainability and to the preservation of the environment to bring environmental lawsuits. After all, they would seem to bring the "concrete adverseness" the Court has identified as central to standing doctrine.[115] But the Case and Controversy Clause, as interpreted by *Morton* and *Lujan,* requires such organizations to identify something more—most notably, a "concrete and particularized" injury that is "actual or imminent."[116] Individual plaintiffs face the same burden.[117] *Massachusetts v. EPA*, though, seems to ease the burden of standing as it applies to States, presenting them with an opportunity to expand their role in environmental litigation.[118] Since the issuance of the *Massachusetts* decision, States have become very frequent litigants in cases challenging federal action or inaction on environmental matters, as well as many other matters. However, two 2023 decisions by the Supreme Court—one on adoption of Indian children, the other on immigration policies—appear to limit the broad standing of states that was declared in the *Massachusetts* decision.[119]

4 The Supremacy Clause: Preemption

The Constitution's "Supremacy Clause", art. VI, cl. 2, states,

> This Constitution, and the laws of the United States which shall be made in pursuance thereof; and all treaties made, or which shall be made, under the authority of the United States, shall be the supreme law of the land; and the judges in every state shall be bound thereby, anything in the Constitution or laws of any State to the contrary notwithstanding.

---

[111] *Mass. V. EPA, 549* U.S. at 522.

[112] *Id.* at 522–23.

[113] *Id.* 524–25.

[114] *Id.* at 526.

[115] *Id.* at 517.

[116] Defs. of Wildlife, 504 U.S. at 560.

[117] *Id.*

[118] 549 U.S. at 518; 520.

[119] *Haaland v. Brackeen.*, __ S.Ct. ___ (June 15, 2023); *United States* v. *Texas*, ___ S.Ct. ___ (June 23, 2023).

As a result, Congressional legislation may preempt, and thus invalidate, state law.[120]

Preemption can take various, though not "rigidly distinct," forms.[121] First, "Congress may withdraw specified powers from the States by enacting a statute containing an express pre-emption provision."[122] Courts might also find "field preemption" where "Congress may have intended 'to foreclose any state regulation in the area,' irrespective of whether state law is consistent or inconsistent with 'federal standards.'"[123] Field preemption may apply where the relevant federal "framework of regulation [is] 'so pervasive … that Congress left no room for the States to supplement it' or where there is a 'federal interest … so dominant that the federal system will be assumed to preclude enforcement of state laws on the same subject.'"[124]

Finally, "state laws are pre-empted when they conflict with federal law."[125] This conflict occurs where either "compliance with both federal and state regulations is a physical impossibility"[126] or "the challenged state law 'stands as an obstacle to the accomplishment and execution of the full purposes and objectives of Congress.'"[127]

There are two further "cornerstones of … pre-emption jurisprudence." "First, 'the purpose of Congress is the ultimate touchstone in every pre-emption case.'" "Second, '[i]n all pre-emption cases, and particularly in those in which Congress has 'legislated … in a field which the States have traditionally occupied,'" there is an assumption against preemption "unless … [preemption] was the clear and manifest purpose of Congress.'"[128]

Given the federal government's large role in environmental regulation, preemption is often relevant. Here, we discuss three examples: the Toxic Substances Control Act (TSCA),[129] the Federal Insecticide, Fungicide and Rodenticide Act (FIFRA),[130] and the Atomic Energy Act (AEA).[131]

As the examples of these statutes will show, preemption is complex, and, because "'the purpose of Congress is the ultimate touchstone in every pre-emption case,'"[132] the analysis generally focuses on the statute at issue as opposed to any more free-standing constitutional doctrine. But that must not distract from preemption's status

---

[120] *See, e.g.*, *Crosby v. Nat'l Foreign Trade Council*, 530 U.S. 363, 372–73 (2000) (collecting cases).

[121] Id. at 372 n.6.

[122] *Arizona v. U.S.*, 567 U.S. 387, 399 (2012).

[123] *Oneok, Inc. v. Learjet, Inc.*, 575 U.S. 373, 377 (2015).

[124] *Arizona v. U.S.*, 567 U.S. at 399 (quoting *Rice v. Santa Fe Elevator Corp.*, 331 U.S. 218, 230 (1947)).

[125] *Id.* at 399.

[126] *Id.* (quoting *Florida Lime & Avocado Growers, Inc. v. Paul*, 373 U.S. 132, 142–143 (1963)).

[127] *Id.* (quoting Hines v. Davidowitz, 312 U.S. 52, 67 (1941)). *See also Oneok*, 575 U.S. at 377. *But see Va. Uranium, Inc. v. Warren*, 139 S. Ct. 1894, 1907–09 (2019) (lead concurring opinion) (questioning the validity of "obstacle preemption").

[128] *Wyeth v. Levine*, 555 U.S. 555, 565 (2009).

[129] 15 U.S.C. §§ 2601 et seq.

[130] 7 U.S.C. §§ 136 et seq.

[131] 42 U.S.C. §§ 2011 et seq.

[132] *Wyeth*, 555 U.S. at 565.

as a fundamental constitutional principle. Under TSCA, FIFRA, the AEA, or any other federal law, the Supremacy Clause and the concept of preemption structure who can or cannot regulate on a given subject.

4.1 Toxic Substances Control Act

TSCA[133] provides an example of fairly detailed explicit preemption. The Act's "primary purpose" is "to assure that" technological "innovation and commerce in" "chemical substances and mixtures do not present an unreasonable risk of injury to health or the environment."[134] Among other things, it provides that EPA must require testing of certain chemical substances under defined circumstances.[135] It also restricts manufacture and use of some chemical substances.[136]

In 2016, Congress overhauled TSCA by passing the Frank R. Lautenberg Chemical Safety for the 21st Century Act, which added significant detail to the Act's preemption provision.[137] It provided that "no State or political subdivision of a State may establish or continue to enforce"[138] "action to require the development of information about a chemical substance or category of chemical substances that is reasonably likely to produce the same information" required by the EPA.[139] States also may not "prohibit or otherwise restrict the manufacture, processing, or distribution in commerce or use of a chemical substance" that the EPA has either (1) found does not present an unreasonable risk or (2) has restricted in such a way to remove unreasonable risk.[140] The statute preempts states from "requiring the notification of a use of a chemical substance" where the EPA has already required notification of said use.[141] States cannot "establish a statute, criminal penalty, or administrative action prohibiting or otherwise restricting the manufacture, processing, distribution in commerce, or use of such chemical substance that" the EPA has designated "a high-priority substance under" 15 U.S.C. § 2605(b)(1)(B)(i).[142] TSCA's preemption provision does, however, provide exceptions[143] and states that EPA can—and in

[133] 15 U.S.C. §§ 2601 et seq.

[134] 15 U.S.C. § 2601(b)(3).

[135] 15 U.S.C. § 2603(a).

[136] 15 U.S.C. § 2604(a).

[137] 15 U.S.C. § 2617.

[138] 15 U.S.C. § 2617(a)(1).

[139] 15 U.S.C. § 2617(a)(1)(A).

[140] 15 U.S.C. § 2617(a)(1)(B); 15 U.S.C. § 2605(i)(1); 15 U.S.C. § 2605(a).

[141] 15 U.S.C. § 2617(a)(1)(C).

[142] 15 U.S.C. § 2617(b)(1).

[143] 15 U.S.C. § 2617(d).

some cases must—waive preemption.[144] It also expressly does not preempt "common law rights" or "statut[es] creating a remedy for civil relief.[145]

4.2 *Federal Insecticide, Fungicide and Rodenticide Act*

The Federal Insecticide, Fungicide and Rodenticide Act (FIFRA)[146] provides a more narrow example of explicit preemption. It is "a comprehensive regulatory statute" governing pesticide use.[147] The Supreme Court has ruled that FIFRA does not generally preclude state or local pesticide regulation.[148]

But FIFRA contains some express preemption provisions. The first essentially replicates impossibility-based conflict preemption: a state may "not permit any sale or use prohibited by this Act."[149] The second, Section 136v(b), is more specific: a state "shall not impose or continue in effect any requirements for labeling or packaging in addition to or different from those required under" the Act.[150]

In *Bates v. Dow Agrosciences L.L.C.*,[151] the Supreme Court held that because Section 136v(b) "appl[ies] only to 'requirements,'" it does not preempt "occurrenc[es] that merely motivat[e] an optional decision."[152] Thus, for FIFRA to preempt a state law, the state law "must be" both "a requirement 'for labeling or packaging'" and "'in addition to or different from" the requirements imposed by the subchapter.[153]

The *Bates* Court ruled that FIFRA does not preempt "[r]ules that require manufacturers to design reasonably safe products, to use due care in conducting appropriate testing of their products, to market products free of manufacturing defects," or "to honor their express warranties or other contractual commitments" because such rules "do not qualify as requirements for 'labeling or packaging.'"[154] But the petitioner's fraud and negligent-failure-to-warn claims *were* labeling or packaging requirements; they "set a standard for a product's labeling that the … label [was] alleged to have violated by containing false statements and inadequate warnings."[155] But because "a state-law labeling requirement is not pre-empted by Section 136v(b)

---

[144] 15 U.S.C. § 2617(f).

[145] 15 U.S.C. § 2617(g)(1)(A).

[146] 7 U.S.C. §§ 136 et seq.

[147] *Wis. Pub. Intervenor v. Mortier*, 501 U.S. 597, 601 (1991).

[148] *Id.* at 600.

[149] 7 U.S.C. § 136v(a).

[150] 7 U.S.C. § 136v(b).

[151] 544 U.S. 431 (2005).

[152] *Bates*, 544 U.S. at 443.

[153] *Id.* at 444.

[154] *Id.*

[155] *Id.* at 446.

if it is equivalent to, and fully consistent with, FIFRA's misbranding provisions,"[156] the Court remanded for the lower court to determine whether that was the case.[157]

4.3 Atomic Energy Act

The Atomic Energy Act (AEA)[158] presents an example of implicit field preemption.[159] Under the AEA, "the Federal Government maintains complete control of the safety and 'nuclear' aspects of energy generation; the States exercise their traditional authority over the need for additional generating capacity, the type of generating facilities to be licensed, land use, ratemaking, and the like."[160] Whether the AEA preempts state law may also depend on "whether there is a nonsafety rationale" for the state law.[161]

Virginia Uranium v. Warren considered whether the AEA preempted a Virginia law banning uranium mining.[162] Virginia Uranium, a company that wanted to mine uranium in the state, "alleged that, under the Constitution's Supremacy Clause, the AEA preempt[ed] state uranium mining laws like Virginia's."[163] A divided Court found that the AEA did not preempt Virginia's ban. [164]

In an opinion for himself and Justices Thomas and Kavanaugh, Justice Gorsuch emphasized the statutory interpretation aspects of preemption analysis.[165] The thrust of his opinion was that "Congress … has spoken directly to the question of uranium mining on private land, and every bit of what it said indicates that state authority remains untouched."[166]

---

[156] *Id.* at 447.

[157] *Id.* at 453.

[158] 42 U.S.C. §§ 2011 et seq.

[159] *Pac. Gas & Elec. Co. v. State Energy Res. Conservation & Dev. Comm'n*, 461 U.S. 190, 212 (1983); *Va. Uranium, Inc. v. Warren*, 139 S. Ct. 1894, 1902 (2019) (lead concurring opinion) ("Unlike many federal statutes, the AEA contains no provision preempting state law in so many words.").

[160] *Pac. Gas*, 461 U.S. 190, 212 (1983).

[161] *Id.* at 213. *But see Va. Uranium*, 139 S. Ct. at 1903 (lead concurring opinion) ("[O]nly state laws that seek to regulate the activities discussed in §2021 without an NRC agreement—activities like the construction of nuclear power plants—may be scrutinized to ensure their purposes aim at something other than regulating nuclear safety."), 1913–14 (Ginsburg, J., concurring) (finding no need to inquire into purpose where a state law "targets an exclusively state-regulated activity.").

[162] *Va. Uranium*, 139 S. Ct. at 1900–01 (lead concurring opinion).

[163] *Id.* at 1901 (lead concurring opinion).

[164] *Id.* at 1900 (lead concurring opinion), 1909 (Ginsburg, J., concurring).

[165] *Id.* at 1901 (lead concurring opinion) ("We examine these arguments about the AEA's preemptive effect much as we would any other about statutory meaning, looking to the text and context of the law in question and guided by the traditional tools of statutory interpretation.").

[166] *Id.* at 1902 (lead concurring opinion).

Justice Ginsburg, writing for herself and Justices Sotomayor and Kagan, concurred in the judgment.[167] She began by stating that Justice Gorsuch's "discussion of the perils of inquiring into legislative motive" and his choice to call obstacle preemption into question were unnecessary to resolve the case.[168] She ultimately agreed that "[e]very indication … is that Congress left private conventional mining unregulated."[169]

5 First Amendment: Freedom of Speech

The First Amendment to the U.S. Constitution states, "Congress shall make no law respecting an establishment of religion, or prohibiting the free exercise thereof; or abridging the freedom of speech, or of the press; or the right of the people peaceably to assemble, and to petition the Government for a redress of grievances."

5.1 Greenwashing

In the field of environmental law, a First Amendment concern that has recently grown in prominence regards "greenwashing", or a "set of deceptive marketing practices in which an entity publicly misrepresents or exaggerates the positive environmental impact of a product, a service, or the entity itself."[170] Under the First Amendment's protection of freedom of speech, the two major threshold questions regarding greenwashing are: (1) whether the First Amendment applies to the marketing speech, and (2) if the First Amendment does cover these practices, what level of scrutiny a court should apply.[171]

According to Professors Shanor and Light, there are five major First Amendment questions at issue in greenwashing.[172] The first three involve the distinction between commercial and political speech; while false or misleading *commercial* speech is not covered by the First Amendment, false or misleading *political* speech is covered and protected by a standard of strict scrutiny. In addition, while commercial assertions of opinions are protected by an intermediate scrutiny standard, commercial assertions of facts are protected by a less stringent standard similar to rational basis

[167] *Id.* at 1909 (Ginsburg, J., concurring).

[168] *Id.* (Ginsburg, J., concurring).

[169] *Id.* at 1912 (Ginsburg, J., concurring).

[170] Shanor and Light (2022), p. 2033.

[171] Shanor and Light (2022), pp. 2075–2076.

[172] Shanor and Light (2022), p. 2076.

review.[173] Thus, the questions can be posed as: (1) whether the marketing practices can be characterized as commercial or political speech; (2) for commercial speech, whether the assertion is fact or opinion; (3) for a factual assertion, whether the statement is false, misleading, or fraudulent. The final two questions regard the evidentiary burden: (4) whether the commercial entity or the regulator bears the burden of proving the truth or falsity of a statement; and (5) what evidence is necessary to fulfill this burden.[174]

5.2 *Securities Regulation and Compelled Disclosures*

On March 21, 2022, the Securities and Exchange Commission (SEC) announced it long-awaited proposal for climate disclosure requirements.[175] The rule would require regulated companies to include a series of climate-related disclosures in their public filings. These disclosures include information regarding climate related risks "reasonably likely" to materially impact a company's business or consolidated financial statements as well as certain greenhouse gas emission metrics. The proposal has faced significant debate over potential First Amendment concerns, identified by the two major threshold questions: (1) whether the First Amendment should apply to these compelled disclosures, and (2) if the First Amendment does cover these disclosures, what level of scrutiny a court should apply. When the rule is issued in final form, it will certainly be challenged in court (several states have already announced they will sue), and First Amendment claims will probably be included in the suits.

5.3 *The First Amendment's Scope of Coverage*

The Supreme Court has implied more than once that securities disclosures are outside the scope of First Amendment coverage. In *Ohralik v. Ohio State Bar Association*, the Court noted that "the exchange of information about securities" is one example of a disclosure that can be regulated without "offending" the First Amendment.[176] More recently, in *Paris Adult Theater I v. Slayton*, the Court stated that "neither the First Amendment nor 'free will' precludes" state blue sky laws that

---

[173] Shanor and Light (2022), p. 2076 (citing *Cent. Hudson Gas & Elec. Corp. v. Pub. Serv. Comm'n*, 447 U.S. 557, 563–64 (1980) for intermediate scrutiny applied to commercial assertions of opinion and *Zauderer v. Off. of Disciplin. Couns. of Sup. Ct. of Ohio*, 471 U.S. 626, 651 (1985) for a lesser standard applied to commercial assertions of fact).

[174] Shanor and Light (2022), pp. 2076–2077.

[175] The Enhancement and Standardization of Climate-Related Disclosures for Investors, 87 Fed. Reg. 21334 (proposed Mar. 21, 2022).

[176] *Ohralik v. Ohio State Bar Ass'n*, 436 U.S. 447, 456 (1978).

regulate what sellers may write or publish about their securities products.[177] However, courts may still distinguish climate-related disclosures from other traditional securities disclosures, whether by analyzing a historical framework to determine what information qualifies as a securities disclosure, reviving an old distinction between "financial" and "non-financial" business information, or some other method.[178] This would thus raise a potential avenue for First Amendment challenges.

5.4 *The Applicable Level of Scrutiny*

Should a court find the First Amendment does apply to the SEC's climate disclosure rules, the next imperative question would be to determine the level of scrutiny that should apply. This is currently a significant topic of debate; some scholars have argued that the climate disclosures constitute commercial speech that should be subjected to a lesser degree of scrutiny.[179] Others, including the Attorneys General of Missouri and West Virginia, have argued that strict scrutiny should apply.[180]

One key case is *Zauderer v. Office of Disciplinary Counsel of Sup. Ct. of Ohio*, where the Supreme Court found that "purely factual and uncontroversial" commercial disclosures would not violate the First Amendment.[181] While some scholars argue that the SEC's proposed climate disclosure rule should qualify under *Zauderer*,[182] others argue the opposite.[183] Upon promulgation of the final SEC climate disclosure rule, these questions are likely to be heavily litigated.

6 Fourth Amendment: Searches and Seizures

The Fourth Amendment of the Constitution protects "[t]he right of the people to be secure in their persons, houses, papers, and effects, against unreasonable searches and seizures… and no Warrants shall issue, but upon probable cause…."[184] Environmental law cases implicating Fourth Amendment concerns generally deal with administrative inspections of commercial property, whether with or without a warrant. While searches without a warrant are typically deemed "unreasonable" under the meaning of the Fourth Amendment, a few exceptions have developed,

---

177 *Paris Adult Theater I v. Slayton*, 413 U.S. 49, 64 (1973).

178 Haan (2022).

179 Law Scholars (2022).

180 Haan (2022), p. 1.

181 *Zauderer v. Off. of Disciplin. Couns. Of Sup. Ct. of Ohio*, 471 U.S. 626, 651 (1985).

182 Law Scholars (2022), pp. 5–9.

183 Haan (2022), pp. 14–15; Berry and Schulp (2022).

184 U.S. Const. amend. IV.

with three exceptions important to environmental law cases in particular: (1) searches pursuant to consent (the most prevalent); (2) searches of "pervasively regulated businesses"; and (3) observations of objects or activities in the "public view".[185]

6.1 *Warrantless Searches Pursuant to Consent*

Even without a warrant, a search is valid if an owner or operator of a property gives consent "freely and voluntarily".[186] Consent can be inferred through a permit that allows sampling by the applicable regulatory agency. Relevant to environmental law cases, the Environmental Appeals Board of EPA considers the "totality of circumstances" to determine whether consent was voluntarily given. In general, if an agency official presents their identification and has an appropriate person sign a Notice of Inspection (as typically done) prior to entering the premises, the consent is considered voluntary.[187]

6.2 *Warrantless Searches of Pervasively Regulated Businesses*

In *Marshall v Barlow's Inc.*, the Supreme Court outlined an exception for allowing warrantless searches of commercial entities subject to "pervasive" regulations.[188] The Court reasoned that "[c]ertain industries have such a history of government oversight that no reasonable expectation of privacy … could exist…."[189]

While the "pervasive" requirement was originally understood to cover only longstanding regulatory schemes, the Supreme Court has since begun to emphasize using instead "a predictable and guided federal regulatory presence" to determine whether the exception applies in a particular situation.[190] In *Donovan v. Dewey*, the Supreme Court focused on the scope and legitimacy of the regulatory scheme as opposed to the length of time the scheme had been in place, and expressly mentioned health and safety programs as examples of recent regulatory schemes that could give rise to the exception. Environmental protections would fall within this scope.[191]

The Supreme Court further clarified the exception in *New York v. Burger*, requiring (1) a "'substantial' government interest that informs the regulatory scheme

---

185 Gerrard (2023), § 11.04(1).

186 Gerrard (2023), § 11.04(1).

187 Gerrard (2023), § 11.04(1).

188 Gerrard (2023), § 11.04(1); *Marshall v. Barlow's Inc.*, 436 U.S. 307, 317 (1978).

189 *Marshall v. Barlow's Inc.*, 436 U.S. 307, 312 (1978).

190 *Donovan v. Dewey*, 452 U.S. 594, 604 (1980).

191 Gerrard (2023), § 11.04(1).

pursuant to which the inspection is made," (2) that the warrantless search be "necessary to further the regulatory scheme," and (3) that "the statute's inspection program, in terms of the certainty and regularity of its application, must provide a constitutionally adequate substitute for a warrant."[192] Such an inspection program must advise the owner or operator that the "search is being made pursuant to the law and has a properly defined scope, and it must limit the discretion of the inspecting officers."[193] While the *Burger* test has not been extensively applied to environmental law contexts, the 10th Circuit in V-*1 Oil Co. v. Wyoming Dep't of Environment Quality* applied it to the Wyoming Environmental Quality Act to deem a warrantless search of a gasoline station unconstitutional. In particular, while the Court acknowledged that environmental concerns could be considered a "substantial government interest", the Act failed the third prong as it did not provide for a constitutionally adequate substitute for a warrant.[194] The Act did not provide an "assurance of regularity" of inspections.[195] In addition, it generally applied to all Wyoming businesses, thereby preventing the owner of any particular business from being on notice that their property would be subject to warrantless inspection.[196]

Based on the holdings of *Burger*, *Wyoming*, and other previous cases regarding the pervasively regulated business exception, courts can be expected to apply the exception to the environmental law context by (1) allowing the exception for environmental statutes that regulate specific industries or segments of commerce (so long as the last two prongs of *Burger* are met), and (2) prohibiting the exception for environmental statutes that regulate a wide range of industries/commercial activity.[197]

6.3 Warrantless Searches of Property in Public View

Warrantless government searches of certain areas open to the "public view" have been found constitutional. Such searches are supported by two theories: (1) the governmental action does not rise to the level of a "search", thereby avoiding a Fourth Amendment concern, and (2) the search is constitutional under "plain view" (or "open fields") principles.[198]

A narrow majority of the Supreme Court found that EPA's aerial surveillance of a plant in Dow Chemical Co. v. United States did not constitute a search under the

[192] *New York v. Burger*, 482 U.S. 691, 702–703 (1987) (quoting *Donovan v. Dewey*, 452 U.S. 594, 600–603 (1980)).

[193] *Id.* at 703.

[194] *V-1 Oil Co. v. Wyo., Dep't of Env't Quality*, 902 F.2d 1482 (10th Cir. 1990).

[195] *Id.* at 1487.

[196] *Id.*

[197] Gerrard (2023), § 11.04(1).

[198] Gerrard (2023), § 11.04(1).

meaning of the Fourth Amendment, and that Dow's "expectations of privacy" were unsupported.[199] While certain actions such as fences are likely sufficient to establish "expectations of privacy", the difficulties of taking action against aerial remote devices and the prevalence of such devices in the modern era make it likely that this exception will be increasingly used by EPA and subsequently tested in courts.[200]

6.4 *Searches Pursuant to Warrant*

In cases where the government chooses to obtain a warrant for a search, typically for longer or more complex inspections, the Fourth Amendment requires a showing of "probable cause". In Marshall v. Barlow's Inc., the Supreme Court held that the showing of "probable cause" in civil or administrative searches is different from the showing required in criminal cases.[201] Particularly relevant to environmental law cases, the Court recognized that administrative agencies need to be able to inspect properties even in cases where they do not suspect a regulatory violation.[202] The Court thus authorized warrants when the target property is identified by the agency pursuant to an inspection plan prepared prior to the warrant application and the plan is derived from "neutral sources".[203] Notably, only the basis must be neutral, not the plan itself.[204]

While EPA has not formally recognized that the *Marshall* principles apply to their inspection plans under federal environmental statutes, they have generally conformed their practices to the principles, as have most state inspection plans.[205]

7 Fifth Amendment

7.1 *Procedural Due Process*

The Fifth Amendment reads, in relevant part, "No person shall … be deprived of life, liberty, or property, without due process of law."[206] Courts have understood the "due process clause" to create both procedural and substantive constraints on the government's ability to take property or place new limits property rights, as well as

---

[199] *Dow Chemical Co. v. United States*, 476 U.S. 227 (1986).

[200] Gerrard (2023), § 11.04(1).

[201] *Marshall v. Barlow's Inc.*, 436 U.S. 307, 320 (1978).

[202] Gerrard (2023), § 11.04(1).

[203] *Marshall v. Barlow's Inc.*, 436 U.S. 307, 321 (1978).

[204] Gerrard (2023), § 11.04(1).

[205] Gerrard (2023), § 11.04(1).

[206] U.S. Const. amend. V.

on incarcerating and executing criminals. While the Fifth Amendment limits the federal government, the Fourteenth Amendment, passed in the aftermath of the Civil War, has a parallel clause which courts have read to place the same due process obligations on states.[207] When depriving a person of their liberty or a property interest, the government must meet the procedural burden of providing notice of the depravation to the rightsholder and an opportunity for them to have any of their objections heard in a meaningful manner before an impartial decision maker.[208]

Notice requirements can vary based on the ability of the government to contact the person being given notice,[209] but also require that a regulation or statute be sufficiently clear on the regulated conduct so as to "provide a person of ordinary intelligence fair notice of what is prohibited."[210] "Void for vagueness" defenses are not a common defense to violations of environmental laws or regulations, and have generally failed as an avenue for avoiding enforcement.[211]

Far more often, procedural due process rights have played a role in litigation in which property owners allege that government imposition of an environmental regulation or a penalty for non-compliance with an environmental statute did not have a mechanism to challenge the regulation's application. The lack of pre-enforcement judicial review is a feature, not a bug, of many of the environmental statutes in which Congress wanted to avoid polluters using litigation to delay remediation costs.[212]

Eldridge established the canonical three-part weighing test to determine if procedural mechanisms in place for a pre-deprivation hearing are sufficient to meet the due process needs. The court will weigh the significance of the private interest of the person being deprived, the public interest, and the extent to which additional

[207] In relevant part, the Fourteenth Amendment reads "No State shall make or enforce any law which shall … deprive any person of life, liberty, or property, without due process of law;" *See* U.S. Const. amend. XIV, § 1.

[208] *Mathews v. Eldridge*, 424 U.S. 319, 333 (1976) (holding that "The fundamental requirement of due process is the opportunity to be heard "at a meaningful time and in a meaningful manner."); *See* also Goldberg v. Kelly, 397 U.S. 254, 271 (1970) (holding that "an impartial decision maker is essential" to due process).

[209] Proper notification to meet due process obligations requires the government to try and notify the rightsholder in such a way as to be "reasonably certain to inform those affected" or, if "conditions do not reasonably permit such notice" to notify the rightsholder in such a way as to not be "substantially less likely to bring home notice than other of the feasible and customary substitutes." *Mullane v. Cent. Hanover Bank & Tr. Co.*, 339 U.S. 306, 315 (1950).

[210] *United States v. Williams*, 553 U.S. 285, 288 (2008).

[211] *See United States v. Sweeney*, 483 F. Supp. 3d 871, 938–39 (E.D. Cal. 2020) (finding CWA regulations on the definition of wetlands sufficiently specific); See also *United States v. Cooper*, 173 F.3d 1192, 1202 (9th Cir. 1999); *Huntress v. United States*, No. 18-CV-2974 (JPO), 2019 U.S. Dist. LEXIS 55156, at *13–14 (S.D.N.Y. Mar. 29, 2019); *United States v. Protex Indus., Inc.*, 874 F.2d 740, 743 (10th Cir. 1989). *But see Nat'l Crushed Stone Asso. v. EPA*, 601 F.2d 111, 120 (4th Cir. 1979) (noting "The fact that the regulations do not define recycling may well make them void for vagueness under our decision in *duPont* where we set aside an EPA regulation because we were "not sure what it means in the context in which it is used.")

[212] *See, e.g., N. Shore Gas Co. v. EPA*, 930 F.2d 1239, 1244 (7th Cir. 1991).

procedural safeguards would decrease the risk of error.[213] Reflecting the policy this test, due process generally does not prevent temporary deprivations in emergency situations before a full adjudicatory process is possible.[214]

Perhaps the most high-profile example of procedural due process interacting with environmental law can be found in *Sackett v. EPA*. Michael and Chantell Sackett, a husband and wife who owned a contracting and excavation business, purchased a residential plot of land just over 90 meters from shore of Priest Lake in Idaho. Part of the plot had been identified as a covered wetland in the online National Wetlands Inventory, which meant that the Sacketts would need to seek a permit under Section 404 of the Clean Water Act (CWA) to fill or dredge the wetlands portion of their property.[215] The Sacketts filled in part of the parcel to make it suitable for further construction without seeking a permit. During the construction process they were asked by EPA inspectors to produce a permit. Seven months after the initial EPA visit, the EPA issued the Sacketts a compliance order which required them to "immediately undertake activities to restore the site, in accordance with [an EPA-created] Restoration Work Plan" or face penalties of up to $75,000 per day.[216] The Sacketts demanded a hearing in which they could challenge EPA's determination that their land contained wetlands covered by the CWA, which EPA denied. The Sacketts then sued, arguing that the compliance order was illegally arbitrary and capricious under the Administrative Procedure Act (the APA is a "super statute" which is the backbone of much of American administrative law),[217] and that the compliance order violated their Fifth Amendment rights by not involving a hearing.

The Ninth Circuit held that the CWA precluded pre-enforcement judicial review of compliance orders, and that issuance of compliance orders without judicial review does not conflict with Fifth Amendment due process rights. The court held that because the CWA required EPA to prove the underlying violation before a court to enforce a compliance order, and because penalties for violation of a compliance order that did correspond to an underlying CWA violation were not enforceable, the Sacketts did not have their Fifth Amendment rights violated.[218] Before any penalties

---

[213] *Mathews v. Eldridge*, 424 U.S. 319, 323 (1976).

[214] *Hodel v. Va. Surface Mining & Reclamation Ass'n*, 452 U.S. 264, 299–301 (1981) (Holding that the Secretary of the Interior can issue immediate cessation orders under the Surface Mining Control and Reclamation Act which are not immediately appealable in an adjudication. Because such orders require a determination that a mine is operating in violation of its permit and that the operation "creates an immediate danger to the health or safety of the public, or … environmental harm" they reflect the longstanding doctrine that a pre-deprivation adjudicative hearing is not required by due process in emergency situations.).

[215] Parts of the property were clearly listed on the National Wetlands Inventory. *See* Robinson (2011). The Sacketts also ultimately objected to the EPA's understanding of CWA applicability to wetlands. The CWA applies to "navigable waters" which is defined in statute as "Waters of the United States." The meaning of the definition is currently being litigated. *See Sackett v. United States EPA*, 8 F.4th 1075, (9th Cir. 2021) (cert. granted).

[216] *Sackett v. EPA*, 566 U.S. 120, 125 (2012).

[217] *See* Kovacs (2015), p. 1209.

[218] *Sackett v. EPA*, 622 F.3d 1139, 1145–46 (9th Cir. 2010).

could be assessed against them, they would have an ability to make their case against the issuance of the compliance order in a court of law. The Supreme Court reversed the Ninth Circuit ruling, but did not reach the Fifth Amendment claims, holding that the Sacketts had a right to judicial review of the order under the APA.[219] Years later the Sacketts exercised their right of judicial review to bring the case that was discussed above in the context of the Commerce Clause.

Overall, there has been a mixed record for property owners who have tried to challenge environmental statutes on procedural due process grounds. Courts have held that the EPA cannot issue administrative compliance orders under the Clean Air Act that carry civil and criminal penalties for noncompliance because due process would require a hearing where the regulated entirety could contest "whether the conduct underlying the issuance of the [administrative compliance order] actually took place and … whether the alleged conduct amounts to a CAA violation," and EPA did not have the power to adjudicate such hearings under the statute.[220] The First Circuit also found a Fifth Amendment violation when EPA issued a lien on a property under CERCLA without pre-enforcement judicial review, in part because there was the possibility of substantial delay before even a post-deprivation hearing where the property owners could challenge the lien.[221]

That said, most courts have found that agencies can issue penalties or regulations through administrative law processes when the penalties can ultimately be disputed in court or must be evaluated in proceedings to enforce the penalties. The Sixth Circuit has held that EPA could issue penalties for not complying with an information request under CERCLA without an adjudicatory hearing because enforcement of the penalties required court proceedings.[222] Courts have also held that EPA could issue corrective action orders with penalties under RCRA without pre-enforcement judicial review so long as EPA held an administrative or civil hearing prior to enforcing the penalties,[223] that notice and comment provided the required procedural due process associated with the reputational deprivation of a site being listed on the National Priorities List under CERCLA,[224] and that EPA could deny a facility the ability to store waste under RCRA from a CERCLA site without an adjudicatory hearing.[225]

[219] *Sackett v. EPA*, 566 U.S. 120 (2012).

[220] *TVA v. Whitman*, 336 F.3d 1236, 1258–59 (11th Cir. 2003).

[221] *Reardon v. United States*, 947 F.2d 1509, 1518–23 (1st Cir. 1991).

[222] *United States v. Gurley*, 384 F.3d 316, 326 (6th Cir. 2004).

[223] *Amoco Oil Co. v. EPA*, 959 F. Supp. 1318, 1324 (D. Colo. 1997).

[224] *SCA Servs. of Ind., Inc. v. Thomas*, 634 F. Supp. 1355, 1367–74 (N.D. Ind. 1986).

[225] *Chem. Waste Mgmt. v. EPA*, 56 F.3d 1434 (D.C. Cir. 1995).

7.2 Substantive Due Process

The Due Process clause, in declaring that no person shall "be deprived of life, liberty, or property, without due process of law" has been understood by the courts to create more than just a right to procedural safeguards under a procedural due process line of jurisprudence, but also to "[protect] individual liberty against 'certain government actions regardless of the fairness of the procedures used to implement them.'"[226] The liberties protected consist of the rights enumerated in the first eight amendments, as well as other, non-enumerated rights "deeply rooted in this Nation's history and tradition ... and implicit in the concept of ordered liberty, such that neither liberty nor justice would exist if they were sacrificed."[227] While the government can still infringe on these liberties in theory, a reviewing court applies a "strict scrutiny" standard of review, which in practice almost always privileges the liberty.[228]

Courts have recognized a number of unenumerated rights protected by the substantive due process clause, including a right to marriage,[229] sexual privacy,[230] and (until recently) abortion.[231] Courts have been fairly consistent however in refusing to recognize environmental due process rights. The most recent effort to recognize environmental rights in the due process clause, the *Juliana* case, has, in line with the trend, proven unsuccessful thus far.

Juliana was brought by 21 youth plaintiffs alleging the federal government violated their due process rights and violated public trust obligations by permitting and encouraging burning fossil fuels. While the Ninth Circuit found the claims non-justiciable, both the district court and the dissent by Judge Staton in the Ninth Circuit found the plaintiffs raised valid equal protection claims, although on slightly different bases. The district court located the due process violation in the assertion that climate change "will cause human deaths, shorten human lifespans, result in widespread damage to property, threaten human food sources, and dramatically alter the planet's ecosystem,[232]" whereas Judge Staton specifically rejected the existence of an environmental due process right, even insofar as it led to deaths, but

[226] *Collins v. City of Harker Heights*, 503 U.S. 115, 125 (1992) (citing *Daniels v. Williams*, 474 U.S. 327, 331 (1986)).

[227] *Washington v. Glucksberg*, 521 U.S. 702, 721 (1997).

[228] *Witt v. Dep't of the Air Force*, 527 F.3d 806, 817 (9th Cir. 2008).

[229] *Loving v. Virginia*, 388 U.S. 1 (1967) (finding a due process right to interracial marriage); *Obergefell v. Hodges*, 576 U.S. 644 (2015) (holding that the due process right to marriage also extends to same sex marriages).

[230] *Lawrence v. Texas*, 539 U.S. 558 (2003).

[231] *Roe v. Wade*, 410 U.S. 113 (1973), *overruled by Dobbs v. Jackson Women's Health Org.*, 142 S. Ct. 2228 (2022).

[232] *Juliana v. United States*, 217 F. Supp. 3d 1224, 1250 (D. Or. 2016).

found a due process right to the perpetuity of the nation and to not be subject to the "dissolution of the Republic" which would ensue if climate change is not mitigated.[233]

Judge Staton's dissent cites to another case in which due process rights were recognized, but not on environmental grounds. The Sixth Circuit recognized a due process right to "bodily integrity" after residents of Flint, Michigan were subject to lead exposure in their drinking water.[234] Flint's water crisis was widely recognized in the United States as an environmental justice crisis.[235]

7.3 Takings

The Fifth Amendment's takings clause declares that "nor shall private property be taken for public use, without just compensation."[236] The takings clause recognizes the government's ability to take private property for the public good, but also constrains the government by requiring that these takings be compensated. How the government can take and limit property rights affects a host of environmental issues, and, as one might expect, the takings clause has featured in cases as diverse as the ability of local ordinances to condition building permits on the property owner creating a bike path without purchasing the land for the bike path,[237] the ability of the government to ban sale of endangered bird feathers without compensating owners of such feathers before the law was passed,[238] and a state's power to prohibit mining coal when doing so damaged neighboring homes without buying out the mining rights.[239]

In evaluating the scope of the takings clause, the courts have wrestled with each of the three questions imbedded in the takings clause: when is the government enacting a taking, what is sufficiently public use, and what constitutes just compensation?[240]

Some instances of takings are clear. When the government expropriates land whole-cloth, there is a clear taking. Less clear however is if a taking occurs when

---

[233] *Juliana v. United States*, 947 F.3d 1159, 1179 (9th Cir. 2020).

[234] *Guertin v. Michigan*, 912 F.3d 907, 916 (6th Cir. 2019).

[235] *See* e.g., Eligon (2016).

[236] U.S. Const. amend. V.

[237] *Dolan v. City of Tigard*, 512 U.S. 374 (1994) (finding an unconstitutional taking).

[238] *Andrus v. Allard*, 444 U.S. 51 (1979) (finding no regulatory taking).

[239] *Pa. Coal Co. v. Mahon*, 260 U.S. 393 (1922) (finding a taking. The case is the canonical source for the proposition that regulations which "[go] too far" constitute a taking).

[240] This chapter will not address the final question, but what constitutes just compensation is a critical question when it comes to the cost of implementing many environmental policies which involve takings. Generally, the rule is that takings must be compensated at market value, but that rule is complicated to apply in takings which are not full expropriations of single parcel of real property. For some modern discussion on how to determine the "denominator" of the asset being taken, see *Murr v. Wisconsin*, 137 S. Ct. 1933 (2017).

the government institutes land use or environmental regulations which restrict some, but not all property rights or requires a property owner to meet certain regulatory standards.

To determine if a so called "regulatory taking" has occurred, the courts have generally stuck to a balancing test articulated in *Penn Central v. City of New York* to determine what regulations implicate the takings clause. In *Penn Central*, New York City had passed a historic preservation act which required listed landmarks to seek approval from a Landmarks Preservation Commission before conducting any alterations on the building's exterior.[241] The owners of Grand Central Station proposed development of a skyscraper which would have been built atop the station. When the Commission disapproved of the development, Penn Central sued, claiming that the restrictions on their rights to develop the airspace above the existing station constituted a taking for which they had not been compensated.[242] The Supreme Court held that the law did not constitute a taking, citing as major factors to consider (1) "The economic impact of the regulation on the claimant," (2) interference with "distinct investment-backed expectations," and (3) "the character of the governmental action," with physical invasions being takings more than "public [programs] adjusting the benefits and burdens of economic life to promote the common good." [243] The Court found that the regulation was not a taking because it specifically ensured that Penn Central would still make a profit on Grand Central Station, Penn Central was still potentially able to build in the airspace above the station if it did so in a way that met the historic building regulation requirements, and Penn Central was given development rights for other buildings they owned as a result of the inability to develop over Grand Central Station.[244] The Court also noted that zoning laws which promote "the health, safety, morals, or general welfare" have never been considered takings even if they "destroyed or adversely affected recognized real property interests."[245]

While the balancing test laid out in *Penn Central* remains good law for figuring out when a regulation might have gone "too far" and constitute a taking, the Court has also articulated at least two areas which constitute *per se* regulatory takings which are categorically prohibited without compensation.

The first such category are so called "total wipeout" regulations, first discussed in *Lucas v. South Carolina Costal Commission.* Lucas purchased two beachfront lots in South Carolina, on which he planned to build homes. Before he began building the homes however, the state legislature passed the Beachfront Management Act, which barred any permanent development on designated beachfronts, including Lucas' parcels, in order to prevent erosion and protect the vulnerable dune

[241] *Penn Cent. Transp. Co. v. New York City*, 438 U.S. 104, 108–15 (1978).

[242] *Id.* at 115–22.

[243] *Id.* at 124.

[244] *Id.* at 135–38.

[245] *Id.* at 125.

ecosystem.[246] The erosion was bad enough that the parcels were in fact underwater for extended periods of time. The land was designated as critical land which should be barred from development "on the basis of 'the best scientific and historical data' available."[247]

The Court held that the Commission's use of the law could constitute a per se taking if it "[denied] all economically beneficial or productive use of [Lucas'] land,"[248] and that such denial, unless imposed by a regulation to prevent property use which could have otherwise been prevented through a common law nuisance claim, required compensation.[249] From an environmental perspective, it is worth highlighting the limited scope of what the Court considered as beneficial property right. Justice Blackmun and Stevens in their dissents both note that Lucas could still enjoy his land for camping, fishing, and recreation.[250]

The second category of per se takings occurs when a regulation requires "a permanent physical occupation." In *Loretto v. Teleprompter Manhattan CATV Corp.* the court found that a New York law which required landlords to allow minimal cable television equipment on their apartment buildings and limited compensation for supporting the equipment to fee set by a government commission.[251] The court ruled that the "permanent physical occupation" enacted by the forcible instillation of cable equipment was a per se taking, regardless of the small footprint of the equipment (a small half-inch diameter cable and two boxes just over one cubic foot) or any public benefit associated with enabling cable access to buildings.[252] While *Loretto* stressed multiple times that the permanence of the occupation was significant in the decision,[253] the court (depending on one's viewpoint) either extended the principles of *Loretto* to temporary occupations or created a new category of per se takings in *Cedar Point Nursery v. Hassid.*

In *Cedar Point*, the Court held that a California regulation which required farms to allow union organizers access to their property for up to three hours per day for up to four 30-day periods per year was a per se taking. The court held that the dividing line between a per se taking and a taking which would be evaluated under *Penn Central* is if there was "a physical taking," no matter the temporal duration or physical size, rather than "a use restriction."[254] While the full implications of this decision have not yet been played out, many scholars have sounded the alarm about the

---

[246] *Lucas v. S.C. Coastal Council*, 505 U.S. 1003, 1007–09 (1992).

[247] *Id.* at 1038.

[248] *Id.* at 1015.

[249] *Id.* at 1029.

[250] *Id.* at 1044, 1065 fn.3.

[251] *Loretto v. Teleprompter Manhattan Catv Corp.*, 458 U.S. 419, 422–24 (1982).

[252] *Id.* at 441.

[253] "Since these early cases, this Court has consistently distinguished between flooding cases involving a permanent physical occupation, on the one hand, and cases involving a more temporary invasion, or government action outside the owner's property that causes consequential damages within, on the other. A taking has always been found only in the former situation." Id. at 428.

[254] *Cedar Point Nursery v. Hassid*, 141 S. Ct. 2063, 2075 (2021).

implications *Cedar Point* could have on laws mandating access to facilities for environmental inspections. [255]

While takings jurisprudence is created by federal judges, the underlying property law that takings clause issues hinge on is often a matter of state law, a dynamic which can be seen in *Stop the Beach Renourishment v. Florida Department of Environmental Protection.*

Florida state common law holds that the state owns the land between the ocean's low-tide line and the mean high-water line. If sediment deposits create new dry land "gradually and imperceptibly", a private landowner adjacent to the ocean automatically takes title to the new land. If, however, there is a "sudden or perceptible loss of or addition to land by the action of the water" then the property lines do not shift.[256] In 1961, Florida established an anti-erosion policy which funds depositing sand and maintaining sand on eroded or eroding beaches. Upon authorization of a beach maintenance program, a state board sets an anti-erosion line at the mean high-water line which then becomes the property line between coastal landowners' property and the state ocean property. This new line is set, and, unlike under the common law rule, not subject to change based on changing coastlines.[257]

In 2003, the city of Destin and Walton County applied for the necessary permits to restore 6.9 miles of beach within their jurisdictions that had been eroded by hurricanes. A group of landowners sued, arguing that the beach restoration amounted to a taking of their right to have their property contact the water, and the right to future property expansion through sediment deposit under the common law rule.[258] The court split 4-4 on the question of whether a judicial order could constitute a taking, but held unanimously that under Florida law the right to deposited land was subordinate to the State's right to fill in the ocean land it clearly owned, and therefore no taking occurred.[259] As sea levels rise, state law will continue to interact with federal constitutional law and impact what compensation states will owe landowners for various adaptation strategies.

Takings jurisprudence has generally threatened to increase the cost or undermine many environmental regulations, but when government is exercising eminent domain powers to take property with compensation, the courts have given legislatures and agencies broad latitude to determine what takes are for a public use. Eminent domain prohibits the government from taking property simply to give it to another private party for private enjoyment, but the court *has* allowed eminent domain to force transfers of property between private parties if there is a claimed public purpose for doing so. In the seminal case *Kelo v. City of New London*, New London exercised its eminent domain powers to take residential homes and transfer the property to a private developer to create retail and recreational spaces as part of

---

[255] *See, e.g.*, Johnson (2022).

[256] *Stop the Beach Renourishment, Inc. v. Fla. Dep't of Envtl. Prot.*, 560 U.S. 702, 708–09 (2010).

[257] *Id.* at 709–10.

[258] *Id.* at 702–03.

[259] *Id.* at 731–32.

an economic development scheme.[260] Homeowners challenged the taking, arguing that the government improperly used eminent domain because economic development was not a public use.[261] The court held that this transfer to a private developer met the standards of a "public purpose" required by the takings clause, and noted a general deference to legislative and executive determinations of public purpose.[262] In the aftermath of Kelo, public backlash led many states to limit use of eminent domain.[263] Still, the broad power of eminent domain enables states, localities, and the federal government to acquire and pay for land for redevelopment into high-density housing, nature parks, creation of non-car transit infrastructure, or siting high-voltage transmission, all critical infrastructure needed to respond to a changing climate.

8 Fourteenth Amendment: Equal Protection

The Equal Protection Clause of the Fourteenth Amendment commands that a state may not "deny to any person within its jurisdiction the equal protection of the laws."[264] In the context of racial discrimination, the Supreme Court has held that "proof of racially discriminatory intent or purpose is required to show a violation of the Equal Protection Clause."[265]

At first blush, the extent to which the intent requirement hampers environmental litigation may seem unclear. The Supreme Court, in *Arlington Heights v. Metro Hous. Dev. Corp.*, observed that state action can violate equal protection even if it does not rest "solely on racially discriminatory purposes."[266] And where "a clear pattern, unexplainable on grounds other than race, emerges from the effect of" a "state action … [t]he evidentiary inquiry is" purportedly "relatively easy."[267] But over time, it has become clear that the discriminatory intent requirement significantly limits the ability of environmental justice plaintiffs to succeed on claims under the Equal Protection clause.[268]

In *Bean v. Southwestern Waste Management Corp.* a judge in the Southern District of Texas denied a temporary restraining order and a preliminary injunction that would have prevented the operation of a solid waste facility near a

260 *Kelo v. City of New London*, 545 U.S. 469, 473–75 (2005).

261 *Id.* at 475–76.

262 *Id.* at 480.

263 Mihaly and Smith (2011).

264 U.S. Const. amend. XIV.

265 *Vill. of Arlington Heights v. Metro. Hous. Dev. Corp.*, 429 U.S. 252, 265 (1977).

266 *Id.* at 265.

267 *Id.* at 266.

268 Wilson (2022), p. 1731.

predominantly Black high school in the Houston area.[269] The plaintiffs specifically challenged the Texas Department of Health (TDH)'s decision to grant a permit to build the facility, which satisfied the Equal Protection Clause's requirement that challenged conduct be "state action."[270] The court nonetheless denied the temporary restraining order and preliminary injunction because it found that the plaintiffs had failed to show likelihood of success on the merits.[271] Specifically, it determined that they had failed to "establish a substantial likelihood of proving that the decision to grant the permit was motivated by purposeful racial discrimination."[272]

Although the opinion acknowledged that "[s]tatistical proof can rise to the level that it, alone, proves discriminatory intent,"[273] it also found that the plaintiffs' evidence did not rise to that level.[274] The court first held that there was insufficient evidence to show a substantial likelihood that TDH had a pattern or practice "of discriminating in the placement of solid waste sites."[275] It then noted that a considerable majority of the sites operating with TDH permits in Houston were in census tracts that were majority white and non-Hispanic.[276] It then analyzed the plaintiffs' remaining data and determined that closer scrutiny rendered the statistics less compelling than they seemed on the surface.[277]

Nonetheless, the judge acknowledged that TDH's decision to grant the permit was "insensitive and illogical."[278] The judge also observed that plaintiffs could still win permanent injunctive relief by showing, for instance, that the more specific locations of waste sites were near predominantly Black communities.[279] But the court, acknowledging that it was "obligated, as all Courts are, to follow the precedent of the United States Supreme Court," denied the preliminary relief.[280]

R.I.S.E., Inc. v. Kay[281] provides another example of the Equal Protection Clause's intent requirement posing a barrier to successful environmental justice litigation. There, Residents Involved in Saving the Environment ("R.I.S.E.") sued the county board of supervisors in opposition to a proposed landfill near their property.[282] Sixty-four percent of those living "within a half-mile radius of the proposed regional landfill site" and twenty-one of the twenty-six families living on the road along

---

[269] 482 F. Supp. 673, 674–75 (S.D. Tex. 1979).

[270] *Id.* at 676.

[271] *Id.* at 677.

[272] *Id.* at 680.

[273] *Id.* at 677.

[274] *Id.* at 680.

[275] *Id.* at 677.

[276] *Id.*

[277] *Id.* at 678.

[278] *Id.* at 681.

[279] *Id.* at 680.

[280] *Id.*

[281] 768 F. Supp. 1144, 1145 (E.D. Va. 1991).

[282] *Id.* at 1145.

which "[m]ost of the landfill-bound traffic" would travel were Black.[283] The court also noted three other landfills built in predominantly Black parts of the county.[284] Given all this, the court concluded that "[t]he placement of landfills in" the county from 1969 to" the date of the decision "had a disproportionate impact on black residents."[285]

But, as in *Bean*, that was not enough. The court noted that "the historical placement of landfills in predominantly black communities" provided "'an important starting point' for the determination of whether official action was motivated by discriminatory intent."[286] But it also asserted that "at worst," the board members "appear[ed] to have been more concerned about the economic and legal plight of the County as a whole than the sentiments of residents who opposed the placement of the landfill in their neighborhood."[287] Since "the Equal Protection Clause does not impose an affirmative duty to equalize the impact of official decisions on different racial groups," the court entered judgment for the defendants.[288]

Professor Jill Evans argued in a 1998 article that these cases were wrongly decided even under the Equal Protection Clause's intent requirement.[289] Regarding *Bean*, a "proposal to place a municipal landfill in the same general area was rejected" just eight years before the *Bean* controversy—at which time the area and school were predominantly white.[290] Moreover, twelve of 13 "disposal facilities owned and operated by the city from 1920 through the early 1970s" "were in black neighborhoods."[291]

The *R.I.S.E.* case may have had similar inconsistencies.[292] For instance, the board there had denied a variance for a landfill in a predominantly white area in part because "permitting the landfill's continued operation would result in a significant decline in property values of the adjacent properties."[293] When the plaintiffs in *R.I.S.E.* raised similar concerns, they did not carry the day.[294]

Nonetheless, both cases demonstrate the potential of the Equal Protection Clause's intent requirement to pose significant barriers to challenging even blatant disparities between the respect state actors give to the environment in different communities.

---

[283] *Id.* at 1144.

[284] *Id.* at 1148.

[285] *Id.* at 1149.

[286] *Id.*

[287] *Id.* at 1150.

[288] *Id.*

[289] Evans (1998), pp. 1184–1186.

[290] Evans (1998), p. 1281.

[291] Evans (1998), p. 1281.

[292] Evans (1998), pp. 1285–1286.

[293] Evans (1998), p. 1285.

[294] Evans (1998), p. 1285.

9 Conclusions

The federal courts have not found the U.S. Constitution to create any rights to a healthful environment for human beings, or to create any rights of nature for its own sake. As interpreted by the Supreme Court, several clauses inhibit the abilities of governments to act on climate change: the Commerce Clause restrains federal actions over isolated wetlands, and state actions that discriminate against other states; the Supremacy Clause limits state actions that are preempted by federal law; and the Fifth Amendment requires the government to provide an opportunity for some regulated parties to be heard, and also requires the government to compensate those whose property it takes. The Case or Controversy Clause restricts who has standing to sue. So far, at least, the First Amendment's right to free speech has not inhibited government action on climate change, though efforts to use it for this purpose are ongoing. The Fourth Amendment prohibition against unreasonable searches and seizes has, for the most part, not inhibited government enforcement. And the Equal Protection Clause has not proven to be useful in environmental justice cases because of the requirement to prove discriminatory intent.

In sum, the U.S. Constitution provides much more assistance to industrial interests than to environmental interests.

References

Belitz H (2017) Environmental law in the age of the Supreme Court Bar. Environs Environ Law Policy J 40:133–184

Berry T, Schulp J (2022) SEC should reject climate rules over first amendment issue. CATO Institute. https://www.cato.org/commentary/sec-should-reject-climate-rules-over-first-amendment-issue. Accessed 9 Mar 2023

Eligon J (2016) A question of environmental racism in flint. New York Times. https://www.nytimes.com/2016/01/22/us/a-question-of-environmental-racism-in-flint.html. Accessed 9 Mar 2023

Evans J (1998) Challenging the racism in environmental racism: redefining the concept of intent. Ariz Law Rev 40:1219–1303

Gerrard M (ed) (2023) Environmental law practice guide 2

Green D (2012) Massachusetts v. EPA without Massachusetts: private party standing in climate change litigation. Environs Environ Law Policy J 36:35–63

Haan S (2022) The First Amendment and the SEC's Proposed Climate Risk Disclosure Rule 3. https://papers.ssrn.com/sol3/papers.cfm?abstract_id=4138712

Johnson O (2022) Let the exceptions do the work: how Florida should approach environmental regulation after *Cedar Point Nursery v. Hassid*. Univ Miami Law Rev 77:258–294

Kovacs K (2015) Superstatute theory and the administrative common law. Ind Law J 90:1207–1260

Law Scholars (2022) Comment Letter on Proposed Rule for the Enhancement and Standardization of Climate-Related Disclosures for Investors. http://democracyforward.org/wp-content/uploads/2022/06/Comment-SEC-Disclosure-Rule-6.17.22.pdf. Accessed 9 Mar 2023

Mihaly M, Smith T (2011) Kelo's trail: a survey of state and federal legislative and judicial activity five years later. Ecol Law Q 38(3):703–729

Robinson J (2011) Idaho couple puts wetlands rules to the test at Supreme Court. National Public Radio. https://www.npr.org/templates/story/story.php?storyId=141887844. Accessed 9 Mar 2023

Shanor A, Light S (2022) Greenwashing and the First Amendment. Colum Law Rev 122:2033–2118

Wilson B (2022) Making me ill: environmental racism and justice as disability. Univ Penn Law Rev 170:1721–1755

The Environmental Rights Provisions of U.S. State Constitutions: A Comparative Analysis

John C. Dernbach

Abstract Nearly every U.S. state has environmental provisions in its constitution. These provisions cover a wide range of issues concerning the manner in which the environment and natural resources are to be governed (e.g., public funding, eminent domain, public trust, access to water, legislative responsibility). But only six declare a right to a quality environment. These are Hawai'i, Illinois, Massachusetts, Montana, New York, and Pennsylvania. This chapter provides a detailed examination of the constitutional text and case law in each of these states, and analyzes the similarities and differences among them. This analysis shows the wide variety of ways in which even similar provisions have been interpreted and applied at the state level. It shows that while four of these states treat their right to a certain quality environment as judicially enforceable, two (Illinois and Massachusetts) do not. Among the states with a judicially enforceable right, Hawai'i and Montana have the most well developed bodies of case law. This analysis also shows similarities and differences in the ways that these constitutional provisions both guide and constrain state authority under statutes and regulations. Perhaps the most basic issue is whether environmental rights have actual parity with other constitutionally protected rights. These states have answered that question in varying ways.

Lindsay Williamson, Widener Law Commonwealth Class of 2023, and Keri Nace, Widener Law Commonwealth Class of 2024, provided excellent research assistance.

J. C. Dernbach (✉)
Widener University, Commonwealth Law School, Harrisburg, PA, USA
e-mail: jcdernbach@widener.edu

G. Antonelli et al. (eds.), *Environmental Law Before the Courts*,
https://doi.org/10.1007/978-3-031-41527-2_2

1 Introduction

Nearly all U.S. state constitutions have environmental or natural resources provisions. Most deal with public funding, eminent domain, public trust, legislative responsibility, public land acquisition and management, water access and development, fishing and hunting, and the like.[1] Recognizing that more needed to be done to protect the environment, many states in the 1960s began to add a policy statement concerning environmental protection or a legislative mandate to protect the environment.[2] And at the beginning of the modern environmental movement in the late 1960s and 1970s, five states adopted different and potentially more far-reaching constitutional provisions, recognizing a public right to a quality environment. These were Hawai'i, Illinois, Massachusetts, Montana, and Pennsylvania. New York joined them in 2021. There is no right to a quality environment in the U.S. Constitution.[3]

To be sure, the vast bulk of the environmental protection work in the U.S. is accomplished under statutes and regulations designed and administered by the federal government as well as state and local governments for that purpose. The various state rights to a quality environment supplement them; they do not replace them.

These state constitutional provisions are nonetheless important, despite their small number, because they provide a polestar or guiding principle for state and local governments; furnish a bulwark against environmental backsliding; give citizens access to courts to vindicate their environmental rights; fill gaps where environmental statutes and regulations are inadequate; and strengthen the legitimacy of, and public support for, protected values and resources. Because these rights are guaranteed equally to all citizens, they can also provide a means of redressing environmental injustice.[4] As constitutional provisions, they supply a basis for invalidating inconsistent statutes and regulations, and also furnish a constitutional defense for environmental protection actions whose lawfulness is challenged on other grounds.

---

[1] May and Daly (2015), p. 218 (explaining that more than 200 such provisions exist in 46 state constitutions).

[2] Zackin (2013), p. 151.

[3] A recent case makes a somewhat similar claim under the U.S. Constitution, but it is not a claim based on a right to a quality environment. In Juliana v. United States, the plaintiffs claim that the federal government has known dangers of climate change for decades, and that it nonetheless has "created and enhanced dangers through fossil fuel extraction, production, consumption, transportation, and exportation." As a result, they argue, the U.S. government has violated the public trust doctrine as well as substantive due process under the U.S. Constitution by interfering with their right to a "climate system capable of sustaining human life." The case was dismissed for lack of standing. Juliana v. U. S., 947 F.3d 1159 (9th Cir. 2020).

Under the various public trust laws, the government is obligated to protect certain natural resources for the benefit of present and future generations. A public trust claim is at the core of the Juliana case. Blumm and Wood (2017). The public trust, however, is not the same as a right to a certain quality environment.

[4] van Rossum and Manahan (2021).

For these reasons, there is a growing movement to add environmental rights to other state constitutions.[5] The recently adopted New York provision is evidence of that movement, and proposals for similar provisions are pending in other states.[6] Around the world, the constitutions of nearly 170 countries, containing three-fourths of the world's population, include some form of environmental protection.[7] And there is growing interest in enforcing existing provisions of national constitutions and strengthening these provisions—in no small part because of disruptive climate change.[8] Indeed, there is at least one recent proposal for an amendment to the U.S. Constitution to address climate change and other environmental issues.[9] At its root, much of this growing activity and advocacy grow out of a normative claim that environmental rights are an integral part of a broader set of human rights, and ought to be given the same level of legal protection as other human rights. In 2022, the U.N. General Assembly strengthened this claim by adopting a landmark resolution recognizing "the right to a clean, healthy and sustainable environment as a human right."[10]

To say that states have constitutional provisions recognizing or declaring a right to a certain level of environmental quality, is not to say that they all recognize this normative claim, or even that state courts have treated these amendments similarly in all ways. There are considerable differences in state court interpretation of these provisions. Some differences are due to differences in constitutional language. Some may relate to the claims made in a particular case, or the procedural posture in which it was decided. And some may simply reflect different judicial sensibilities about the appropriate way to decide such cases. But whatever the reason, the similarities and differences are worth examining in some detail. They provide an understanding of how judicial interpretation of these provisions could evolve over time. Nearly all of these provisions, after all, are at most half a century old—far newer than nearly all of the other provisions in state constitutions or the U.S. Constitution. They thus provide an understanding of the possible trajectories of future decisions.

In addition, because most of these cases are decided strictly with reference to a particular state's constitutional provision and relevant cases, there has been little opportunity to compare or understand alternative provisions and cases in other states to see what can be learned from them. A comparative approach can suggest better (and poorer) ways of approaching particular questions. This is true not only for litigation strategy and judicial interpretation after constitutional provisions have been adopted; it is also useful in drafting such provisions. As climate change and other environmental problems intensify, such constitutional provisions will likely

---

[5] van Rossum (2022).

[6] Joselow (2021).

[7] Dernbach et al. (2018), pp. 844–845.

[8] May and Daly (2015), pp. 255–280.

[9] Kury (2021), pp. 107–120.

[10] G.A. Res. A/76/L.75, The Human Right to a Clean, Healthy and Sustainable Environment (July 26, 2022).

be an increasingly important part of the legal and judicial toolbox for addressing them.

This chapter first shows and describes the six constitutional provisions in question, and reviews some similarities and differences in the text of these provisions. The rest of the chapter addresses key issues in judicial interpretation. A recurring question in these cases is whether these provisions are self-executing—that is, whether courts can enforce them as is, in the absence of additional legislation. Because courts answer this question in different ways, this question is addressed throughout the sections summarized below.

Section 3 examines the role of both the courts and the legislature in defining the scope of these provisions—what the right to a certain quality environment means. Not surprisingly, differences in the text of these provisions between states lead courts to different results. This section also looks at a closely related question--the role of the legislature in defining the right to a quality environment. Two state constitutions (Hawai'i and Illinois) specifically authorize the legislature to define what their specified level of environmental quality means. These states have approached this issue in contrasting ways, one enlarging its meaning and the other essentially foreclosing review altogether.

Section 4 looks at the impact of these provisions on standing requirements. While some states treat plaintiffs under these provisions in the same way as they do other plaintiffs, two states have slightly relaxed standing requirements under these provisions.

Section 5 reviews who can be sued under these provisions. It first examines case law under these provisions to see whether they authorize a right of action against the state for violating them. While four states authorize a right of action against the state, two (Illinois, Massachusetts) do not. This section then looks at the question of whether these provisions authorize a cause of action against private parties. Except for Montana, the answer is no. In the first major case under any state's environmental rights provision, Pennsylvania courts said yes; they have since appeared to backtrack from that decision.

Finally, Sect. 6 explains the different legal tests for determining the constitutionality of challenged actions, mostly in Hawai'i and Montana. These are the two states with the most well developed case law overall on the right to a quality environment.

2 Right to Quality Environment: Similarities and Differences in Text

Six states have constitutional provisions stating that their citizens have a right to a quality environment. The right to a certain quality environment (often expressed as the right to a clean environment) is different from the right to use the environment in various ways, such as hunting, fishing, or in otherwise using or extracting resources. It is also different from the right to have the government hold, manage, and protect certain public resources in trust for the benefit of present and future generations.[11] These states, the text of the relevant constitutional provision, the date when each provision was adopted, and its location in the state constitution are shown in Table 1.

As Table 1 shows, these provisions are all relatively recent in origin. The oldest, (Illinois in 1970) is new when measured against the age of most state and national constitutional provisions. The recent adoption of these provisions is a direct result of growing public concern about environmental conditions and their relation to public health.

These six states express this right in different ways. Hawai'i, Montana, and New York all declare, in somewhat similar words, a public right to a clean and healthful environment. Illinois simply declares a right to a "healthful environment" without describing other desired environmental qualities, such as clean.

Pennsylvania, by contrast, recognizes "a right to clean air, pure water, and to the preservation of the natural, scenic, historic and esthetic values of the environment." The Massachusetts provision, adopted one year later, states this right in nearly identical language, and adds "freedom from excessive noise" and a right to the "conservation, development and utilization" of certain natural resources.[12] While neither state uses "healthful," these provisions can easily be read that way; clean air, pure water, and the preservation of environmental values are unquestionably healthful to humans.

These differences are based on considered decisions by their drafters. The committee that drafted the Illinois provision, for example, was interested in protecting the environment insofar as it affected human health, but its interest did not extend beyond that. "The Committee…explained that it selected the word 'healthful' as best describing the kind of environment that ought to be the objective."[13] The drafters of the Montana provision, by contrast, used "clean and healthful" because they wanted to maintain a high quality environment and "did not intend to merely prohibit that degree of environmental degradation which can be conclusively linked to ill health or physical endangerment."[14]

Three states (Montana, New York, Pennsylvania) have these provisions in their declaration or bill of rights—their analogue to the Bill of Rights in the U.S. Constitution. This is no small thing, for it emphasizes that environmental rights are of the same status as other recognized rights. The preamble to Article I of the Pennsylvania constitution, where that state's environmental rights provision is

[11] Two states—Hawai'i and Pennsylvania—also provide broad constitutional public trust protection for certain natural resources. Pa. Const. Art. I, § 27; Haw. Const. Art. XI, § 1. While beneficiaries of a constitutional public trust also have rights, these provisions are outside the scope of this chapter.

[12] The Massachusetts provision also includes a right to the "conservation and utilization" of certain natural resources. As explained earlier, this is different from the right to a certain quality of environment.

[13] Glisson v. City of Marion, 20 N.E.2d 1034, 1044 (Ill. 1999).

[14] Mont. Env't Info. Ctr. v. Dep't of Env't Quality, 988 P.2d 1236, 1249 (Mont. 1999).

Table 1 State constitutional provisions recognizing a right to a quality environment

State	Text	Year adopted	Location in state constitution
Hawai'i	Each person has the right to a clean and healthful environment, as defined by laws relating to environmental quality, including control of pollution and conservation, protection and enhancement of natural resources. Any person may enforce this right against any party, public or private, through appropriate legal proceedings, subject to reasonable limitations and regulation as provided by law.	1978	Art. XI (Conservation, Control, and Development of Resources), § 9
Illinois	Each person has the right to a healthful environment. Each person may enforce this right against any party, governmental or private, through appropriate legal proceedings subject to reasonable limitation and regulation as the General Assembly may provide by law.	1970	Art. XI (Environment), § 2
Massachusetts	The people shall have the right to clean air and water, freedom from excessive and unnecessary noise, and the natural, scenic, historic, and esthetic qualities of their environment; and the protection of the people in their right to the conservation, development and utilization of the agricultural, mineral, forest, water, air and other natural resources is hereby declared to be a public purpose. The general court shall have the power to enact legislation necessary or expedient to protect such rights.[a]	1972	Art. 97
Montana	All persons are born free and have certain inalienable rights. They include the right to a clean and healthful environment and the rights of pursuing life's basic necessities, enjoying and defending their lives and liberties, acquiring, possessing and protecting property, and seeking their safety, health and happiness in all lawful ways. In enjoying these rights, all persons recognize corresponding responsibilities.	1972	Article II (Declaration of Rights), § 3
New York	Each person shall have a right to clean air and water, and a healthful environment.	2021	Art. I (Bill of Rights), § 19
Pennsylvania	The people have a right to clean air, pure water, and to the preservation of the natural, scenic, historic and esthetic values of the environment.[b]	1971	Article I (Declaration of Rights), § 27

[a] Article 97 further provides: "In the furtherance of the foregoing powers, the general court shall have the power to provide for the taking, upon payment of just compensation therefor, or for the acquisition by purchase or otherwise, of lands and easements or such other interests therein as may be deemed necessary to accomplish these purposes. Lands and easements taken or acquired for such purposes shall not be used for other purposes or otherwise disposed of except by laws enacted by a two thirds vote, taken by yeas and nays, of each branch of the general court."

(continued)

Table 1 (continued)

The effect of these additional provisions is to provide a form of public trust protection to lands acquired in this manner. Most of the reported appellate cases under Article 97 involve the acquisition, use, and disposition of public land thus acquired. *See, e.g.*, Smith v. City of Westfield, 82 N.E.3d 390 (Mass. 2017)

[b] Article I, Section 27 reads in full: "The people have a right to clean air, pure water, and to the preservation of the natural, scenic, historic and esthetic values of the environment. Pennsylvania's public natural resources are the common property of all the people, including generations yet to come. As trustee of these resources, the Commonwealth shall conserve and maintain them for the benefit of all the people."

The latter two sentences recognize that the Commonwealth has a constitutional public trust duty to conserve and maintain public natural resources for the benefit of present and future generations. It uses different words and concepts than the first sentence, and was intended to be understood as a separate constitutional right. Dernbach (2020), pp. 795–799. Most of the recent litigation under this provision involves its public trust component. Dernbach (2020), pp. 807–818. *See also* Pa. Env't Defense Found. v. Commonwealth, 279 A.3d 1194 (Pa. 2022); Pa. Env't Defense Found. v. Commonwealth, 255 A.3d 289 (Pa. 2021)

located, makes that point emphatically by declaring that the following rights represent "the general, great and essential principles of liberty and free government."[15] As will be seen, the courts in all three of these states now take these provisions seriously.

Of the other three states, two (Hawai'i and Illinois) place their provisions in an environmental section to the constitution, while one (Massachusetts) has its provision in a general list of amendments. Of these three, only Hawai'i permits judicial enforcement of a right to a quality environment.

In two states (Massachusetts and Pennsylvania), the same provision includes a right to a quality environment and language recognizing a public trust for certain natural resources. In both states, the majority of litigation involves the public trust part of the provision.[16]

While each of the six states recognizes, in different language, a right to a quality environment, Montana's provision adds: "In enjoying these rights, all persons recognize corresponding responsibilities." A separate provision (Art. IX, § 1) states the responsibility of each person to "maintain and improve" Montana's environment for "present and future generations."

In that same provision, Montana assigns the state with the same responsibility, and requires the state to adopt appropriate legislation and provide adequate remedies to prevent violations:

---

[15] Robinson Twp. v. Commonwealth, 83 A.3d 901, 953–54 (Pa. 2013) ("The right delineated in the first clause of Section 27 presumptively is on par with, and enforceable to the same extent as, any other right reserved to the people in Article I [Declaration of Rights].") (plurality). The first clause of Section 27 recognizes the environmental right addressed in this chapter.

See also p. 948: ("The Declaration of Rights assumes that the rights of the people articulated in Article I of our Constitution—vis-à-vis the government created by the people—are inherent in man's nature and preserved rather than created by the Pennsylvania Constitution.").

[16] *See* footnotes a and b to Table 1.

(1) The state and each person shall maintain and improve a clean and healthful environment in Montana for present and future generations.
(2) The legislature shall provide for the administration and enforcement of this duty.
(3) The legislature shall provide adequate remedies for the protection of the environmental life support system from degradation and provide adequate remedies to prevent unreasonable depletion and degradation of natural resources.

Somewhat similar language exists in the constitutions of three other states. The Massachusetts provision authorizes the legislature (known in Massachusetts as the General Court) to adopt legislation to protect constitutional environmental rights. Separate provisions in the Hawai'i Constitution (Art. IX, § 8) and Illinois Constitution (Art. XI, § 1) authorize or require similar legislation. By contrast, Pennsylvania and New York are silent on the role of the legislature in implementing their provisions.

Two states (Hawai'i and Illinois) use nearly identical language to authorize any person to bring an action for their enforcement. However, they add that this authorization is subject to "reasonable limitation" as the legislature "may provide by law."

3 Scope

Courts and legislatures can have a role in interpreting the scope of the meaning of a quality environment, depending on how the provisions are written.

3.1 Role of Courts in Defining Right to a Quality Environment

Textual differences in how the right to a quality environment is expressed have affected judicial decisions. For example, the use of "healthful environment" in the Illinois provision, instead of "clean and healthful," or words like that, has limited the scope of that state's provision. In 1999, the Illinois Supreme Court held that protecting species was not a legally cognizable interest for standing purposes under that state's constitution because it is not included in the plaintiff's right to a "healthful environment."[17] The plaintiff sued to prevent a city's planned construction of a dam and reservoir on a creek, claiming that the project would destroy the habitat for two endangered and threatened species (least brook lamprey and Indiana crayfish) protected under the state's Endangered Species Act. Although that act did not provide a private cause of action, the plaintiff alleged that the Illinois Constitution specifically authorized citizens to sue to protect their right to a "healthful environment." While the Court agreed that the constitution authorizes lawsuits for that purpose, it held that the word "healthful" was specifically intended to focus on the

[17] Glisson v. City of Marion, 720 N.E.2d 1034 (Ill. 1999).

human health effects of environmental degradation, and did not extend to endangered and threatened species. The drafting history of the provision, the Court explained, showed that "healthful" was chosen, instead of words like clean, to limit the provision to human health. A dissenting opinion argued that the text provided the plaintiff with standing, and that resort to the drafting history to limit the reach of the provision was improper. "Altering a watercourse in a way that will eradicate species of threatened or endangered wildlife," the dissent argued, "clearly has obvious and important implications for the healthfulness of the environment."

Somewhat similarly, a federal district court in Montana held that the right to a "clean and healthful environment" did not provide a basis for challenging a 2021 statute generally prohibiting persons and entities from withholding goods, services, or employment based on a person's vaccination status or whether the person has an immunity passport.[18] The text and legislative history of the amendment, the court explained, showed that the right was intended to apply to the natural environment, and not to institutional settings.

A recurring question in recent years is whether constitutional environmental provisions should be interpreted and drafted to include a right to be protected against human-induced climate change.[19] In Hawai'i, where the right to a clean and healthful environment is to be defined by the legislature, that state's Supreme Court has concluded that the right "subsumes a right to a life-sustaining climate system" because the legislature has adopted greenhouse gas reduction goals and taken other measures to reduce greenhouse gas emissions.[20]

Higher courts in other states may be asked to decide the same question in the absence of legislation. In Montana, appellate courts may decide whether the right to a "clean and healthful environment" extends to a right to be free from human-induced climate change. In *Held v. Montana*, 16 youth plaintiffs sued the state, claiming that the state violated their constitutional right by adopting a fossil-fuel based State Energy Plan and by creating a legislative exception for climate change to Montana's Environmental Policy Act. Of course, these claims bring climate change under the scope of that state's environmental right. The trial court held that the plaintiffs had standing, limited their claim to declaratory relief, and scheduled a trial for 2023.[21] For Pennsylvania, where courts also have the power to decide the scope of the right to "clean air, pure water, and the preservation of the natural, scenic, historic, and esthetic values of the environment," Robert McKinstry and the

---

[18] Mont. Med. Ass'n v. Knudsen, 581 F.Supp.3d 1232 (D. Mont. 2022).

[19] Kury (2021), pp. 107–120.

[20] Matter of Maui Electric Co., Ltd., 506 P.3d 192, 202-03 n. 15 (Haw. 2022). *See also* In re Application of Maui Elec. Co., 141 P.3d. 1, 5 (Haw. 2017), (holding that Haw. Const. art. XI, § 9, includes the right to be protected "from the effect of greenhouse gas emissions").

[21] Held v. Montana, Dist. Ct. Lewis & Clark County (Cause No. CDV-2020-307) (Aug. 4, 2021), http://climatecasechart.com/wp-content/uploads/sites/16/case-documents/2021/20210804_docket-CDV-2020-307_order.pdf.

author have argued that this language recognizes a right to be free from human-induced climate change.[22]

3.2 *Role of Legislature in Defining or Limiting Right to Quality Environment*

A question closely related to the scope of these provisions is the role that the legislature is allowed to play in defining what the right to a quality environment means. This is a different question than whether these provisions are self-executing. This question assumes that courts have authority to enforce them without further legislation, but asks instead whether that authority can be defined or limited by the legislature.

To be sure, legislative authority to protect the environment and public health existed before these provisions were adopted, and continues afterward. Indeed, environmental statutes and regulations generally make it unnecessary (or at least less necessary) for plaintiffs to invoke constitutional provisions. Many states adopted these amendments at the dawn of the modern environmental movement, when environmental laws and their enforcement were much weaker than they are now. Still, a continuing justification for constitutional provisions like these is to fill gaps in current levels of statutory and regulatory protection.

The Hawai'i constitution (Art. XI, § 9) explicitly authorizes the legislature to define the meaning of its right to a quality environment: "Each person has the right to a clean and healthful environment, as defined by laws relating to environmental quality, including control of pollution and conservation, protection and enhancement of natural resources." The drafters of this provision believed that authorizing courts to decide the boundaries of the right "could lead to confusion and inconsistencies."[23] This provision, the state's Supreme Court has explained, gives "flexibility to the definition of the right over time," enabling it to be "reshaped and redefined through statute, ordinance and administrative rule-making procedures."[24] As already noted, the court has concluded that the many state statutes for reducing greenhouse gas emissions mean that the right to a clean and healthful environment "subsumes a right to a life-sustaining climate system."[25]

Two states (Illinois and Hawai'i) authorize the legislature to impose "reasonable limitation and regulation" on a person's right to sue to enforce their constitutional environmental rights. As a practical matter, these provisions authorize the

[22] McKinstry and Dernbach (2018), pp. 69–70.

[23] Cty. of Haw. v. Ala Loop Homeowners, 235 P.3d 1103, 1121 n.24 (Haw. 2010).

[24] In re Application of Maui Elec. Co., Ltd., 408 P.3d 1, 13 (Haw. 2017). "By contrast, the privacy right is to be implemented, not defined, by the legislature." Org. of Police Officers v. City and Cnty. of Honolulu, 494 P.3d 1225, 1243 n. 21 (Haw. 2021) (emphasis omitted).

[25] Matter of Maui Electric Co., Ltd., 506 P.3d 192, 202-03 n. 15 (Haw. 2022).

legislature to limit the reach of these rights. In *People v. Pollution Control Board*,[26] the Illinois legislature exempted motor car racing events from the noise regulation authority of the state's Pollution Control Board. The legislature did so after the state's attorney general brought an action against a corporation that owned a car racing facility for exceeding the noise limits established by the Board. The attorney general argued that the legislative exemption violated the constitutional right to a "healthful environment." The Illinois Court of Appeals disagreed, holding that the legislature has the authority to impose reasonable limits on a person's right to sue under the constitutional provision, and that the exemption was reasonable. The court said it lacked the authority to challenge "the wisdom and propriety" of the legislature's exemption decision.

The Hawai'i Supreme Court reached a different result on the meaning of the same limiting language. In a 2010 case, *County of Hawai'i v. Ala Loop Homeowners*, a group of homeowners used the environmental rights provision of the state constitution to challenge a proposed development. The defendant argued that the action was barred by statute, saying that the Hawai'i constitution (Article XI, § 9) authorized "reasonable limitations and regulations," and that this statute appropriately precluded judicial review.[27] The court held that this statute did not mean what the defendant claimed. Even if the statute did mean that, the court reasoned, the drafting history of this constitutional language indicated that it was intended to "encompass such matters as statutes of limitations or procedural or jurisdictional limitations," and not "to eliminate private enforcement altogether."

New York and Pennsylvania are silent on the role of the legislature in defining the right to a quality environment. At a minimum, constitutionally stated rights are rights that an individual has against the government. That is, a court in an appropriate case may hold that the government violated these rights and award relief. Thus, the duty of the government, including the legislature, to protect these rights may be implied. In Pennsylvania, the Supreme Court has held that the state's police power extends to protection of the right to clean air, pure water, and the preservation of certain values in the environment.[28]

4 Standing

Many states apply their ordinary standing test in environmental rights cases, but two states relax them somewhat for these cases. The general rule for standing in Pennsylvania is that a petitioner must establish "a substantial, direct and immediate

---

[26] People v. Pollution Control Bd., 473 N.E.2d 452 (Ill. App. 1984).

[27] Cnty. of Hawai'i v. Ala Loop Homeowners, 235 P.3d 1103 (Haw. 2010).

[28] *See, e.g.*, United Artists' Theater Cir. v. City of Phila., 635 A.2d 612 (Pa. 1993). *See also* Dernbach (1999), pp. 150–156.

interest in the outcome of the litigation," and that rule is applied in environmental rights cases.[29] Montana also applies its ordinary standing rules to these cases.[30]

Illinois and Hawai'i provide somewhat more relaxed standing requirements for suits claiming the right to a certain quality environment. The Illinois provision "does away with the 'special injury' requirement" for standing "typically employed in environmental [public] nuisance cases."[31] As a result, "a plaintiff need not allege a special injury to bring an environmental claim" under that state's environmental rights provision. Hawai'i employs a three-part standing test, requiring that the plaintiff must have suffered an actual or threatened injury; the injury must be fairly traceable to the defendant's actions; and a favorable decision would likely provide relief for the plaintiff's injury. But since 1975 (three years prior to the adoption of its environmental rights provision, it has not required plaintiffs to assert an injury that is different in kind from the injury suffered by the public.[32] Thus, the Hawai'i Supreme Court has held that an environmental organization has standing to vindicate its right to a clean and healthful environment even though its members did not use or own the land at issue.[33]

5 Possible Defendants

A reader who has even a passing familiarity with U.S. constitutional law involving the Bill of Rights might reasonably assume that these state provisions uniformly provide a cause of action against the government and almost certainly do not provide a cause of action against private parties. Neither assumption, it turns out, is fully accurate.

5.1 *Right of Action Against State?*

The enforceability of the right to a quality environment against the government varies dramatically. The three states that declare a right in their bill of rights or declaration of rights (Montana, New York, Pennsylvania) see this right as akin to other rights. That means, among other things, that the constitutional right authorizes lawsuits against the state to redress violations. While that is how constitutional rights

[29] Robinson Twp. v. Commonwealth, 83 A. 901, 917 (Pa. 2013).

[30] Mont. Env't Info. Ctr. v. Dep't of Env't Quality, 988 P.2d 1236, 1242 (Mont. 1999) ("(1) the complaining party must clearly allege past, present, or threatened injury to a property or civil right; and (2) the alleged injury must be distinguishable from the injury to the public generally, but the injury need not be exclusive to the complaining party.").

[31] City of Elgin v. Cnty. of Cook, 660 N.E.2d 875, 891 (1995).

[32] In re Application of Maui Elec. Co., 408 P.3d 1, 22 (Haw. 2017).

[33] Life of the Land v. Land Use Comm'n, 623 P.2d 431, 441 (Haw. 1981).

are understood, environmental rights sometimes have a legal status similar to those of other rights (e.g., free speech, freedom of assembly), and sometimes they do not.

Montana has repeatedly heard cases where individuals and organizations have challenged state action for violation of the state's right to a "clean and healthful environment," and awarded relief against the state in many of them.[34] In doing so, the Montana Supreme Court has described this right as "fundamental."[35] This court has even held that, as an agent of the state, it lacks the authority to issue an order that would violate the right to a clean and healthful environment.[36]

A 2022 trial court decision in New York, *Fresh Air for the Eastside v. State of New York*, suggests that the New York provision will be similarly applied in that state.[37] A citizen group claimed that the operations of a solid waste landfill violated the "right to clean air and water, and a healthful environment," which went into effect on January 1, 2022. It complained, among other things, of odors, fugitive emissions, and greenhouse gas emissions from the landfill, many of which violate state law. Defendants included the state, which issued a permit for the landfill and has statutory responsibility for regulating its operation, and the company that operates the landfill. The trial court refused to dismiss the case against the state, holding that constitutional rights limit the government, and explaining that the state lacks the authority or discretion to violate the constitutional environmental rights of citizens.

Pennsylvania appellate courts have rendered one decision in a lawsuit against the state concerning the constitutional right "to clean air, pure water, and to the preservation of the natural, scenic, historic and esthetic values of the environment." As previously explained, the Pennsylvania provision also recognizes a public trust in certain natural resources. Judicial interpretation of each part of Pennsylvania's entire environmental provision was interrupted in 1973 by a Commonwealth Court decision that substituted a judicially invented three-part test for the text of the amendment.[38] (The Commonwealth Court is an intermediate appellate court in Pennsylvania.) As a result, there was no meaningful judicial development of either part of the provision for more than four decades; instead, nearly all of the case law concerned this three-part test.[39] The state's Supreme Court overruled that decision in 2017, reinstating the text of the amendment as the law to be applied, and thus recognizing that each of the two parts needs to be analyzed separately.[40] In doing so, it explained how the environmental rights part of the provision should be applied:

[34] *See, e.g.,* Park Cnty. Env't Council v. Mont. Dep't of Environmental Quality, 477 P.3d 288 (Mont. 2020); Mont. Env't Info. Ctr. v. Dep't of Env't Quality 988 P. 2d 1236 (Mont. 1999).

[35] Clark Fork Coal. v. Mont. Dep't of Nat. Res. and Conservation, 481 P.3d 198, 217-18 (Mont. 2021).

[36] Cape-France Enter. v. Est. of Peed, 29 P.3d 1011, 1017 (Mont. 2001).

[37] Fresh Air for the Eastside v. State of N. Y., Index No. E2022000699 (Monroe Cnty. Sup. Ct., Dec. 7, 2022).

[38] Payne v. Kassab, 312 A.2d 86 (Pa. Commw. Ct. 1973), a*ffirmed*, 361 A.2d 263 (Pa. 1976).

[39] Dernbach and Prokopchak (2015).

[40] Pa. Env't Defense Found. v. Commonwealth, 161 A.3d 911 (Pa. 2017).

"This clause places a limitation on the state's power to act contrary to this right, and while the subject of this right may be amenable to regulation, any laws that unreasonably impair the right are unconstitutional." This right, in other words, is a self-executing right against the government. Since that time, the Commonwealth Court has heard and decided one case under this provision against the government—a claim that a local government's zoning ordinance and a permit issued under that ordinance were inconsistent with the right to a quality environment.[41]

But not all state courts have read their amendment as authorizing suits against the state. While the Illinois constitution declares a right to a healthful environment, the state's Supreme Court has held that this provision "does not create any new causes of action."[42] It does not provide a right of action against the government. "Thus, while a plaintiff need not allege a special injury to bring an environmental claim, there must nevertheless still exist a cognizable cause of action" that is separate from, and in addition to, the constitutionally declared right.[43]

In Massachusetts, there is, at best, substantial doubt whether the constitutional right to a quality environment provides a right of action against the government. There are no state court decisions saying that there is an enforceable right to clean air and clean water.[44] In a case where the plaintiff asked a federal district court for relief based on a violation of the "right to clean air and water," the court declined, saying that it is "emphatically not the role of the federal courts to develop and expand upon state law."[45]

5.2 Right of Action Against Private Parties?

Typically, constitutional rights are rarely enforceable against private parties. For the most part, that is how the provisions in these six states have worked. For example, while the trial court in *Fresh Air for the Eastside v. State of New York* allowed the case against the state to proceed, it nonetheless dismissed the case against the private company operating the landfill. It held that there is nothing in the text of the new constitutional provision authorizing suits against private parties.[46]

In contrast, Montana appears to recognize two different types of actions against private parties under its environmental rights provision. The first is derived not only from the right to a clean and healthful environment but also from accompanying language in the same provision (Art. IX, § 1) stating that "the state and each person

[41] Frederick v. Allegheny Twp. Zoning Hearing Bd., 196 A.3d 677 (Pa. Commw. Ct. 2018).

[42] City of Elgin v. Cnty. of Cook, 660 N.E.2d 875, 891 (1995). *See also* NBD Bank v. Krueger Ringier, Inc., 686 N.E.2d 704, (Ill. App.1997) (Illinois constitution does not authorize "private right of action in tort against a seller of contaminated property.").

[43] City of Elgin, 660 N.E.2d at 891.

[44] Hootstein v. Amherst-Pelham Reg'l Sch. Comm., 361 F.Supp.3d 94, 113-14 (D. Mass. 2019).

[45] *Id.* at 115, quoting Pimentel v. City of Metheun, 323 F.Supp.3d 255, 274 (D. Mass. 2018).

[46] Fresh Air for the Eastside v. State of N. Y., slip. op. at 10–12.

shall maintain and improve a clean and healthful environment in Montana for present and future generations…." In *Cape-France Enterprises v. Estate of Peed,*[47] Cape-France, which owned a large tract of land, entered a buy-sell agreement with Lola Peed and her granddaughter (Peed) for five acres of that tract so that they could build a motel or hotel. To subdivide the property, Peed was required to drill a well for water. But it turned out that a groundwater contamination plume from a nearby drycleaner was spreading underneath the Cape-France property, and drilling the necessary well would likely spread the contamination and create potential cleanup liability for Cape-France. Cape-France sued to rescind the agreement on the grounds of impossibility or impracticability, and the trial court granted that relief.

On appeal, Peed argued, among other things, that potential liability is an insufficient reason to rescind a contract. The supreme court said that argument "ignores an important—and, in fact, a decisive—point." Because of what the Montana constitution provides about the right to a clean and healthful environment and the duty of each person to ensure that quality environment, "it would be unlawful for Cape–France, a private business entity, to drill a well on its property in the face of substantial evidence that doing so may cause significant degradation of uncontaminated aquifers and pose serious public health risks." By reasoning that a private party could violate the state's right to a clean and healthful environment, the court indicated that the constitutional provision could provide a basis for lawsuits against such parties.

Second, the Montana Supreme Court recognizes that the state's right to a clean and healthful environment authorizes lawsuits for damages against private parties, but only where no common law cause of action is available to provide a remedy.[48] The Court had previously recognized tort actions for damages under other provisions of the state constitution. Thus, it affirmed the dismissal of an action brought by downstream landowners against a mining company, claiming contamination of their property and diminished water flows, and requesting damages for violation of their constitutional right to a clean and healthful environment.[49] In that case, the court held, the plaintiffs failed to demonstrate that traditional tort remedies would not redress the damage to them.

Pennsylvania's first major case under its provision, and the earliest major case under any of these environmental provisions, also read the provision to authorize actions against private parties. The case created doubt about whether the provision, Article I, § 27, is self-executing, because many dissenting judges argued that there should be legislative authorization for the provision to be used against private parties. Since then, the Pennsylvania Supreme Court appears to have backed away from this decision.

---

[47] Cape-France Enter. v. Est. of Peed, 29 P.3d 1011 (Mont. 2001).

[48] Sunburst Sch. Dist. No. 2 v. Texaco, Inc., 165 P.3d 1079 (Mont. 2007). The court said it was important to "avoid constitutional issues wherever possible." *Id.* at 1093.

[49] Shammel v. Canyon Res. Corp., 167 P.3d 886 (Mont. 2007).

In *Commonwealth v. National Gettysburg Battlefield Tower, Inc.*,[50] the attorney general challenged the proposed construction of an observation tower on private land outside of Gettysburg Battlefield National Park. No local or state governmental approval was required to construct the tower. The attorney general argued that the tower's visibility throughout the Gettysburg Battlefield would interfere with the public right to preservation of the natural, scenic, historic, and esthetic values of that environment. The public's right to the preservation of those values, the attorney general claimed, imposed a substantive limitation on such private development.

The trial court decided that article I, section 27 is self-executing because, among other reasons, other provisions in the state's Declaration of Rights had previously been held to be self-executing.[51] The trial court also denied the requested injunction, ruling that the state "failed to show by clear and convincing evidence that the natural, scenic, historic, and esthetic values of the Gettysburg area will be irreparably harmed by the construction of the proposed tower on the proposed site."

The attorney general lost on appeal to both the Commonwealth Court and the Pennsylvania Supreme Court.[52] Still, the Commonwealth Court held that section 27 is self-executing against a private party.[53] While the Pennsylvania Supreme Court affirmed the Commonwealth Court's decision, there was no majority opinion on whether section 27 is self-executing.[54] This decision established the Commonwealth Court's opinion as binding precedent on the question of whether the Amendment is self-executing against a private party. For reasons that appear to be outside the realm of precedent, that point was lost on subsequent courts, which held that section 27 is not self-executing.[55] The most obvious explanation is that the case led lawyers and judges to view section 27 as entirely a grant of governmental authority, and not as a limitation on that authority.

There is a substantial question about whether *Gettysburg Tower* is precedent that lawyers can rely on, even though it has not been overruled. No Pennsylvania appellate court has subsequently authorized a lawsuit against a private party based on the right to a quality environment. In more recent decisions, as explained above, the Pennsylvania Supreme Court has stated that the right to a quality environment is self-executing against the government, and has been silent on the applicability of that right against private parties.[56]

---

[50] Commonwealth v. Nat'l Gettysburg Battlefield Tower, Inc., 302 A.2d 886 (Pa. Commw. Ct. 1973), *affirmed*, 311 A.2d 588 (Pa. 1973).

[51] Commonwealth v. Nat'l Gettysburg Battlefield Tower, Inc., 13 Adams Cnty. L.J. 75, 79–80 (citing Erdman v. Mitchell, 207 Pa. 79 (1903)).

[52] Commonwealth v. Nat'l Gettysburg Battlefield Tower, Inc., 302 A.2d at 894–95 (Pa. Commw. Ct.); 311 A.2d at 595 (Pa.).

[53] Gettysburg Tower, 302 A.2d at 892.

[54] Gettysburg Tower, 311 A.2d at 595.

[55] *See, e.g.*, Robinson Twp. v. Commonwealth, 52 A.3d 463, 488–89 (Pa. Commw. Ct. 2012), *affirmed in part, reversed in part*, 83 A.3d 901 (Pa. 2013).

[56] Pa. Env't Defense Found. v. Commonwealth, 161 A.3d 911, 931 (Pa. 2017).

6 Standard for Determining Constitutionality of Challenged Actions

The standards that courts employ for determining whether there is a violation of the constitutional right to a quality environment also vary from state to state. Montana, Hawai'i, and Pennsylvania employ significantly different approaches.

Montana employs a strict scrutiny test that is similar to that used in cases involving other constitutional rights. In *Montana Environmental Information Center v. Department of Environmental Quality*,[57] the state legislature exempted arsenic discharges from 23 ground-water pumping tests from review under the state's water quality nondegradation statute. The plaintiffs demonstrated that the tests would increase the concentration of arsenic in the receiving water in excess of the concentration that would have required review under the nondegradation statute. They argued that the exemption violated their right to a "clean and healthful" environment. The state's Supreme Court held that this exemption should be subject to strict scrutiny, the most demanding level of scrutiny under that state's constitution. The court remanded the case to the trial court for a strict scrutiny determination of its constitutionality: whether (1) there was a compelling state interest for enactment of the exemption, (2) the exemption was "closely tailored to effectuate that interest," and (3) the exemption represented "the least onerous path that can be taken to achieve the state's objective."

In a subsequent case, the Montana Supreme court held a statute unconstitutional without a remand. In that case, the court addressed a legislative amendment to the Montana Environmental Policy Act (MEPA), which requires an assessment of environmental impact prior to government actions that may significantly affect the environment.[58] The amendment barred courts from issuing equitable relief to prevent violations of MEPA. In this case, the Department of Environmental Quality approved an exploration permit for mining in the Yellowstone River watershed 15 miles from the national park, even though its MEPA analysis did not adequately analyze wildlife and water quality effects. Under the statutory amendment, the company would have been able to continue conducting exploration activities while the MEPA defects were being corrected. In addition to vacating the permit, the court held the amendment unconstitutional on its face, using strict scrutiny analysis, because it removed "the Plaintiffs' only available remedy adequate to prevent potential constitutionally-proscribed environmental harms."[59]

---

[57] Mont, Env't Info. Ctr. v. Dept. of Env't Quality 988 P.2d 1236 (Mont. 1999).

[58] Park Cnty. Env't Council v. Montana Dep't of Env't Quality, 477 P.3d 288 (Mont. 2020).

[59] Park Cnty., 477 P.3d at 309. The Montana Supreme Court has held certain exemptions to be consistent with the constitutional right to a "clean and healthful environment," but only because the exemption did not authorize any environmental degradation. In both cases, the project or activity was still subject to MEPA and required the issuance of specific permits before it could proceed. Clark Fork Coal. v. Mont. Dep't of Nat. Res. and Conservation, 481 P.3d 198 (Mont. 2021); N. Plains Res. Council, Inc. v. Mont. Bd. of Land Comm'rs., 288 P.3d 169 (Mont. 2012).

In Hawai'i, where the legislature defines the right to a "clean and healthful environment," private litigants nonetheless have two opportunities to challenge the constitutionality of government actions. Both are procedural.

First, the Hawai'i provision (Article XI, § 9) authorizes state court actions for its enforcement. After recognizing a right to a clean and healthful environment that may be defined by the legislature through laws "relating to environmental quality," the provision states: "Any person may enforce this right against any party, public or private, through appropriate legal proceedings, subject to reasonable limitations and regulation as provided by law."[60] In the *Ala Loop Homeowners* case, discussed earlier, the Hawai'i Supreme Court held that this provision creates a right of action to enforce Chapter 205 of Hawai'i's statutes, which concerns land use.[61]

In that case, a county exempted a proposed charter school project on a 28-acre parcel from an otherwise applicable special permit process under Chapter 205 designed to review health and safety issues. The Supreme Court first decided that Chapter 205 is a law "relating to environmental quality," which means that is a law defining the meaning of a "clean and healthful environment" that can be privately enforced. Among other things, the court explained, Chapter 205 requires "consideration of issues relating to the preservation or conservation of natural resources."

The court then held that the right of action under Article XI, § 9 does not require further legislation to be judicially enforceable; it is self-executing. The text indicates that it can be enforced without legislation, and its drafting history indicates it was meant to be treated that way. In addition, the court noted, a 1986 statute authorizes attorney fees for environmental rights litigation under Chapter 205 because the legislature found that "the public has rarely used" the right to a clean and healthful environment. Finally, as explained above, the court held that another statute did not bar the constitutional right of action. Thus, the Hawai'i constitution creates a private right of action to enforce laws relating to environmental quality.

Second, the Hawai'i Supreme Court has held several times that this right includes a constitutionally protected property right to an opportunity to be heard prior to state action that may adversely affect this right. A 2017 decision, for example, *In re Application of Maui Electric Company* (MECO), involved a challenge by the Sierra Club to a power purchase agreement between MECO and another company.[62] Under that agreement, MECO would purchase electricity generated by a facility burning sugar processing residue and other fuels, including coal. Public utilities such as MECO are regulated monopolies; they have an exclusive right to supply electricity in a particular region, and the Public Utilities Commission (PUC) is supposed to ensure that they receive a reasonable rate of return on their investment, but no more than a reasonable return. Under Hawai'i law, public utilities such as MECO are allowed to recover power purchase costs from customers if the PUC approves. The

[60] This provision is limited to actions in state court; it does not enlarge the jurisdiction of federal courts. Fiedler v. Clark, 714 F.2d 77, 79–80 (9th Cir. 1983).

[61] Cnty. of Haw. v. Ala Loop Homeowners, 235 P.3d 1103 (Haw. 2010).

[62] In re Application of Maui Electric Company, 408 P.3d 1 (Haw. 2017).

Sierra Club sought to intervene in the PUC proceeding concerning that decision, challenging, among other things, the use of coal under the agreement, and injury to members living near the facility. The PUC denied its motion to intervene and approved the power purchase agreement.

The Sierra Club argued on appeal that it was entitled to a hearing under the due process clause of the state constitution (Art. I, § 5), which provides that "[n]o person shall be deprived of life, liberty or property without due process of law." In general, persons are entitled to prior notice and an opportunity to be heard before the government can deprive them of life, liberty, or property. The right to a "clean and healthful environment," the Club argued, is a property right protected under the due process clause, and the state's Supreme Court agreed. Under state law, the court reasoned, a property interest does not need to be real estate, chattels, or money to be protected under the due process clause; the property interest can also be a "legitimate claim of entitlement" to a particular benefit. The right to a clean and healthful environment, the court held, "is a legitimate entitlement stemming from and shaped by independent sources of state law, and is thus a property interest protected by due process."

Hawai'i legislation requires the PUC "when exercising its duties to recognize the 'need' to reduce reliance on fossil fuels and to 'explicitly consider' the levels and effect of greenhouse gas emissions." Because this is a law "relating to environmental quality" under the Hawai'i constitution, the court said, the Sierra Club has a protectable property interest in having the PUC perform this analysis. As a result, the court held that the PUC was required to hold a hearing on "the impacts of approving the Agreement on Sierra Club's members' right to a clean and healthful environment, including the release of harmful greenhouse gases…."[63] The state Supreme Court has found a similar procedural due process right to be heard in three subsequent cases involving the right to a clean and healthful environment.[64]

The development of Pennsylvania's case law has been complicated by a tendency of both litigants and courts to treat the two separate provisions of its

[63] A dissenting opinion challenged, among other things, the conclusion holding that the Sierra Club had a protected property interest in this case.

[64] Matter of Haw. Elec. Light Co., Inc., 445 P.3d 673 (Haw. 2019) (in proceeding involving rate recovery for power purchase agreement, court held PUC violated environmental and due process rights of nonprofit organization by refusing to provide the organization with an opportunity to be heard on the impacts of greenhouse gas emissions caused by agreement, vacated PUC approval of the agreement, and remanded case to PUC for consideration of these impacts); Matter of Gas Co., LLC, 465 P.3d 633 (Haw. 2020) (in rate increase proceeding based on two recently established liquid natural gas projects, court held PUC violated environmental and due process rights of non-governmental organizations by refusing to allow them to be heard on impact of these projects on climate change, vacated PUC decision, and remanded case to PUC); Protect and Preserve Kahoma Ahupua'a Ass'n v. Maui Plan. Comm'n, 489 P.3d 408 (Haw. 2021) (court vacated Planning Commission's decision to approve affordable housing project, holding that Coastal Zone Management act is law relating to environmental quality under Article XI, § 9, that association was denied due process to protect its right to a clean and healthful environment, and that Commission was required to make findings on consistency of proposed project under Coastal Zone Management Act).

constitutional provision (right to a quality environment and public trust) as an undifferentiated whole. While the state's supreme court has clearly stated that each part is subject to separate rules and need to be analyzed separately,[65] plaintiffs and lower courts often argue and analyze the two parts without distinguishing them.[66] Still, Pennsylvania's Commonwealth Court has decided one case on the right to clean air, pure water, and the preservation of certain values in the environment since the courts began honoring the text of this provision. That case, *Frederick v. Allegheny Township Zoning Hearing Board*,[67] involved a challenge to a local government ordinance authorizing shale gas development and a subsequently issued permit. The court employed a test different from that in Montana and Hawai'i for determining the constitutionality of these actions.

The challenged ordinance allowed "oil and gas well operations in all zoning districts so long as they satisfy enumerated standards designed to protect the public health, safety and welfare." The township used that ordinance to issue a permit for an unconventional gas well in a district that was zoned residential/agricultural (R-2). Under the permit, a drilling rig at a particular location, the "Porter Pad," would drill thousands of feet underground and enable other actions needed to recover natural gas from shale. Objectors, who were adjacent landowners, challenged the ordinance as, among other things, a violation of their right to clean air, pure water, and the preservation of certain values in the environment. The Commonwealth Court upheld the ordinance and permit.

The court described the process and standard for reviewing the objectors' claims as follows: "Judicial review of the government's action requires an evidentiary hearing to determine, first, whether the values in the first clause of the Environmental Rights Amendment [the right to a quality environment] are implicated and, second, whether the governmental action unreasonably impairs those values."[68] Zoning, the court said, implicates the natural, scenic, historic, and esthetic values of the environment, thus satisfying the first requirement. "It does so by placing compatible uses in the same zoning district; by establishing minimum lot sizes and dimensional requirements; providing parking and signage controls; and requiring landscape and screening controls."[69] But the court found that the objectors did not prove unreasonable impairment of those values. The permit issued under the amended ordinance, the court held, would not result in any cognizable harm:

> [T]he Zoning Board found that oil and gas development and agricultural uses "have long safely coexisted within rural communities." The only feature of the Porter Pad that will be visible from any of Objectors' homes is the portion of the drilling rig that rises over the treetops. Once drilling operations cease, the rig will be removed during the pumping phase.

---

[65] Pa. Env't Defense Found. v. Commonwealth, 161 A.3 at 931 (explaining legal test for determining violation of right to a quality environment) and 933 (explaining legal test for determining violation of public trust).

[66] Dernbach (2021), pp. 194–196.

[67] Frederick v. Allegheny Twp. Zoning Hearing Bd., 196 A.3d 677 (Pa. Commw. Ct. 2018).

[68] *Id.* at 695.

[69] *Id.*

> When pumping ends, the land can be returned to its original state. In the meantime, oil and gas drilling will support the agricultural use of land in the R-2 Zoning District.[70]

The court also held that the local government was not required to conduct a pre-decision analysis on the impact of its ordinance on protected environmental rights before adopting the amendment.

7 Conclusion

While six U.S. states have provisions declaring a right to a certain quality environment, they vary dramatically. Two states (Hawai'i and Montana) have reasonably well-developed case law and appear to play a significant role in protecting the environment in those states. The environmental rights provisions in two state constitutions (Illinois and Massachusetts) appear to play only a small role in environmental protection. While Pennsylvania is developing a significant body of law on a constitutional public trust,[71] its environmental rights provision is hobbled by an early decision that it provides a general right of action against private parties. And New York is just beginning to implement its environmental rights provision.

As this chapter shows, environmental rights provisions in state constitutions raise a multitude of issues. Perhaps the most basic issue, as identified in the introduction, is whether environmental rights have actual parity with other constitutionally protected rights. Thus far, these states have decided this issue in different ways.

References

Blumm M, Wood M (2017) "No ordinary lawsuit": climate change, due process, and the public trust doctrine. Am Univ Law Rev 67:1–87

Dernbach J (1999) Taking the Pennsylvania constitution seriously when it protects the environment: Part II—Environmental Rights and Public Trust. Dickinson Law Rev 104:97–164

Dernbach J (2020) Natural resources and the public estate. In: Gormley K, McNally J (eds) The Pennsylvania constitution: a treatise on rights and liberties. Bisel, Philadelphia, pp 793–830

Dernbach J (2021) Thinking anew about the environmental rights amendment: an analysis of recent Commonwealth Court decisions. Widener Commonwealth Law Rev 30:147–199

Dernbach J, Prokopchak M (2015) Recognition of environmental rights for Pennsylvania citizens: a tribute to Chief Justice Castille. Duquesne Law Rev 53:335–369

Dernbach J, Kristl K, May J (2018) Recognition of environmental rights for Pennsylvania citizens: Pennsylvania Environmental Defense Foundation v. Commonwealth of Pennsylvania. Rutgers Law Rev 70:803–856

[70] *Id.* at 697 (citations omitted).

[71] Dernbach (2020), pp. 807–818. *See also* Pa. Env't Defense Found. v. Commonwealth, 279 A.3d 1194 (Pa. 2022); Pa. Env't Defense Found. v. Commonwealth, 255 A.3d 289 (Pa. 2021).

Joselow M (2021) Environmental rights push could boost youth climate cases. ClimateWire (March 3, 2021). https://www.eenews.net/articles/environmental-rights-push-could-boost-youth-climate-cases/

Kury F (2021) The constitutional question to save the planet: the people's right to a healthy environment. Environmental Law Institute Press, Washington, D.C.

May J, Daly E (2015) Global environmental constitutionalism. CUP, Cambridge

McKinstry R, Dernbach J (2018) Applying the Pennsylvania environmental rights amendment meaningfully to climate disruption. Mich J Environ Adm Law 9:50–114

van Rossum M (2022) The green amendment: the people's right to a clean, safe, and healthy environment. Disruption Books, Austin, NY

van Rossum M, Manahan K (2021) Constitutional green amendments; making environmental justice a reality. Nat Resour Environ 36:1–5

Zackin E (2013) Looking for rights in all the wrong places: why state constitutions contain America's positive rights. Princeton University Press, Princeton & Oxford

John C. Dernbach is Emeritus Professor of Environmental Law and Sustainability at Widener University Commonwealth Law School. He can be reached at jcdernbach@widener.edu.

Environmental Justice Before U.S. Courts

Sara A. Colangelo and Abigail E. André

Abstract The history of environmental justice litigation in federal, state, and administrative courts illustrates how difficult it is to remedy intersectional harm using a single legal tool. In the United States, there is no federal "environmental justice law" that litigants can wield in court. The movement has instead attempted to harness constitutional and civil rights claims, environmental statutes, and common law to address the disproportionate environmental harm suffered by communities of color and under-resourced communities. However, the architects of those laws did not design them to combat environmental injustice, and litigation thereunder rarely provides communities with complete redress. While advocates have pursued multi-faceted approaches with some success, the arc of environmental justice litigation highlights the need for an expansion of existing law or new approaches to address the entrenched harms that characterize environmental racism and injustice.

This Chapter introduces readers to environmental justice and explains why litigation has been a lever of limited utility to advance the movement's goals in the last several decades. It surveys the historical application of different types of laws that advocates have employed in cases seeking environmental justice. The Chapter also highlights the non-litigation value lawsuits can provide from the perspective of movement organizing strategies. Finally, it identifies recent successful efforts as well as new setbacks and explores how novel legal developments could reshape the horizon of environmental justice litigation.

The authors thank Scott Badenoch, Amy Laura Cahn, and Sheila Foster for insightful feedback on early drafts of this chapter, as well as Emma Schwartz (Georgetown J.D. 2023) and Margaret McCallister (Georgetown J.D. 2024) for excellent research assistance.

S. A. Colangelo (✉)
Georgetown University, Washington, DC, USA
e-mail: sara.colangelo@law.georgetown.edu

A. E. André
The Ohio State University, Moritz College of Law, Columbus, OH, USA
e-mail: andre.16@osu.edu

G. Antonelli et al. (eds.), *Environmental Law Before the Courts*,
https://doi.org/10.1007/978-3-031-41527-2_3

1 Introduction: Mobilizing the Environmental Justice Movement

Despite America's progress in ratcheting down net amounts of air, water, and land-based pollution, "significant racial and socioeconomic disparities persist" in the distribution of environmental burdens.[1] These inequities manifest across several aspects of lived experience.[2] Two decades ago, Luke Cole and Sheila Foster observed that "numerous environmental hazards: garbage dumps, air pollution, lead poisoning, toxic waste production and disposal, pesticide poisoning, noise pollution, occupational hazards, and rat bites... are inequitably distributed by income or race."[3] Further, the studies Cole and Foster surveyed demonstrated race to be the best "predictor of exposure" to such hazards.[4] Expert analyses repeatedly confirmed these findings.[5] Further, the cumulative impacts of both chemical and non-chemical stressors (*i.e.* pollutants and structural racism) have compounded and amplified the burden of pollution over time.[6]

A web of long-standing societal and governmental practices gave rise to this inequitable distribution of burdens, and a significant body of scholarship details these roots.[7] The development of land use and zoning laws is a frequent origin for environmental justice analyses.[8] Scholars such as Shalanda Baker also emphasize the fossil-fuel-based energy system's role in "ongoing structural inequality because

[1] Bullard et al. (2007), p. xi. *See, e.g.*, Spiller et al. (2021), pp. 127,004-3 to 127,004-11; Mikati et al. (2018), p. 480.

[2] This Chapter addresses the lived experience of others. Accordingly, the authors intend our nomenclature to be as circumstance-specific as possible related to individuals with non-white racial identities, acknowledging there are preferences among terminology including "communities of color." Where source material uses the term "minority," the reader may assume it refers to racial minorities. Finally, we do not use the term "environmental justice communities" to avoid the implication that communities with environmental justice concerns are homogenous.

[3] Cole and Foster (2000), pp. 54–55.

[4] *Id.*

[5] *See, e.g.*, Mohai and Saha (2015), p. 7 (examining disparities in the location of pollution sources by race and economic status); Taylor (2014) (same).

[6] EPA defines cumulative impacts as "the total burden – positive, neutral, or negative – from chemical and nonchemical stressors and their interactions that affect the health, well-being, and quality of life of an individual, community, or population…." U.S. Env't Prot. Agency, EPA 600/R-22/014a, Cumulative Impacts: Recommendations for EPA's Office of Research and Development (2022).

[7] For a summary of the complex factors contributing to environmental injustice, *see, e.g.*, Rothstein (2017); Foster (2004), p. 10.

[8] Scholars cite "market-based forces that make [areas that] are already heavily industrial particularly appealing for the siting of new [locally unwanted land uses], as well as entrenched political and economic disempowerment of …affected communities." Ball (2006), p. 871. *See also* Foster (2017), pp. 136–137 (collecting sources on land use and zoning influences on environmental disparity, and noting counter arguments). Professor Vicki Been has been a consistent and prominent contributor to the scholarship on siting of undesireable land uses and environmental justice. *See, e.g.*, Vicki Been, *Locally Undesirable Land Uses in Minority Neighborhoods: Disproportionate Siting or Market Dynamics?*, 103 YALE L.J. 1383 (1994).

it places disproportionate burdens on poor communities and communities of color."[9] Others point to disparate enforcement of environmental protections and investment in critical infrastructure.[10] Today, such inequities have become entrenched, and are grimly confirmed by COVID-19 morbidity and mortality rates, and vulnerability to climate change-driven health and safety harms.[11]

The environmental justice movement responds to these disparities. The movement "concern[s]" itself with "the 'environmental racism' embedded in decisions to disproportionately dispose of hazardous wastes in predominantly Black communities… [and] with redistributing the negative externalities associated with other environmentally harmful industries…."[12] It draws upon strands of the civil rights, social justice, labor, and environmental movements; it intersects policy, economics, as well as federal, state, and local laws.[13] Legal efforts are one strand of broader movement building strategies. Further, "the environmental justice movement is []dynamic," and not static—diverse, and not a monolith in terms of its advocates, concerns, or goals.[14]

Seminal activism formally launched the movement in the latter half of the twentieth century. This included, most prominently, the 1968 Memphis Sanitation Strike and the 1982 weeks-long protest against the siting of carcinogen-contaminated soil in a predominantly Black community in Warren County, North Carolina.[15] The movement's leaders later drafted 17 Principles of Environmental Justice in 1991, affirming the right to "self determination" and to be "free from environmental hazards," and demanding "the right to participate as equal partners at every level of decision-making," among others.[16]

The definition of environmental justice by the U.S. Environmental Protection Agency (EPA) enshrines some, but not all, of those principles. EPA defines environmental justice as "the fair treatment and meaningful involvement of all people regardless of race, color, national origin, or income, with respect to the development, implementation, and enforcement of environmental laws, regulations, and

[9] Baker (2019), p. 6. Baker details scholarship suggesting that "[f]ossil fuels… led to 'the westward expansion, growth of urban centers, rise of monopoly capitalism, concentration of wealth, migration and immigration of working-class people and people of color, segregation [and] impoverishment.'" *Id.* (citing Energy Democracy: Advancing Equity in Clean Energy Solutions 8 (Denise Fairchild & Al Weinrub eds., 2017)).

[10] *See, e.g.*, Kuehn (1994), p. 625.

[11] *See, e.g.*, Magesh et al. (2021); Chen et al. (2022), p. 440.

[12] Baker (2019), p. 14.

[13] *See, e.g.*, Bryant (1995), pp. 5–6.

[14] Bullard (2007), p. viii.

[15] *See, e.g.*, *id.* at 1–5.

[16] *Principles of Environmental Justice*, United Church of Christ, https://www.ucc.org/what-we-do/justice-local-church-ministries/justice/faithful-action-ministries/environmental-justice/principles_of_environmental_justice/ (last visited Jan. 24, 2023).

policies."[17] Others discuss environmental justice as encompassing "distributive justice, or the equitable distribution of environmental burdens and benefits; procedural justice, or the fairness of the decision-making process itself; and corrective justice, or the fairness of punishment and compensation."[18]

2 Litigation: A Tool of Limited But Continuing Utility

Predating most early environmental justice activism, advocates sought legal solutions to pollution-based harms on behalf of communities. In 1970, residents from a predominantly Black neighborhood sought to enjoin the construction of two highways through their local park by alleging violations of the Fourteenth Amendment and Title VIII of the Fair Housing Act.[19] That same year, the newly formed Natural Resources Defense Council joined the California Rural Legal Assistance to ban DDT in a lawsuit on behalf of female farmworkers.[20]

In the case most frequently credited as the first environmental justice lawsuit, *Bean v. Southwestern Waste Management Corp.*, a community led by Linda McKeever Bullard challenged the Texas Department of Health's issuance of a solid waste landfill permit in a predominantly Black neighborhood in Houston.[21] The vast majority of Houston's solid waste sites were already in predominantly Black and Latino neighborhoods.[22] Yet the Court held those facts insufficient to "show…that the decision to grant the permit [wa]s attributable to an intent to discriminate on the basis of race."[23]

The loss in *Bean* is emblematic of much environmental justice litigation: unsatisfactory from a legal relief perspective, but still valuable to advance other goals within the movement. Although the plaintiffs lost their civil rights suit, the matter altered Houston's siting practices, solidified the law as a tool for change within the movement's toolkit, and inspired sociologist Dr. Robert Bullard—known as the father of environmental justice—to devote his life to advancing the movement.[24]

Though critical of litigation's utility for and alignment with environmental justice, Luke Cole nonetheless credited the capacity of litigation to foster

[17] *Environmental Justice*, EPA.gov, https://www.epa.gov/environmentaljustice (last visited Jan. 24, 2023).

[18] Failed Promises: Evaluating the Federal Government's Response to Environmental Justice (David M. Konisky ed. 2015).

[19] Harrisburg Coal. Against Ruining the Env't v. Volpe, 330 F. Supp. 918, 926 (M.D. Pa. 1971) (denying claims).

[20] Env't Def. Fund, Inc. v. Hardin, 428 F.2d 1093 (D.C. Cir. 1970) *sub. history omitted.*

[21] 482 F. Supp. 673 (S.D. Tex. 1979).

[22] *Id.* at 678–679.

[23] *Id.*

[24] *See* Bullard (2000), p. 53.

accountability, garner media attention, and inspire alliances.[25] He observed that even pursuing civil rights suits, which have almost universally failed as legal or administrative claims, "can be useful in building morale, raising the profile of a community's struggle and educating the public and government officials about environmental racism."[26] Others similarly remark on the leverage lawsuits can provide to "forc[e] federal and state agencies to grapple[] with [environmental justice] in the assessment and approval of projects" and to inspire creative evolutions in legal strategy.[27]

Still, environmental justice litigation suffers significant shortcomings from both normative and strategic perspectives. Litigation may risk "disempowerment of the client community," which violates a key tenet of environmental justice that communities speak for themselves.[28] The litigation forum can also reproduce power asymmetries that contribute to the marginalization of non-white populations in the environmental decision-making process.[29] Even a "victory,"—such as the $600 million payment from settling defendants in the Flint water crisis—is often partial compensation at best for individuals' harm and trauma, leaving communities disillusioned with the law.[30] Further, piecemeal lawsuits are inefficient and some argue inherently deficient to remedy what can be considered "political and economic" struggles.[31] Scholars critique the "ill-fit between…federal civil rights and environmental laws…and [harms] caused by an agglomeration of private and public action."[32] But it is not merely an "ill fit" that stymies progress. The persistence of environmental injustice in the U.S. has been fueled by broader, more fundamental *gaps* and *barriers* in various legal regimes, many of which this Chapter addresses below.

2.1 Antidiscrimination Cases: Constitution and Civil Rights Act-Based Claims

Equal Protection Clause

Because the environmental justice movement grew from the civil rights movement, early litigation efforts attempted to leverage civil rights laws and constitution-based

[25] Cole (1994), pp. 525–526, 541–542 [hereinafter *Environmental Justice Litigation*].

[26] *Id.*

[27] Henderson et al. (2021), p. 17.

[28] Cole (1994), p. 525; *see also* Drury and Chu (1994/1995), p. 52.

[29] *See, e.g.*, Cole (1995), pp. 702–704 (arguing for lawyers to empower and provide legal expertise to attack environmental injustice before it begins rather than through litigation, which removes the matter from the hands of the community).

[30] Paul Egan, *Federal Judge Gives Final Approval to $626.25M Settlement in Flint Water Crisis*, Detroit Free Press (Nov. 10, 2021), https://www.freep.com/story/news/local/michigan/flint-water-crisis/2021/11/10/federal-judge-approves-settlement-flint-lead-poisoning-case/5556131001/.

[31] Cole (1994), p. 524.

[32] Daly (2002), p. 18.

claims to remedy disparate pollution burdens. Cases challenging discriminatory land use practices relied heavily on the equal protection clause of the Constitution, under which neither states nor the federal government may deny persons "equal protection of the laws."[33] Despite sweeping constitutional language, these cases tended to fail because of the need to prove discriminatory *intent*, not merely discriminatory *impact*.[34] Plaintiffs must establish that race was "a motivating factor in the decision" and that decision-makers were motivated by the action's adverse effects on a group.[35] Discriminatory intent is vastly more difficult to establish than discriminatory impact because the "processes that result in the concentration of polluting sources in poor, minority communities are neutral on their face."[36]

The discriminatory intent requirement often proves too high a hurdle in constitutional environmental justice cases, even where explicit bias appears obvious.[37] For example, in *R.I.S.E., Inc. v. Robert A. Kay*, residents challenged plans to site a garbage dump in a predominantly Black neighborhood.[38] Evidence of discriminatory intent was overwhelming: every government run dump in the county was in a majority-Black neighborhood; similar proposals in white neighborhoods had failed; and the county only solicited input from wealthy white homeowners.[39] Though the court agreed that the siting had a disparate impact, it found no discriminatory intent because the government's decision had "balanced the economic, environmental, and cultural needs of the County in a responsible and conscientious manner."[40]

Either a single discriminatory act[41] or a clear pattern "unexplainable on grounds other than race" can establish discriminatory intent.[42] Single act cases are typically proven by showing unequal enforcement or unequal access to government services. In a rare success for corrective justice, residents established that 30% of predominantly Black neighborhoods were unpaved compared to 18% of white neighborhoods and that while 50% of Black homes had no sewer connection, only 28% of

[33] U.S. Const. amend. XIV, § 1 (applies to states); U.S. Const. amend V (applies to the federal government); *see, e.g.*, Cole (1994), p. 538.

[34] Cole (1994), p. 539; *see also* Washington v. Davis, 426 U.S. 229, 242 (1976) ("[A] law … is [not] invalid under the Equal Protection Clause simply because it may affect a greater proportion of one race than another.").

[35] Arlington Heights v. Metro. Hous. Dev. Corp., 429 U.S. 252, 264–66 (1977). Discriminatory intent exists when someone "treats some people less favorably than others because of their race, color, religion, sex, or national origin." Int'l Bhd. Of Teamsters v. United States, 431 U.S. 324, 388 n.15 (1977). Discriminatory impact more broadly includes "practices that are facially neutral … [yet] in fact fall more harshly on one group than another…." *Id.*

[36] Foster (2017), p. 136.

[37] *See, e.g.*, Northern (1997), p. 538.

[38] 768 F. Supp. 1144 (E.D. Va. 1991).

[39] *Id.* at 1148–1149.

[40] *Id.* at 1149–1150.

[41] *See, e.g.*, Gomillion v. Lightfoot, 364 U.S. 339 (1960); Yick Wo v. Hopkins, 118 U.S. 356 (1886).

[42] Arlington Heights v. Metro. Hous. Dev. Corp., 429 U.S. 252 (1977).

white households lacked this service.[43] This provision has also been used to successfully challenge discriminatory practices related to public services that are connected with the sale or rental of a dwelling, including policing, fire protection, and garbage collection.[44]

To succeed under the "clear pattern" standard, however, communities must not only identify a series of racially motivated decisions but must also rule out alternative non-race-based justifications.[45] For example, in *East Bibb Twiggs Neighborhood Association v. Macon-Bibb County Planning and Zoning Commission*, a community objected to the permitting of a landfill in a predominantly Black neighborhood.[46] The plaintiffs relied on census data, a history of local racism, procedural irregularities, and the commission's prior proceedings as evidence of a "clear pattern" of racial discrimination.[47] In rebuttal, the zoning commission proffered the county's urgent need for additional waste disposal.[48] Ultimately, the court dismissed the plaintiff's constitutional claims based on the government's "reasonable" nondiscriminatory motives.[49]

In the wake of a series of failed Equal Protection Clause cases, environmental justice advocates turned to civil rights laws for redress.[50] The Civil Rights Act (CRA) Title VI and Title VII of the Fair Housing Act (FHA) are commonly used vehicles in environmental justice cases. And yet, like constitutional claims, an exceptionally high evidentiary burden persists.

Title VI

CRA Title VI Section 601 provides a pathway to sue federally funded programs for discrimination on the basis of race, color, and national origin.[51] This provision can be used to challenge state programs, policies, and decisions, including pollution and

---

[43] Ammons v. Dade City, Fla., 594 F. Supp. 1274 (M.D. Fla. 1984), *aff'd*, 783 F. 2d 982 (11th Cir. 1986); *see also* Hawkins v. Town of Shaw, 437 F.2d 1286 (5th Cir. 1971).

[44] *See, e.g.*, Southend Neighborhood Improvement Ass'n v. Cnty. of St. Clair, 743 F.2d 1207, 1210 (7th Cir. 1984). Discriminatory municipal practices that are not closely linked with housing are not covered under Title VIII. *See* Cox v. City of Dallas, 430 F.3d 734, 746 (5th Cir. 2005).

[45] *Arlington Heights*, 429 U.S. at 266.

[46] 706 F. Supp. 880 (M.D. Ga.), *aff'd sub nom.*, 888 F.2d 1573 (11th Cir. 1989), *sub. history omitted.*

[47] The plaintiffs attempted to satisfy factors articulated in *Arlington Heights*, 429 U.S. at 269–270.

[48] 706 F. Supp. at 885.

[49] *Id.* at 887.

[50] Cole (1994), p. 530. Scholars are also exploring whether Section 504 of the Americans with Disabilities Act could provide redress for environmental injustice, particularly where cumulative exposures have resulted in serious medical conditions and disability. *See e.g.*, Wilson (2022), p. 1722. The ADA has no intent requirement and imposes an affirmative duty on states and local governments not to discriminate. *Id.* at 1738. It remains unclear whether a government's "decision to subject people with disabilities to environmental conditions that exacerbate their health conditions constitutes" a violation of the ADA's reasonable accommodation requirement, "but courts have suggested that the answer could be yes." *Id.* at 1743.

[51] 42 U.S.C. § 2000d (1988).

siting permits.[52] At first blush, the CRA seemed to provide an opportunity to supplement facts relevant to traditional environmental law claims with more detail about community impacts.[53] However, in 1983, the Supreme Court severely limited the utility of this provision by holding that Section 601 of the CRA requires a showing of discriminatory intent.[54]

Advocates pivoted to Section 602 of Title VI,[55] which authorizes federal agencies to implement Section 601 through regulation.[56] Most agencies' regulations originally codified a discriminatory effect standard rather than one requiring proof of intent.[57] Despite some early successes,[58] in 2001 the Supreme Court's decision in *Alexander v. Sandoval* held that Title VI does not provide a "freestanding private right of action to enforce regulations promulgated under 602" because the CRA itself does not clearly establish such a right.[59] The ruling flew in the face of precedent and legislative history,[60] and effectively gutted Section 602 disparate impact regulations.[61]

In practice, the success of administrative and judicial Title VI 602 claims has varied greatly among agencies.[62] For example, while communities can file Title VI administrative complaints with the EPA directly, they often either languish for years[63] or are never considered.[64] Additional deficiencies include that: complainants are frequently excluded from investigations; there is no way to challenge decisions of the EPA itself; the relevant EPA office is historically understaffed; and the agency has failed to finalize formal guidance about its approach to Title VI.[65] In contrast, Title VI complaints at the Department of Housing and Urban Development and Department of Transportation are much more effectively managed and have resulted

---

[52] Mank (2008), p. 23; Outka (2005), p. 223.

[53] Cole (1994), p. 530.

[54] Guardians Ass'n v. Civil Serv. Comm'n, 463 U.S. 582, 591–604 (2d Cir. 1980).

[55] 42 U.S.C. § 2000d-1 (1994); Lado (2019), p. 281; Outka (2005), p. 224.

[56] 42 U.S.C. § 2000d-1 (1994).

[57] *See* Cole (1994), p. 532 (summarizing agency regulations).

[58] *See also* S. Camden Citizens in Action v. New Jersey Dep't of Env't Prot., 274 F.3d 771 (3d Cir. 2001); N. Carolina Dep't of Transp. v. Crest St. Cmty. Council, Inc., 479 U.S. 6 (1986).

[59] Alexander v. Sandoval, 532 U.S. 275 (2001).

[60] *Id.* at 295 & n.1 (Stevens, J. dissenting).

[61] The Court did not explicitly invalidate the regulations because they were not challenged by the petitioners, but the ruling had that effect. *Id.* at 281–282.

[62] *See, e.g.*, Dana and Tuerkheimer (2017), p. 97 (deeming EPA's approach to Title VI "ineffectual").

[63] Rosemere Neighborhood Ass'n v. EPA, 581 F.3d 1169, 1175 (9th Cir. 2009) (criticizing the EPA for "fail[ing] to process a *single* complaint … in accordance with its regulatory deadlines."); Californians for Renewable Energy v. EPA, No. 4:15-CV-03292, 2018 WL 158211 (N.D. Cal. July 15, 2015) (same); *see also* Lado (2019), p. 296.

[64] *See* Lado (2019), p. 297; Dana and Tuerkheimer (2017), p. 97.

[65] Lado (2019), p. 319.

in the withdrawal of federal funds.[66] In pursuit of progress in this uphill battle, communities have also brought Section 602 claims using 42 U.S.C. § 1983, which generally forbids the deprivation of federally guaranteed rights.[67] However, courts have held that it cannot be used to create a right of action if the underlying statute does not provide one.[68]

Title VIII

In 1968, Congress passed the FHA to eradicate discrimination within the housing sector.[69] While FHA Title VIII only addresses housing-related discrimination, it reaches private defendants[70] and does not require proof of discriminatory intent.[71]

Title VIII bars refusal to sell or rent housing to any person on the basis of race[72] and prohibits discriminatory municipal practices if they relate to housing.[73] Title VIII has been used to invalidate a variety of discriminatory zoning laws,[74] including housing or land use policies with "segregative effect."[75] As a result, Title VIII provides "a tool with which to challenge government rezoning of residential neighborhoods in communities of color to allow noxious facilities or other undesirable [sic] land uses"[76] as well as discriminatory municipal services.[77] Defendants can overcome claims of housing discrimination if their conduct is justified by a legitimate interest and no feasible alternative would result in a less discriminatory result.[78] Even if legitimate business or public interests exist, plaintiffs can still minimize

---

[66] *See, e.g.*, Civil Rights Case Decisions, Department of Transportation Federal Highway Administration, (Feb. 1, 2023), https://www.fhwa.dot.gov/civilrights/programs/title_vi/case_decisions.cfm; File a Complaint, U.S. Department of Housing and Urban Development (Feb. 1, 2023), https://www.hud.gov/fairhousing/fileacomplaint%20.

[67] 42 U.S.C. § 1983 (1974).

[68] *See S. Camden*, 274 F.3d at 790; Blessing v. Freestone, 520 U.S. 329, 340–31 (1997) (limiting what constitutes a "federal right" as those clearly defined by statute).

[69] 42 U.S.C. § 3601 (1976); Texas Dep't of Hous. & Cmty. Affs. v. Inclusive Communities Project, Inc., 576 U.S. 519, 521 (2015).

[70] *See* Lazarus (1992), p. 840.

[71] *See, e.g.*, 576 U.S. at 545–46; United States v. City of Black Jack, 508 F.2d 1179, 1184–85 (8th Cir. 1974).

[72] 42 U.S.C. § 3604(a) (1970).

[73] 42 U.S.C. § 3604(b) (1970).

[74] Town of Huntington, N.Y. v. Huntington Branch, N.A.A.C.P., 488 U.S. 15, 16–18 (1988) (striking down zoning laws forbidding multifamily unit construction); *Greater New Orleans Fair Hous. Action Center v. St. Bernard Par.,* 641 F.Supp.2d 563, 569, 577–578 (E.D. La. 2009) (invalidating zoning law that restricted rental to blood relatives in predominantly white neighborhood).

[75] *See* United States v. City of Parma, 494 F. Supp. 1049, 1055 (N.D. Ohio 1980), *aff'd*, 661 F.2d 562 (6th Cir. 1981), cert. denied, 456 U.S. 926 (1982); *see also* Been (1994), pp. 1403–1404.

[76] Cole (1994), p. 535.

[77] *See, e.g.*, 743 F.2d at 1210; *see also* Rajotte, see footnote 46, at 179.

[78] *See Ricci v. DeStefano,* 557 U.S. 557, 578 (2009); *City of Parma*, 494 F. Supp. at 1055; *Inclusive Communities Project, Inc.*, 576 U.S. at 533.

disparate impacts by identifying a less harmful alternative that still serves the regulated entity's needs.[79]

Ultimately, civil rights and constitutional claims have been insufficient to provide wide scale remedies for environmental injustice.[80] Yet these claims can be leveraged as a rallying cry to strengthen a community group, raise public awareness, create political power, and increase pressure on decision-makers.[81]

2.2 *Substantive and Procedural Environmental Law Claims*

Even statutes designed to tackle pollution across air, water, and land have failed to achieve robust success for the environmental justice movement. Their failures are well-documented by scholars. They include significant issues of regulatory design and enforcement priorities, such as focusing on "protecting the collective"[82] or more affluent areas from environmental violations, rather than vulnerable communities from intersectional hotspots of harm.[83] Nonetheless, advocates have pursued judicial remedies using citizen suit provisions available in most major environmental statutes,[84] and challenging governmental decision-making.

Substantive "Citizen" Enforcement

Federal and state environmental law in the U.S. envisages two types of enforcement cases: those brought by the government (public enforcement) and those brought by private individuals, groups, or coalitions (private enforcement, termed "citizen

[79] *Ricci,* 557 U.S. at 578.

[80] Cole (1994), p. 525 (observing that "we as a movement are not winning civil rights cases").

[81] *See, e.g.*, Cole (1994), p. 541. For an example of litigation successfully producing pressure to settle environmental justice claims, *see* Holt v. Scovill, Inc., No. 3:07- CV-00727 (M.D. Tenn. 2008), *aff'd sub nom.* Holt v. City of Dickson, 2011 WL 134249 (M.D. Tenn. Jan. 14, 2011). In *Holt*, a Black family sued government officials and private companies for polluting its well water with carcinogenic chemicals. Over the course of fifteen years, the defendants reassured the Holts that their water was safe to drink while simultaneously warning their white neighbors of possible contamination. Ultimately, every member of the Holt family fell seriously ill. After filing a complaint and successfully defending several motions to dismiss, the Holts settled their case for more than two million dollars. *See* Legal Defense Fund, *Case: Holt v. Scovill*, Economic Justice, https://www.naacpldf.org/case-issue/holt-v-scovill/) (last visited Jan. 23, 2023).

[82] Yang (2002), p. 14.

[83] *See e.g.*, CWA, 33 U.S.C. §§ 1251–1387 (2000); NEPA, 42 U.S.C. §§ 4321–4347 (1994); CAA, 42 U.S.C. §§ 7401–7671q (1994). Additionally, when major environmental statutes were written in the 1970s, lawmakers held a conventional view of the environment as wilderness rather than the place where we live. *See* Biermann (2021), p. 61; *see generally* The First National People of Color Environmental Leadership Summit: The Washington Court on Capitol Hill, Washington, D.C., October 24–27, 1991: Proceedings xiii (Charles Lee ed., 1991).

[84] *See, e.g.*, TSCA, 15 U.S.C. § 2619 (1994); CWA, 33 U.S.C. § 1910 (2000); CAA, 42 U.S.C. § 7604 (1994); RCRA, 42 U.S.C. § 6972 (1994); SWDA, 42 U.S.C. § 300j-8 (1994).

suits").[85] Congress' inclusion of citizen enforcement provisions is borne of its awareness that government under-enforcement is likely due to agency capacity and/or agency inaction.[86] With ever growing administrative responsibility and lagging resources, the EPA simply "lacks the ability to enforce all environmental laws to the maximum extent possible."[87]

Citizen suit provisions empower citizens to prosecute regulated entities for violations of some environmental laws.[88] Individuals and organizations can combat air pollution through claims against operating facilities for Clean Air Act permit violations,[89] force remediation of toxic substances at Superfund sites,[90] or stem discharges of pollutants from permitted animal feeding operations under the Clean Water Act.[91] Communities have sought to control pollution not only by suing polluters, but also government entities responsible for regulating them.[92] Plaintiffs have challenged a variety of agency actions (or inactions), including the issuance of permits,[93] failure to regulate,[94] lack of federal oversight,[95] or the sufficiency of a superfund cleanup.[96]

Still, strict procedural and substantive limitations "control private enforcement in a manner that complements rather than supplants public enforcement."[97] For example, individuals bringing private polluter cases must generally comply with strict

[85] Guana (1995), p. 4. The vast majority of environmental enforcement takes place at the state and administrative levels. *See, e.g.*, Heisler (2011), p. 4.

[86] Guana (1995), p. 40.

[87] *Id.*

[88] *See, e.g.*, CWA, 33 U.S.C. § 1910 (2000); CAA 42 U.S.C. § 7604 (1994); RCRA, 42 U.S.C. § 6972 (1994). Note that some environmental statutes do not contain citizen suit provisions.

[89] *See* Friends of Buckingham v. Virginia Air Pollution Control Bd., 947 F.3d 68 (4th Cir. 2020).

[90] CERCLA, 42 U.S.C. §§ 9601–9675 (1994); 40 CFR §§ 300–311 (2003); *see also* Office of Land and Emergency Management, EPA 502/P-21/001, EJ Action Plan: Building Up Environmental Justice in EPA's Land Protection and Cleanup Program (2022).

[91] Goeke v. Nat'l Farms, Inc., 512 N.W.2d 626 (Neb. 1994).

[92] Guana (1995), p. 41.

[93] *See, e.g.*, Pine Bluff for Safe Disposal v. Arkansas Pollution Control & Ecology Comm'n, 127 S.W.3d 509, 515 (Ark. 2003) (challenging the issuance of a permit for a chemical weapons incinerator); Chester Residents Concerned for Quality Living v. Com., Dep't of Env't Res., 668 A.2d 110 (Pa. 1995) (challenging a permit for the construction and operation of an infectious waste sterilization facility).

[94] Citizens for a Better Env't v. Costle, 515 F. Supp. 264, 270–271 (N.D. Ill. 1981) (requiring EPA to promulgate regulations under the Clean Air Act); Natural Resources Defense Council, Inc. v. Callaway, 392 F. Supp. 685, 686 (D.D.C. 1975) (holding that EPA has a duty to adopt a definition of "navigable waters" under the Clean Water Act).

[95] *Costle*, 515 F. Supp. 271 (finding that EPA has a duty to determine whether state SIPs comply with the Clean Air Act).

[96] *See generally*, Ferris (1994), pp. 678–679.

[97] Guana (1995), p. 43.

notification requirements[98] and may only bring cases when there is no preexisting diligent government prosecution.[99] Though the definition of "diligent" government action varies by circuit and statute,[100] citizen suits may not duplicate the work of public officials.[101] Further, citizen suits can also only force an agency to perform "nondiscretionary duties," such as meeting statutory deadlines and promulgating regulations.[102] Courts have generally refused to intervene when litigants target the *substance* of a decision—and therefore agency interpretation and expertise.[103]

Limited remedy provisions in the citizen suit provisions of environmental laws also render plaintiffs only able to obtain injunctive relief; in some cases, penalties are also only paid to the U.S Treasury.[104] These limitations not only control the type of remedy available, but also the scope of harm considered: citizen suits generally cannot offset past, cumulative, recurring, or potential harms.[105] In addition to these procedural and substantive hurdles, the Supreme Court has limited who can bring citizen suits to those who can prove "injury in fact" under Article III of the U.S. Constitution.[106]

Of course, successful suits against polluters and inactive government officials are also dependent upon community awareness that a violation has occurred and an ability to identify the violator.[107] This prerequisite is a fundamental barrier for com-

---

[98] *See, e.g.*, CERCLA § 9659(d)(1) (1988) (requiring 60-day notice for CERCLA enforcement suits); Hallstrom v. Tillamook County, 493 U.S. 20, 31 (1989) (holding that notice requirements should be strictly enforced).

[99] Significant caselaw exists surrounding what constitutes "diligent" prosecution. *Compare* N. & S. Rivers Watershed Ass'n v. Scituate, 949 F.2d 552, 558 (1st Cir. 1991) (precluding a citizen suit because agency proceedings had begun), *with* Friends of the Earth v. Consol. Rail Corp., 768 F.2d 57, 63 (2d Cir. 1985) (requiring a preexisting government-filed judicial proceedings to preclude a citizen suit).

[100] *Compare N. & S. Rivers*, 949 F.2d at 558 (holding that an administrative compliance action bars a citizen suit), *with Consol. Rail Corp.*, 768 F.2d at 62 (holding that an administrative consent decree does not preclude suit).

[101] For example, CERLCA citizen suits cannot proceed before the government has identified potentially responsible parties and initiated cleanup. *See generally* Gaba and Kelly (1990), p. 929; *see* 42 U.S.C. § 9607.

[102] *See, e.g.*, Env't Def. Fund v. Thomas, 870 F.2d 892, 896–900 (2d Cir. 1989), *cert. denied sub nom.* Ala. Power Co. v. Env't Def. Fund, 493 U.S. 991 (1989) (commenting on the nondiscretionary duty to make *some* decision regarding revision of sulfur dioxide NAAQS).

[103] *See, e.g.*, Wisconsin's Env't Decade v. Wisconsin Power & Light Co., 395 F. Supp. 313, 323 (W.D. Wis. 1975) (holding that EPA's identification of violations is mandatory).

[104] Clean Water Act, 33 U.S.C. § 1365(a) (1988); Clean Air Act, 42 U.S.C. § 7604(a) (1988 & Supp. V 1993); RCRA, 42 U.S.C. § 6972(a) (1988); *see also* Friends of the Earth v. Archer Daniel Midland Co., 780 F. Supp. 95, 101–02 (N.D.N.Y. 1992) (refusing to approve CWA citizen suit settlement including money payments to private environmental groups).

[105] *See, e.g.*, Gwatlney of Smithfield, Ltd. v. Chesapeake Bay Found., Inc., 484 U.S. 49, 64 (1987) (holding that NPDES does not permit citizen suits for "wholly past violations").

[106] Lujan v. Defs. of Wildlife, 504 U.S. 555, 560 (1992); *see also* TransUnion LLC v. Ramirez, 141 S. Ct. 2190 (2021).

[107] Guana (1995), p. 72.

munities with limited resources and vast information asymmetries compared to regulators or regulated entities.[108] Given the substantive, procedural, and practical hurdles, scholars critique the citizen suit framework as "plac[ing] too great a burden on local communities to self-protect against environmental harm."[109]

Public Participation & Process Cases

Most environmental laws require agency decision-makers to invite public participation before certain types of decisions are made. For example, legally mandated public participation provisions may require agencies to host open meetings or hold hearings, provide information, or invite and respond to public comments.[110] Unlike citizen enforcement suits, public participation cases attack the adequacy of an agency's decision-making *process* rather than the decision itself.

The main statute invoked for such litigation is the National Environmental Protection Act (NEPA). Under NEPA public participation is required when federal actions may result in significant environmental impacts.[111] While Congress designed NEPA to encourage government agencies to consider the environmental impacts of its actions, NEPA *does not mandate* that agencies prioritize the most environmentally friendly approach to projects.[112] It covers actions "which are potentially subject to Federal control and responsibility."[113] Therefore, the statute reaches not only government led projects, permitting, policy, and regulatory implementation, but also federally funded actions.

As a result, NEPA provides several pathways for assessing possible environmental impacts. If the proposed action is not likely to result in significant environmental impacts and falls under agency-defined categorical exclusion, no further NEPA analysis is required.[114] The vast majority of federal actions fall under categorical exclusions or are found to have no significant impact.[115]

But if the extent of a project's environmental impact is unclear, then decision-makers must perform an Environmental Assessment (EA).[116] An EA includes a high-level description of the proposed action, possible alternatives, and environmental impacts.[117] Depending on the EA's findings, agencies will either make a

---

[108] *Id.* at 46, n. 159.

[109] Outka (2005), p. 235.

[110] *See, e.g.*, FIFRA, 7 U.S.C. § 136 (2000); CWA, 33 U.S.C. §§ 1365, 1344(o), 1342(j) (2004); NEPA, 42 U.S.C. §§ 4332(c), 4368 (1994); RCRA, 42 U.S.C. § 6974.

[111] 42 U.S.C. § 4331(a). Note that all states have their own versions of NEPA, which generally mirror the federal standards. *See, e.g.*, California Environmental Quality Act (CEQA), Cal. Pub. Res. Code § 21002.

[112] Council on Env't Quality, A Citizen's Guide to NEPA: Having your Voice Heard 4 (2021).

[113] 40 C.F.R. § 1508.18 (2019).

[114] 40 CFR §1508.1(d) (2019).

[115] *See, e.g.*, Fleischman et al. (2020), p. 403 (finding that, of 33,976 decisions made by the U.S. Forest Service between 2005 and 2018, 82.3% fell under categorical exclusions, 15.8% received EAs, and for only 1.9% were EISs written).

[116] 40 C.F.R. § 1501.3 (2019).

[117] 40 C.F.R. § 1501.5(c)(2) (2019).

"finding of no significant impact"[118] or embark on a more comprehensive study, called an Environmental Impact Statement (EIS).[119] An EIS must contain significant analysis of the project's potential direct, indirect, and cumulative environmental impacts, as well as descriptions of the project's scope, schedule, and detailed alternatives.[120]

The amount of legally mandated public participation in the NEPA process increases with the intensity of the environmental study required under the Act. For example, the development of an EA calls for public involvement to the extent practicable,[121] whereas agencies preparing an EIS must publish notices, provide reliance materials and drafts, and must also accept and respond to comments at multiple points during the project's development.[122]

While NEPA cases target process, they can still achieve lasting substantive impacts.[123] Every NEPA-related decision an agency makes—including the initial categorization of impacts as "significant," as well as the sufficiency of scoping, analysis, choice of alternatives, and public participation opportunities[124]—can be challenged as "arbitrary, capricious, an abuse of discretion, or otherwise not in accordance with the law."[125] In *Standing Rock Sioux Tribe v. U.S. Army Corps of Engineers*, Tribal members challenged an Army Corps of Engineers' conclusion that the construction of a crude oil pipeline under a significant tribal water source would have no significant environmental impact.[126] Plaintiffs presented evidence that the pipeline had been moved onto tribal land out of concerns about impacts on the Bismark community's water supply. Bismark is 90% white.[127] After years of litigation and massive activism campaigns, the court finally required that the Corps perform an EIS and ultimately vacated the agency's approval of the pipeline.[128]

---

[118] 40 C.F.R. § 1501.4 (2019).

[119] 40 C.F.R. § 1502.1–1502.25 (2019).

[120] *Id.*

[121] 40 C.F.R. § 1501.5(e) (2019).

[122] *See* 40 CFR §1502 (2019).

[123] Outka (2005), p. 235.

[124] *See, e.g.*, Center for Cmty. Action & Env't Just. v. Fed. Aviation Admin., 18 F.4th 592, 597–98 (9th Cir. 2021) (challenging FAA's decision to forgo an EIS before constructing large air cargo facility); Ctr. for Biological Diversity v. U.S. Forest Serv., 349 F.3d 1157, 1168 (9th Cir. 2003) (challenging agency's failure to discuss opposing views in EIS); Nat'l Wildlife Fed'n v. Andrus, 440 F. Supp. 1245, 1253 (D.D.C. 1977) (finding cursory admission of impacts insufficient); Kentucky Riverkeeper, Inc. v. Rowlette, 714 F.3d 402, 411 (6th Cir. 2013) (finding that past actions and impacts must be considered); Am. Marine Rail, LLC, 2000 WL 1299571 (N.Y. Dep't of Env't Conservation) (Aug. 25, 2000) (requiring an EIS for permitting a solid waste transfer program).

[125] NEPA provides no independent right of action, these suits are brought under the Administrative Procedure Act, 5 U.S.C § 706(2)(A) (2000).

[126] 205 F. Supp. 3d 4, 13 (D.D.C. 2016); *see also* Standing Rock Sioux Tribe v. U.S. Army Corps of Eng'rs, 255 F. Supp. 3d 101, 140 (D.D.C. 2017); Standing Rock Sioux Tribe v. U.S. Army Corps of Eng'rs, 440 F. Supp. 3d 1, 9 (D.C. Cir. 2020).

[127] *See* Fredericks et al. (2019), p. 569.

[128] Standing Rock Sioux Tribe v. U.S. Army Corps of Eng'rs, 985 F.3d 1032 (D.C. Cir. 2021).

The sufficiency of public participation opportunities can also be challenged, including opportunities to comment and the quality of agency response.[129] For example, in *El Pueblo Para el Aire y Agua Limpio v. County of Kings*, community members of a 95% Spanish speaking county challenged the siting of a toxic waste incinerator by arguing that the county violated public participation provisions when it failed to publish its decision-making documents in Spanish.[130] The court overturned the county's approval of the incinerator.[131]

Because NEPA requires consideration of impacts on the "human environment," communities may also object to agency decision-making for its failure to consider economic, social, cultural impacts, and cumulative impacts. While non-environmental effects alone cannot trigger an EIS requirement,[132] once an EIS must be prepared based on natural or physical environmental impacts, all *interrelated* and *foreseeable* economic, social, and cultural effects must also be analyzed.[133] Courts have construed "human environment" broadly to include an array impacts, including consideration of how projects may change a community's character[134] and quality of life,[135] effects on culturally significant wildlife,[136] as well as job loss and lost revenue.[137] However, the extent to which NEPA analysis must consider how a project contributes to the cumulative impacts felt by overburdened communities is unsettled.[138]

[129] *See, e.g.*, State of Cal. v. Block, 690 F.2d 753, 773 (9th Cir. 1982) (finding that be "good faith" response to comments); Pyramid Co. of Watertown v. Plan. Bd. of Town of Watertown, 24 A.D.3d 1312, 1315 (N.Y. App. Div. 2005) (EIS inadequate for failing to include supporting data in response to comment about non-environmental impacts).

[130] 22 Envtl. L. Rep. (Envtl. Law Inst.) 20,537 (Cal. Sup. Ct. Dec. 30, 1991).

[131] *Id.*

[132] 40 C.F.R. § 1508.14 (2019); *see also* One Thousand Friends of Iowa v. Mineta, 250 F. Supp. 2d 1064, 1072 (S.D. Iowa 2002) (finding that environmental justice "does not alone give rise to judicial review by this or any other court.").

[133] 40 C.F.R. § 1508.14 (2019); 40 C.F.R. § 1508.1(g)(1) (2020); Nat'l Ass'n of Gov't Emp. v. Rumsfeld, 418 F. Supp. 1302, 1306 (E.D. Pa. 1976) ("[W]hen a federal action does have a significant environmental impact, social and economic impacts must also be considered...."); *see, e.g.*, Hammond v. Norton, 370 F. Supp. 2d 226, 243 (D.D.C. 2005) (defining interrelatedness under NEPA); Sierra Club v. Marsh, 976 F.2d 763, 767 (1st Cir. 1992) (defining foreseeability under NEPA).

[134] Pyramid Co. of Watertown v. Plan. Bd. of Town of Watertown, 24 A.D.3d 1312, 1315 (N.Y. App. Div. 2005).

[135] Hanly v. Mitchell, 460 F.2d 640, 647 (2d Cir. 1972).

[136] *See, e.g.*, Protect Our Cmty. Found. v. LaCounte, 939 F.3d 1029, 1040–41 (9th Cir. 2019).

[137] *See, e.g.*, Barrie v. Kitsap Cnty., 613 P.2d 1148, 1157 (Wash. 1980).

[138] *But see infra* Sect. 3.3. Under traditional cumulative impact analysis, agencies were required to consider the addition of its project to the "aggregate effects of past actions" in the same area. Memorandum from James L. Connaughton, Chairman, Council on Environmental Quality on Guidance on the Consideration of Past Actions in Cumulative Effects Analysis to Heads of Federal Agencies (June 24, 2005), http://energy.gov/nepa/downloads/guidance-consideration-past-actions-cumulative-effects-analysis.

In 1994, Executive Order (EO) 12898 strengthened NEPA's "human environment" inquiry.[139] The Order directs each federal agency to incorporate environmental justice into its mission and consider the impact of its "programs, policies, and activities on minority populations and low-income populations…."[140] Cases challenging the sufficiency of an agency's environmental justice analysis under NEPA commonly cite EO 12,898.[141]

As a threshold matter, courts require that agency consideration of environmental justice impacts include demographic analysis to identify the presence of under-resourced communities or communities of color.[142] When such communities are present, agencies must also consider the "interrelated cultural, social, occupational, historical, or economic factors that may amplify" the environmental effect of a proposed action[143] and heighten "attention to alternatives (including alternative sites), mitigation strategies, monitoring needs, and preferences expressed by the affected community or population."[144] Unfortunately, once an agency has performed this analysis, it is difficult to challenge the results: courts generally defer to agency conclusions regarding the extent of environmental justice impacts.[145]

For all its virtues, public participation and robust environmental analysis alone cannot cure environmental injustice. NEPA and E.O. 12898, as well as many other participation-oriented regulations, prescribe *process, not outcome*. Instead, they require fair consideration of relevant facts, leaving communities to hope government officials will make environmentally responsible decisions. These claims also place immense responsibility on the public to protect themselves.[146]

As with citizen enforcement cases, claims under NEPA require time, resources, education, and mobilization to succeed, all of which raise significant barriers for historically marginalized and overburdened populations. In addition, under facially neutral environmental laws, "[g]overnment action with significant discriminatory

---

[139] Federal Actions to Address Environmental Justice in Minority and Low-Income Populations, Exec. Order No. 12, 898, 59 Fed. Reg. 7629 (Feb. 11, 1994), amended by Exec. Order No. 12,948, 60 Fed. Reg. 6381 (Jan. 30, 1995); *see also* Tackling the Climate Crisis at Home and Abroad, Exec. Order No. 14,008, 86 Fed. Reg. 7619, 7629–30 (Jan. 27, 2021) (amending E.O. 12,898 to create a "White House Environmental Justice Interagency Council").

[140] *Id.* § 1-101; *see also* Villa (2022), p. 163.

[141] *See* Hausrath v. U.S. Dep't of the Air Force, 491 F. Supp. 3d 770, 797 (D. Idaho 2020) (concluding agency's consideration of environmental justice impacts "too cursory.").

[142] Mid States Coal. for Progress v. Surface Transp. Bd., 345 F.3d 520, 541 (8th Cir. 2003).

[143] Council on Env't Quality, Environmental Justice: Guidance Under the National Environmental Policy Act 9 (1997) [hereinafter CEQ Guidance].

[144] *Id.* at 10.

[145] *See, e.g.*, Communities Against Runway Expansion, Inc. v. FAA, 355 F.3d 678, 689–90 (D.C. Cir. 2004) (upholding the sufficiency of agency's demographic analysis even though the size of the area considered diluted potential impacts on communities of color closest to the project); In re Annova LNG Common Infrastructure, LLC, 2020 WL 865088, at *10 (FERC Feb. 21, 2020) (upholding facility's citing where "all project-affected populations are minority or low-income populations, or both" over challenge under NEPA).

[146] Outka (2005), p. 237.

effects may often be in legal compliance with environmental standards."[147] The inability of environmental laws to address cumulative environmental burdens suffered by disadvantaged communities reflects gaps within the laws' scope, a problematically narrow idea of what constitutes the "environment," as well as procedural deficiencies.

2.3 *Tort Claims*

American tort law is borne of a common law system that originated in England and predates the United States' founding. Before the 1970s, state tort actions were the primary legal vehicle available to remedy environmental wrongs in the United States.[148] Tort laws were written to protect individuals' bodies and land from harm and restore "parties to their original condition" following injury caused by an outside party.[149] Tort is facially appealing to the environmental justice movement because it "has long recognized the preeminence given to the right of an individual, regardless of race or socio-economic status, to be free of uninvited, non-consensual, harmful, or offensive conduct."[150]

Today, tort law in the environmental context is often seen as a "gap filler"[151] that is needed in the absence of sufficient regulatory remedies.[152] However, in the remedial sphere for environmental injustice, regulatory gaps abound. As a result, scholars and practitioners alike test the boundaries of tort law to determine to what extent it can be used as a tool for overburdened communities.[153] Though trespass, negligence, and battery have all been discussed as potential avenues for environmental justice litigation,[154] nuisance claims have proven most successful.[155]

Nuisance law forbids "unreasonable interferences" with the use, enjoyment, or value of real property and applies to private property as well as public rights.[156] Private nuisance claims revolve around interference with the use and enjoyment of privately owned land and can be used to remedy a variety of environmental harms including air pollution, noise, foul odor, or even exposure from nearby pollutants

---

[147] Outka (2005), p. 233.

[148] *See* Latham et al. (2011), p. 737.

[149] Schwartz (2010), pp. 1–2.

[150] Northern (1997), p. 492.

[151] *See* Connecticut v. Am. Elec. Power Co., 582 F.3d 309, 330 (2d Cir. 2009), *rev'd*, 564 U.S. 410 (2011).

[152] *See* Latham et al. (2011), p. 739.

[153] *See, e.g.*, Northern (1997), pp. 557–577 (suggesting the expansion of environmental torts to include battery); Brennan (1993), pp. 44–48 (considering the development of toxic torts); *but see* Latham et al. (2011), p. 746 (discussing the limitations of tort in the environmental context).

[154] Latham et al. (2011), pp. 750–751.

[155] *See e.g.*, Baptiste v. Bethlehem Landfill Co., 965 F.3d 214 (3d Cir. 2020).

[156] Restatement (Second) of Torts §§ 821B, 821D (1979).

that are not on a claimant's land.[157] Public nuisance results from an "unreasonable interference" with a public right, which can include the right to unpolluted air and waterways.[158] Unreasonable interference broadly includes "conduct or omissions that offend, interfere with or cause damage to the public in the exercise of rights common to all in a manner such as to offend public morals, interfere with use by the public of a public place or endanger or injure the property, health, safety or comfort of a considerable number of persons."[159]

Recent environmental justice lawsuits demonstrate the potential utility of these claims. For example, in 2020 the Third Circuit reversed and remanded a motion to dismiss regarding a coalition's (including communities of color) private and public nuisance claims based on the release of noxious odors, pollutants, and other contaminants from a landfill.[160] A decade earlier, the Bronx-based environmental justice organization Mothers on the Move, leveraged nuisance claims to shutter a sludge fertilizer plant after years of activism.[161]

Remedies for public nuisance include damages as well as equitable relief. However, the award of damages for nuisance claims can be limited to cases where the gravity of the injury outweighs the utility of the conduct that caused the injury.[162] To establish the utility of polluting conduct, defendants can point to the social value of their business, its importance to local character, or even the "impracticality of preventing or avoiding the invasion."[163]

Torts law's ability to handle the complexity of environmental justice is limited.[164] Nuisance law is not able to adequately remedy problems of complex causation, including background hazards present in the community, workplaces, and homes. Torts are also strictly retrospective: they cannot be used to prevent future harm or improve environmental conditions.[165] Nonetheless, tort law offers a litigation tool in

[157] *See e.g.,* Restatement (Second) of Torts § 822 cmt. a (1979) (defining types of private nuisance); Baptiste, 965 F.3d at 222–24 (successful environmental justice suit against landfill operator for air pollution, contaminants, and odor); Scribner v. Summers, 84 F.3d 554 (2d Cir. 1996) (barium leak from adjacent property); Boomer v. Atl. Cement Co., 257 N.E.2d 870, 874 (N.Y. 1970) (airborne cement dust); Moon v. N. Idaho Farmers Ass'n, 96 P.3d 637, 640 (Idaho 2004) (grass smoke). *See also* Rendleman (2018), p. 1859.

[158] Restatement (Second) of Torts § 821B (1979); Northern (1997), p. 551 (noting that some states consider pollution of public waters or open air a public nuisance).

[159] State v. Schenectady Chems., Inc., 459 N.Y.S.2d 971, 976 (N.Y. Sup. Ct. 1983) (citations omitted), *aff'd as modified,* 479 N.Y.S.2d 1010 (N.Y. App. Div. 1984).

[160] Baptiste, 965 F.3d at 224 (finding petitioners allegations were sufficient to state claims for nuisance).

[161] Patrick Rocchio, *MOM files lawsuit against HP businesses*, Bronx Times (July 18, 2008), https://www.bxtimes.com/mom-files-lawsuit-against-hp-businesses/.

[162] Restatement (Second) of Torts § 826 (1979). There are exceptions to this rule if the injury suffered is unusually serious and damages will not bankrupt the offending company. *Id.* at § 827.

[163] *See* Restatement (Second) of Torts §§ 828 a-c (1979).

[164] Todd (2020), p. 181.

[165] Latham (2011), p. 754.

straightforward cases[166] and is an important component of the legal patchwork available to pursue environmental justice.

3 Promise and Peril: Emerging Tools and Challenges for Environmental Justice Litigation

The sporadic success of past environmental justice claims may be poised for transformation. A confluence of factors has endowed the mission of environmental justice with an overdue sense of urgency.[167] Violence against Black and Asian Americans, the continued disempowerment of Indigenous people, and the ravaging of communities of color by COVID-19 spurred a policy transformation that decades of work within the movement had well positioned for success.[168] President Biden's definitive commitment to environmental justice further amplified its rise in social and political discourse. Novel initiatives and authorities to support environmental justice litigation are emerging across federal and subnational dimensions. Yet, threats to this progress loom from recent Supreme Court jurisprudence on administrative agency authority and race-conscious decision-making.

3.1 Federal Progress Under the Biden Administration

Under the Biden administration, the Federal government has elevated the pursuit of environmental justice by establishing new initiatives and revamping existing—and in some cases, languishing—authorities. The output is singularly ambitious among presidential administrations and this Chapter highlights only some of the policies likely to influence environmental justice litigation. In his first days in office, President Biden issued Executive Orders 13985 (Advancing Racial Equity and Support for Underserved Communities Through the Federal Government) and 14008 (Tackling the Climate Crisis at Home and Abroad). Together, these Orders direct federal agencies to: "work toward proactively achieving environmental justice[,] …develop and implement policies and strategies that strengthen compliance and enforcement,…increase community engagement, and ensure that… [40% of] benefits from federal investments in climate and clean energy flow to underserved

166 Latham (2011), pp. 753–754.

167 *See, e.g.*, Roberts et al. (2022); Bullard (2021), p. 243.

168 *See, e.g.*, Sprainer (2022), p. 123; Politico, *An 'urgent' crisis: City's study of environmental inequities gains new momentum amid Covid*, https://www.politico.com/states/new-york/albany/story/2021/11/08/an-urgent-crisis-citys-study-of-environmental-inequities-gains-new-momentum-amid-covid-1392421 (11/08/21).

communities."[169] EPA has also publicly identified and updated a comprehensive set of legal authorities to pursue environmental justice goals ("EPA Legal Tools to Advance Environmental Justice" (May 2022))[170] and for the first time issued guidance on cumulative impacts ("Cumulative Impacts Addendum" (January 2023)).[171]

Title VI

Notable among these efforts is the creation of EPA's new Office of Environmental Justice and External Civil Rights, which merges previously disparate offices and consolidates more than 200 staff in service of advancing environmental justice.[172] This Office will not only handle but is expected to revitalize, Title VI enforcement.[173] In the first two years of the Biden administration, DOJ and EPA have accepted multiple environmental justice-related civil rights matters. Examples include investigations into the Alabama Department of Health in support of historically Black Lowndes County communities suffering unconscionable wastewater and public health crises;[174] and Louisiana's state air and water permitting agencies.[175]

In August 2022, EPA also released interim guidance on "Environmental Justice and Civil Rights in Permitting" for state offices. The guidance, while not legally binding, indicates that "[i]f [a] permitting action will have a disparate impact on the basis of race, color, or national origin…" without a "legally sufficient justification

---

[169] Press Release, Environmental Protection Agency, EPA Finalizes Environmental Action Plan for Land Protection Cleanup Programs (Sept. 30, 2022) (available at https://www.epa.gov/newsreleases/epa-finalizes-environmental-justice-action-plan-land-protection-and-cleanup-programs) (detailing the new Office of Land and Emergency Management EJ Action Plan and situating it within the broader environmental justice efforts across EPA and DOJ in response to EO's 13,985 and 14,008, including the Biden administration's Justice40 Initiative).

[170] Office of General Counsel, No. 360R22001, EPA Legal Tools to Advance Environmental Justice (2022), https://www.epa.gov/ogc/epa-legal-tools-advance-environmental-justice.

[171] Office of General Counsel, No. 360R22002, EPA Legal Tools to Advance Environmental Justice: Cumulative Impacts Addendum (2023), https://www.epa.gov/system/files/documents/2022-12/bh508-Cumulative%20Impacts%20Addendum%20Final%202022-11-28.pdf.

[172] Press Release, Environmental Protection Agency, EPA Launches New National Office Dedicated to Advancing Environmental Justice and Civil Rights (Sept. 24, 2022) (available at https://www.epa.gov/newsreleases/epa-launches-new-national-office-dedicated-advancing-environmental-justice-and-civil).

[173] The necessity of overhaul to the Title VI program is clear from Section II herein, and is well-detailed in scholarship. *See, e.g.*, Lado (2019), p. 281; *see also* Glenn (2017), p. 45.

[174] *See* Press Release, Department of Justice, Justice Department Announces Environmental Justice Investigation into Alabama Department of Public Health and Lowndes County Health Department (Nov. 9, 2021) (available at https://www.justice.gov/opa/pr/justice-department-announces-environmental-justice-investigation-alabama-department-public) To understand the scale of the health crises fueled by the wastewater sanitation issues in Alabama, *see* Flowers (2020).

[175] Letter from Lilian S. Dorka, Deputy Assistant Administrator for External Civil Rights, Environmental Protection Agency to Dr. Chuck Carr Brown, Secretary, Louisiana Department of Environmental Quality & Dr. Courtney N. Phillips, Secretary, Louisiana Department of Health (October 12, 2022) (available at https://www.epa.gov/system/files/documents/2022-10/2022%2010%2012%20Final%20Letter%20LDEQ%20LDH%2001R-22-R6%2C%2002R-22-R6%2C%2004R-22-R6.pdf).

for the disparate impacts, denial of the permit may be the only way to avoid a Title VI violation."[176] For all its long-standing deficiencies at EPA, Title VI may be at the precipice of renewed utility.

Environmental Enforcement and Public Participation

In a parallel effort to further institutionalize considerations of environmental justice, DOJ has drafted a Comprehensive Environmental Justice Enforcement Strategy. The strategy deploys enforcement resources towards disparate pollution burdens and aims to "provide timely remedies for systemic environmental violations."[177] Pursuant to this strategy, enforcement-initiating events (such as inspections) and enforcement matters (administrative, civil judicial, and criminal) will be more frequently pursued in overburdened communities.[178]

Now, government enforcers must attempt to include types of relief that provide redress to communities harmed from violations of law, such as supplemental environmental projects (SEPs) and mitigation.[179] SEPs are projects that a defendant voluntarily agrees to undertake through settlement. SEPs "obtain environmental and public health … benefits that may not otherwise have occurred in the settlement of an enforcement action."[180] Mitigation is restorative equitable relief that offsets or remedies harm from past violations and is available in litigation and settlements.[181] DOJ and EPA included SEPs and mitigation in enforcement matters for decades, but under President Obama, EPA first memorialized their utility for communities with environmental justice concerns.[182] However, their use was optional, underutilized, and included no mandate for community consultation.[183] The Environmental Justice Enforcement Strategy, however, mandates that attorneys develop and implement a "community outreach plan."[184] Enforcement matters informed by community

[176] Office of General Counsel, Interim Environmental Justice and Civil Rights in Permitting Frequently Asked Questions at 14–15 (2022).

[177] Memorandum from the Associate Att'y Gen. to Heads of Dep't Components & U.S. Att'ys (May 5, 2022) (available at https://www.justice.gov/d9/pages/attachments/2022/05/05/02._asg_strategy_memorandum.pdf) [hereinafter Environmental Justice Enforcement Strategy Memo]. *See also* Exec. Order No. 14,008 § 222(c)(ii), 86 Fed. Reg. 7,619, 7,631 (Jan. 27, 2021).

[178] Environmental Justice Enforcement Strategy Memo at 2.

[179] *Id.* at 5. *See also* Colangelo (2021), p. 4. Citizen enforcement suits have also secured these types of relief. *See, e.g.*, *Holding Polluters Accountable*, Conservation Law Foundation (2022) https://www.clf.org/strategies/environmental-enforcement/ (describing effort to incorporate SEPs into enforcement); United States v. DTE Energy Co., 845 F.3d 735 (6th Cir. 2017), *cert. denied* 138 S.Ct 555 (2017).

[180] Env't Protection Agency, Supplemental Environmental Projects Policy 2015 Update 1 (2015).

[181] Memorandum from Susan Shinkman, Director, Office of Civil Enforcement to Regional Counsels, Environmental Protection Agency (Nov. 14, 2012) (available at https://www.epa.gov/sites/default/files/2016-08/documents/2ndeditionsecuringmitigationemo.pdf).

[182] *Id.*

[183] Environmental Justice Enforcement Strategy Memo at 4–5.

[184] *Id.* at 6–7.

narratives and data will facilitate a more transparent process, and one more respectful of community expertise, interests, and concerns.[185]

Finally, the participation process for environmental justice claims described above (Sect. 3.2) are also likely to evolve with recent and forthcoming changes to NEPA. In 2022, the Council on Environmental Quality (CEQ) published a "Phase I" final rule that restored much of the language to NEPA regulations previously revised by the Trump administration.[186] Experts view the Phase I rule as a path towards reintegrating and amplifying the role of environmental justice and climate change in environmental analyses.[187]

3.2 *State Environmental Justice Statutes and State Law-Based Claims*

While much attention is focused on new federal environmental justice efforts, subnational progress continues. In the absence of comprehensive federal action on environmental justice, particularly during former President Trump's administration, states and municipalities acted and innovated across legislative, regulatory, and enforcement arenas.[188]

Novel and Emerging State Authorities

Recent years have seen a marked uptick in states proposing and enacting legislation with environmental justice goals or standards.[189] According to one report, by October 2021 only 14 states remained without "environmental justice initiatives or

---

[185] *See, e.g.*, Colangelo (2022), pp. 356–360 (urging more significant involvement of communities in government enforcement matters and providing public engagement strategies). *See generally* Hoffer (2003), pp. 1000–1002 (describing the benefits of public participation in environmental decision-making).

[186] 87 Fed. Reg. 23453 (Apr. 20, 2022); The White House, *CEQ Restores Three Key Community Safeguards during Federal Environmental Reviews*, https://www.whitehouse.gov/ceq/news-updates/2022/04/19/ceq-restores-three-key-community-safeguards-during-federal-environmental-reviews/ (Apr. 19, 2022).

[187] *Id.*

[188] *See, e.g.*, Press Release, State of New Jersey Office of the Attorney General, DEP File 12 New Environmental Justice Lawsuits Targeting Polluters in New Jerseys Lower-Income and Minority Communities (August 27, 2020) (available at https://www.njoag.gov/attorney-general-dep-file-12-new-aeoeenvironmental-justiceae%C2%9D-lawsuits-targeting-polluters-in-new-jerseys-lower-income-and-minority-communities/) ("[I]n the last year, New Jersey filed 12 lawsuits targeting polluters across New Jersey whose actions threaten the health and safety of residents in minority and lower-income communities….")

[189] *See, e.g.*, Colorado (HB19-1266, 2021); Washington (SB 5141, HEAL Act 2022); Vermont (S.148, 2022); Dylan Bruce, *ANALYSIS: State Laws are Codifying Environmental Justice*, Bloomberg Law (March 9, 2021), https://news.bloomberglaw.com/bloomberg-law-analysis/analysis-state-laws-are-codifying-environmental-justice. States with relevant Executive Orders include Wisconsin (E.O. 161, 2022) and North Carolina (E.O. 271, 2022).

regulations."[190] Novel or revised laws contain provisions addressing permitting, public participation, cumulative impacts, and/or climate justice[191]—with many now providing concrete mechanisms to enforce these laws and challenge decisions such as permitting.

For example, New Jersey's recent environmental justice law is likely now the nation's most protective regulatory regime. It mandates denial of a permit if cumulative impacts on communities with environmental justice concerns are identified: the cumulative impact assessment is not only for newly proposed facilities, but also for permit renewals. [192] This expands the New Jersey Department of Environmental Protection's ability to regulate aggregate risks posed by multiple facilities. [193] New York, Massachusetts, Vermont, and Washington State also now require consideration of cumulative impacts, with the laws already affecting permits and projects.[194]

Progressive State Law-Based Litigation

State-based legal authorities have proven critical to recent advances in environmental justice litigation. In a widely hailed environmental justice victory, residents and environmental advocates sued the Virginia State Air Pollution Control Board (the Board) over a permit issued for a gas pipeline compressor station set for construction in a historically Black community. The Union Hill community is largely inhabited by descendants of former slaves.[195] Petitioners based their arguments on the 2006 Commonwealth Energy Policy, which mandates that the "development of new, or expansion of existing, energy resources or facilities [] not have a disproportionate adverse impact on economically disadvantaged or minority communities."[196] Petitioners also leveraged the Virginia air permitting rules that require the Board to consider the suitability of the compressor station to the Union Hill neighborhood, as

---

[190] Integral Consulting, Compendium of State Regulatory Activities on Environmental Justice (2021), Integral-State-EJ-Compendium.pdf.

[191] Environmental justice organizations view climate justice as "addressing the climate crisis in a just and equitable way," typically focusing on a just transition to a low carbon economy and considering the needs of communities on the frontlines of greenhouse gas producing industry. *Climate Justice Working Groups*, WeAct for Environmental Justice (2023), weact.org/home-3-2-2-2/getinvolved/mermership/cjwg/.

[192] *See* A.B. 2212, 219th Leg. (N.J. 2020); S.B. 232, 219th Leg. (N.J. 2020).

[193] *See id.* Importantly, the law also requires public hearings throughout the permitting assessment process.

[194] *E.g.*, Michael P. Norton, *DEP official cites 'societal context' in East Springfield biomass plant ruling*, WBUR Local Coverage (Dec. 1, 2022), https://www.wbur.org/news/2022/12/01/palmer-renewable-energy-permit-revoked-upheld; Renata Stiehl, *NYS DEC Denies Air Permit Renewal for Greenridge Generation*, WENY News (June 30, 2022), https://www.weny.com/story/46797720/nys-dec-denies-air-permit-renewal-for-greenidge-generation. California, Rhode Island, and Illinois are also currently considering cumulative impacts legislation as additions to their current environmental justice efforts. *See* A.B. 649, Reg. Sess. (Cal. 2022); S.B. 2087, Reg. Sess. (R.I. 2022); H.B. 3090, 102nd Gen. Assemb. (Ill. 2022).

[195] *See* Amicus Brief for Petitioners, Friends of Buckingham v. Virginia Air Pollution Control Bd., 947 F.3d 68 (4th Cir. 2020) (No. 19-1152), 2019 WL 2462720, at 9–18.

[196] Va. Code Ann. § 67-102(A)(8).

well as potential health harms.[197] They argued that the Board failed to assess the project's potentially disproportionate health impacts on Union Hill and failed to independently evaluate the suitability of the site for the compression station.[198] The Board countered that its reliance on EPA's demographic and environmental data visualization tool to identify the character of the affected community was sufficient.[199] It further argued that it had determined there would not be significant adverse impacts because air pollutants in the community would remain below a set of nation-wide air pollution standards.[200]

But the unanimous Fourth Circuit decision admonished that "environmental justice is not merely a box to be checked."[201] The court found that the Board's analysis was deficient because it failed to make formal findings regarding whether Union Hill was an environmental justice community.[202] The Board's reliance on the national air quality standards—which did not analyze the risk that the specific emissions from the compressor station would present to the Union Hill community—was also deemed insufficient.[203]

Advocates are also testing other litigation strategies, including state constitutional claims, to challenge siting and permits with mixed results.[204] In 2020, the environmental justice organization Rise St. James used language from the Louisiana Constitution to challenge an air operating permit issued by the Louisiana Department of Environmental Quality for proposed chemical plants. The locations sit between two historically Black communities within a stretch of land referred to as "Cancer Alley," the densest concentration of pollution-generating industrial facilities in the country.[205] The organization won in the lower court, but lost on appeal.[206]

Further, there have also been recent matters that illustrate the effect of heightened social and political will around environmental justice. For example, in 2020 a scrap metal shredding facility (General Iron) in a predominantly white, affluent area of Chicago attempted to relocate to a predominantly Latino neighborhood. The

---

[197] Va. Code Ann. § 10.1-1307(E).

[198] *Friends of Buckingham*, at 4.

[199] *Id.* at 38.

[200] *Id.* at 41.

[201] *Id.* at 34.

[202] *Id.* at 92. *But see* Town of Weymouth v. Mass. Dep't of Env't Prot., 961 F.3d 34, 54–55 (1st Cir. 2020) (interpreting the Massachusetts Department of Environmental Protection's environmental justice analysis as going beyond what was required by the now-replaced state Environmental Justice Policy over challenge from disadvantaged community).

[203] *Friends of Buckingham*, 86, 90–91.

[204] States with explicit environmental constitutional protections include New York (N.Y. Const., Art. 1, Sec. 19.), Pennsylvania (P.A. Const., Art 1, Sec. 27), Montana (M.T. Const., Art. 2, Sec. 3), and Hawaii (H.I. Const., Art. XI, Sec. 9).

[205] *See also* Rise St. James v. La. Dep't of Evn't Quality, No. C-694029 (La. Dist. Ct. Dec. 14, 2020), *overruled* Rise St. James v. La. Dep't of Env't Quality, No. 2021 CW 0032, 2021 CW 0037 (*consolidated*) (La. Ct. App. Mar. 15, 2021).

[206] *Id.*

company met with an onslaught of environmental activism after it applied for an air pollution control construction permit from the Illinois Environmental Protection Agency (IEPA). After the community highlighted the company's poor compliance track record to IEPA without success,[207] hunger strikes ensued.[208] Several organizations filed a complaint with HUD's Fair Housing and Equal Opportunity Office, alleging that General Iron's proposed relocation was an example of Chicago's zoning and land-use practices that discriminate against people of color.[209] The plaintiffs also filed a Title VI complaint with EPA against the IEPA. Following investigations by HUD and EPA, along with potent social and political focus, the Chicago Department of Public Health ultimately denied General Iron a permit.

3.3 *Other Emerging Influences on Environmental Justice Litigation*

These recent developments at the federal and state levels implicate several key questions. How will the permit process evolve to address not only the siting of new facilities but also more "routine administrative extension" and the "reissuance of environmental permits[?]"[210] How will the process provide transparency and engagement to stakeholders? How will threat multipliers such as climate change and cumulative impacts shape litigation and broader movement strategies? The answers to some of these questions will be influenced by other legal developments discussed below.

Climate Litigation

The recent proliferation of climate-related lawsuits may influence the arc of environmental justice. One vein of climate suits involves states and municipalities seeking relief from major energy (and greenhouse gas) producers.[211] Though these cases are not styled as environmental justice claims, certain remedies would benefit communities on the frontlines of the public health harms from greenhouse gas

---

[207] The facility was previously the subject of numerous complaints and was already under an Administrative Consent Order with EPA by 2019. Still, the Illinois Environmental Protection Agency approved the permit. *See* Maxwell Evans, *General Iron's Move to East Side Approved by State Regulators Despite Community Outrage*, Block Club Chicago (June 25, 2020, 5:00 PM), https://blockclubchicago.org/2020/06/25/state-environment-regulators-approve-general-irons-move-to-east-side/.

[208] Adam Mahoney, *Pollution is so bad in this Chicago neighborhood, people are on hunger strike to stop it*, Grist (Mar. 1, 2021), https://grist.org/justice/chicago-hunger-strike-scrapyard-pollution-environmental-justice/.

[209] *See, e.g.*, Complaint, Southeast Environmental Task Force, et al. v. City of Chicago, HUD File No. 05-20-0419-6/8/9.

[210] Henderson et al. (2021) *21.

[211] Burkett (2019–2020), pp. 146–149.

producing industries.[212] Multiple suits have demanded funds for repairing or preparing for climate-driven infrastructure harms such as storm surge and flooding. Under-resourced communities, tribes, and communities of color are disproportionately concentrated in flood prone areas and face significant health and safety risks during increasingly common flood events.[213] As U.S.-based climate litigation proceeds, it remains to be seen whether litigants will pursue strategies that conceive of the climate crisis "not merely as a scientific quandary," but as a "civil and human rights crisis."[214]

Lawsuits associated with forced climate displacement of Indigenous communities may also be on the horizon.[215] Claims stemming from physical displacement of tribal members, economic loss related to land and livelihood, and other damages of cultural and religious significance might arise under several types of law.[216] Members of federally recognized and non-recognized tribes might pursue the Federal government, state governments and connected permitting agencies, and/or energy and climate change-inducing greenhouse gas producers. They are also leveraging human rights-oriented forums to highlight these issues and seek government accountability.[217]

Environmental Justice Information Influx

The marked influx in environmental justice information through community science and screening tools, among other streams, will also play an increasingly important role in most types of environmental justice litigation cataloged herein. Community science (sometimes termed citizen science or participatory science) "engages the

[212] *Id.*

[213] Tate et al. (2021), p. 453 (finding that communities of color are overrepresented in flood-prone areas); US Water Alliance, Water Rising: Equitable Approaches to Urban Flooding 15–16 (2020), http://www.uswateralliance.org/sites/uswateralliance.org/files/publications/Water%20Rising%20paper.pdf.

[214] Kim (2021), p. 416.

[215] *See, e.g.*, U.S. Government Accountability Office, GAO-04-142, Alaska Native Villages: Most Are Affected by Flooding and Erosion, but Few Qualify for Federal Assistance 3 (2003); *Complaint on Rights of Indigenous Peoples in Addressing Climate-Forced Displacement*, UN Special Rapporteurs *complaint submitted* (Jan. 15, 2020) (available at http://climatecasechart.com/wp-content/uploads/sites/16/non-us-case-documents/2020/20200116_USA-162020_complaint.pdf).

[216] Theories of liability might sound in tort, property, and various administrative claims. For example, over the last several decades oyster lessees in Louisiana have successfully brought claims against oil and gas companies for damaging oyster beds and oyster harvests using theories of trespass. *See, e.g.*, Alexis v. Hilcorp Energy Co., 493 F. Supp. 3d 497 (E.D. La. 2020). Notably, the Ninth Circuit has held that the Clean Air Act displaced a native Alaskan tribe's public nuisance claims under federal common-law against energy companies for climate change-related damages. *See* Native Vill. of Kivalina v. ExxonMobil Corp., 696 F.3d 849, 857–58 (2012).

[217] *See* Written Submission of Inter-American Commission on Human Rights, *In Support of Thematic Hearing Addressing Forcible Climate Displacement of Indigenous Communities in the United States* (October 28, 2022) (on file with authors).

public in advancing scientific knowledge by formulating research questions, collecting data, and interpreting results."[218]

Community science has the potential to be a powerful driver of change. It empowers residents of overburdened communities to share their expertise and define their environmental concerns.[219] By validating a community's lived experience, these data can also create leverage to persuade decision-makers to act.[220] As the tools used to gather information become more accessible from a financial standpoint, and their technological reliability improves from an evidentiary standpoint, the use of community science will likely explode across environmental justice claims and activism.[221]

In addition, environmental justice screening tools that collect, sort, and visualize environmental, health, and social indicators are also undergoing exponential growth. Early versions of these tools included Cal Enviro Screen and EPA's EJScreen. Now, both have proceeded through significant revisions.[222] States across the country are establishing similar tools or are mandating their creation through legislation or executive order.[223] Further, the CEQ released a beta version of its novel Economic and Climate Justice Screening Tool to assist agencies in identifying and

---

[218] *Participatory Science for Environmental Protection,* EPA.gov (June 16, 2022), https://www.epa.gov/participatory-science.

[219] *See, e.g., id.*; Davis and Ramírez-Andreotta (2021); Ristroph (2012), p. 81.

[220] A high-profile example is the Flint water crisis. Community science including pervasive water testing raised awareness, informed healthcare, and propelled lawsuits towards success. *See, e.g.,* U.S. Water Study Research Team, *U.S. Water Study: Practicing Science as a Public Good,* YouTube. (Oct. 9, 2016), https://www.youtube.com/watch?v=j8Y2Q7WPLOE; *see also* Co-Lead Class Counsel, *Ongoing Litigation Update,* Flint Water Justice (2021), https://www.flintwaterjustice.com/ongoing-litigation-updates (detailing the Flint water class action suits filed 2016–17).

[221] For a comprehensive manual on how community science might be used to support litigation efforts, *see* Harvard Law School Emmett Environmental Law & Policy Clinic, Using Citizen Science Data in Litigation (available at http://clinics.law.harvard.edu/environment/files/2019/03/Supplement-2-Using-Citizen-Science-Data-in-Litigation-March-2019.pdf).

[222] *EJSCREEN: Environmental Justice Screening and Mapping Tool,* U.S. Env't Prot. Agency, https://www.epa.gov/ejscreen.

[223] Examples from states not previously discussed herein include: Connecticut, *see* Press Release, Connecticut Department of Energy & Env't, DEEP and UConn CIRCA Partner to Develop Mapping Tool for Environmental Justice Communities (Oct. 27, 2021) (available at https://portal.ct.gov/DEEP/News-Releases/News-Releases%2D%2D-2021/DEEP-and-UConn-CIRCA-Partner-to-Develop-Mapping-Tool-for-Environmental-Justice-Communities); Michigan, *see Environmental Mapper*, Dept. of Env't, Great Lakes, & Energy, https://www.mcgi.state.mi.us/environmentalmapper/; *MiEJScreen: Environmental Justice Screening Tool*, Dept. of Env't, Great Lakes, & Energy, https://www.michigan.gov/egle/maps-data/miejscreen; New Mexico, see H.B. 51, 55th Leg., 1st Sess. (N.M. 2021); Illinois, *see Illinois EPA EJ Start*, Il. Env't Prot. Agency, https://www.arcgis.com/apps/webappviewer/index.html?id=f154845da68a4a3f837cd3b880b0233c; Pennsylvania. *see Environmental Justice Areas Viewer*, Pa. Dept. of Env't Prot., https://padep-1.maps.arcgis.com/apps/webappviewer/index.html?id=f31a188de122467691cae93c3339469c.

understanding the communities that the administration's environmental justice initiatives are intended to benefit.[224]

These publicly available tools inform environmental decision-makers (federal, state, and local) as well as communities and corporations about key characteristics of a location—by census tract or zip code—likely to be affected by a certain decision. Screening tools will aid attempts to address systemic burdens and cumulative impacts through regulatory processes and litigation. Indeed, practitioners remark that "[t]he standards and best practices for spatial social analysis [will] remain important… evidentiary skirmishes in EJ litigation."[225]

Supreme Court Decisions

In contrast to these developments likely to facilitate advances in environmental justice is recent U.S. Supreme Court jurisprudence. The Court's 2022 decisions circumscribed agency authority across areas of environmental and health protection with the judicially-created "major questions doctrine."[226] According to the conservative super-majority of justices, the doctrine requires Congress to "speak clearly" if authorizing agency action in "extraordinary" cases.[227] The vague standard announced in a trio of opinions invites capacious anti-regulatory interpretation, and amasses power to judges to decide what actions are "extraordinary."[228] Lisa Heinzerling observes, "in the aftermath of these decisions, motivated parties have had little trouble characterizing agency decisions as 'major'—and thus illegitimate" in lawsuits across the country, including matters related to climate change and storage of nuclear waste.[229]

As of this writing, the Court is also drafting a decision on race-conscious admissions practices in higher education.[230] "Depending how broadly it is written, [the] decision… could limit the Biden administration from explicitly mentioning or considering race when crafting environmental justice policy."[231] The judicial interpre-

---

[224] For critiques of the tool, including its lack of racial data and its inability to distinguish between disadvantaged communities *see* Jean Chemnick, *Experts to White House: EJ Screening Tool Should Consider Race,* E&E News: Climate Wire (June 1, 2022 6:40 AM) www.eenews.net/articles/experts-to-white-house-ej-screening-tool-should-consider-race/.

[225] Henderson et al. (2021) *20.

[226] *See, e.g.*, W. Virginia v. Env't Prot. Agency, 142 S. Ct. 2587 (2022); Nat'l Fed. of Indep. Bus. v. Dep't of Lab., Occupational Safety & Health Admin., 142 S. Ct. 661, 665 (2022) (quoting Alabama Ass'n of Realtors v. Dep't of Health & Hum. Servs., 141 S. Ct. 2485, 2489 (2021)) ("We expect Congress to speak clearly when authorizing an agency to exercise powers of vast economic and political significance.").

[227] *Id.*

[228] *See, e.g.*, Heinzerling (2022) https://www.bostonreview.net/articles/how-government-ends/ (describing the Court's "roll[] out" of the major questions doctrine over three cases and why the doctrine limits both agency and Congress' power).

[229] *Id.*

[230] *See* Students for Fair Admissions Inc. v. President & Fellows of Harvard Coll., (Dkt. 20-1199).

[231] Pamela King, *7 Cases that Reshaped Environmental Law in 2022*, E&E News Greenwire (December 22, 2022 1:40 PM), https://www.eenews.net/articles/7-cases-that-reshaped-environmental-law-in-2022/.

tive playing field is tilting. And progressive executive agency actions, overdue to advance environmental justice and explicitly acknowledge the systemic role of race in environmental disparities, are at risk in the courts.

These decisions limiting federal authority, coupled with the potential for a presidential administration change in 2024, warrant apprehension from environmental justice advocates. Future retrenchment looms over the federal progress described above. This specter underscores the importance of continued and expanded use of state law-based tort claims, and progressive state environmental justice efforts including statutes, constitutional amendments, and executive orders.

4 Conclusion

Environmental injustice is a pattern replicated across the globe, though concerns and goals are specific to individuals and cohorts within each jurisdiction.[232] People of ethnic minorities, Indigenous populations, and low-wage earners are more likely to suffer disproportionate environmental burdens wherever they live.[233] While environmental justice litigation in the United States has been embroiled in an uphill battle over the last several decades, there is reason to believe progress is on the horizon.[234] Key cases and innovations demonstrate the capacity of litigation as a tool within the movement. They evince that certain laws can be effectively harnessed for environmental justice, especially as part of a larger strategy that includes non-judicial advocacy.

Still, significant gaps and barriers in our laws persist. And the unprecedented breadth of federal action is vulnerable to restrictions from the Supreme Court or future presidential administrations less committed or outright hostile to the pursuit of environmental justice. State constitutional, statutory, and tort claims can serve as a bulwark against federal retrenchment. Or, perhaps the progressive state-based theories described above may be harbingers of the potential for litigation to drive environmental justice advancements, especially in an atmosphere of engaged federal actors. Certain, though, is that systemic change demands not only time, but also intersectional solutions, of which litigation is but one strand.

---

[232] *See, e.g.*, Perera (2017), pp. 142–146 (discussing disparities in pollution and climate change burdens by race and economic status worldwide).

[233] *See, e.g.*, Hajat et al. (2015); Inter-agency Support Group on Indigenous Peoples: *Issues Paper on Indigenous Peoples and Climate Change*, 2008, available at http://www.cbd.int/doc/meetings/tk/emccilc-01/other/emccilc-01-ipcc-en.pdf.

[234] The same limitations on constitutional claims, for example, are not present in all jurisdictions. *See* Oppenheimer et al. (2012).

References

Baker SH (2019) Anti-resilience: a roadmap for transformational justice within the energy system. Harv Civ Rights-Civ Lib Law Rev 54:1

Ball CA (2006) The curious intersection of nuisance and takings law. Boston Univ Law Rev 86:819

Been V (1994) Locally undesirable land uses in minority neighborhoods: disproportionate siting or market dynamics? Yale Law J 103:1383

Biermann F (2021) The future of 'environmental' policy in the anthropocene: time for a paradigm shift. Environ Polit 30:1–2

Brennan TA (1993) Environmental torts. Vand Law Rev 46(1):44–48

Bryant B (1995) Environmental justice: issues, policies, and solutions, pp 5–6

Bullard RD (2000) Dumping in Dixie: race, class, and environmental quality, p 53

Bullard RD (2021) Introduction: environmental justice—once a footnote, now a headline. Harv Environ Law Rev 45:243

Bullard R et al (2007) Toxic wastes and race at twenty 1987–2007: a report prepared for the United Church of Christ Justice & Witness Ministries, pp 1–14

Burkett M (2019–2020) Litigating separate and equal: climate justice and the fourth branch. Stan Law Rev Online 72:145

Chen Z et al (2022) Ambient air pollutant exposures and COVID-19 severity and mortality in a cohort of patients with COVID-19 in Southern California. Am J Respir Crit Care Med 206:440. https://doi.org/10.1164/rccm.202108-1909OC

Colangelo SA (2021) Environmental enforcement 2021: the likely resurgence of tools targeting environmental justice. Trends 52:4

Colangelo SA (2022) Forging complete justice: equitable relief in environmental enforcement. Harv Environ Law Rev 46:315

Cole LW (1994) Environmental justice litigation: another stone in David's Sling. Fordham Urb Law J 21:523

Cole LW (1995) Community initiatives: macho law brains, public citizens, and grassroots activists: three models of environmental advocacy. Va Environ Law J 14:687

Cole LW, Foster SR (2000) From the ground up: environmental racism and the rise of the environmental justice movement, pp 54–55

Daly E (2002) New hurdles for environmental justice plaintiffs. Nat Resour Environ 17:18

Dana DA, Tuerkheimer D (2017) After Flint: environmental justice as equal protection. Northwest Univ Law Rev Online 111:93

Davis LF, Ramírez-Andreotta MD (2021) Participatory research for environmental justice: a critical interpretive synthesis. https://doi.org/10.1289/EHP6274

Drury RT, Chu F (Fall 1994/Winter 1995) From White Knight lawyers to community organizing: citizens for a better environment — California. Race Poverty Environ 5:52

Ferris D (1994) Communities of color and hazardous waste cleanup: expanding public participation in superfund cleanups. Fordham Urb Law J 21:671

Fleischman F et al (2020) US forest service implementation of the national environmental policy act: fast, variable, rarely litigated, and declining. J For 118:403

Flowers CC (2020) Waste: One Woman's Fight Against America's Dirty Secret

Foster S (2004) The challenge of environmental justice. Rutgers J Law Urb Policy 1:1

Foster SR (2017) Vulnerability, equality and environmental justice: the potential and limits of law. In: Holifield R, Chakraborty J, Walker G (eds) The Routledge handbook of environmental justice, pp 136–137

Fredericks CF et al (2019) Social cost and material loss: the Dakota access pipeline. N Y Univ J Legis Pub Policy 22:563

Gaba JM, Kelly ME (1990) The citizen suit provision of CERCLA: a sheep in wolf's clothing? Sw Law J 43:929

Glenn C (2017) Upholding civil rights in environmental law: the case for Ex Ante Title VI Regulation and Enforcement. N Y Univ Rev Law Soc Change 41:45

Guana E (1995) Federal environmental citizen provisions: obstacles and incentives on the road to environmental justice. Ecol Law Q 22:1
Hajat A et al (2015) Socioeconomic disparities and air pollution exposure: a global review. Air Pollution and Health
Heinzerling L (2022) How government ends. Boston Review (Sept. 28, 2022)
Heisler KR (2011) Understanding environmental enforcement. In: Environmental law enforcement and compliance: leading lawyers on communicating with enforcement agencies, overcoming compliance challenges, and developing response strategies (Inside the Minds), p 4
Henderson D et al (2021) Environmental justice litigation: few wins, still effective. Nat Resour Environ 36:17
Hoffer MA (2003) Closing the door on private enforcement of Title VI and EPA's discriminatory effects regulations: strategies for environmental justice stakeholders after Sandoval and Gonzaga. New Eng Law Rev 38:971
Kim JJ (2021) Leave no one behind: realizing environmental justice through climate litigation remedies. Ecol Law Q 48:409
Kuehn RR (1994) Remedying the unequal enforcement of environmental laws. J CR Econ Dev 9:625
Lado ME (2019) No more excuses: building a new vision of civil rights enforcement in the context of environmental justice. Pa J Law Soc Change 22:281
Latham M et al (2011) The intersection of tort and environmental law: where the twains should meet and depart. Fordham Law Rev 80:737
Lazarus RJ (1992) Pursuing "environmental justice:" the distributional effects of environmental protection. Northwest Univ Law Rev 87:787
Magesh S et al (2021) Disparities in COVID-19 outcomes by race, ethnicity, and socioeconomic status: a systematic review and meta-analysis. 4 JAMA Network Open, at https://jamanetwork.com/journals/jamanetworkopen/fullarticle/2785980
Mank BC (2008) Title VI. In: Gerrard MB, Foster SR (eds) The law of environmental justice: theories and procedures to address disproportionate risks, p 23
Mikati I et al (2018) Disparities in distribution of particulate matter emissions sources by race and poverty status. Am J Pub Health 108:480
Mohai P, Saha R (2015) Which came first, people or pollution? A review of theory and evidence from longitudinal environmental justice studies. Environ Res Lett 10:125,011
Northern KS (1997) Battery and beyond: a tort law response to environmental racism. Wm Mary Environ Law Policy Rev 21:485
Oppenheimer D et al (2012) Comparative equality and antidiscrimination law: cases, codes, constitutions, and commentary
Outka U (2005) Environmental injustice and the problem of the law. Me Law Rev 57:209
Perera FP (2017) Multiple threats to child health from fossil fuel combustion: impacts of air pollution and climate change. Environ Health Perspect 125:141
Rendleman D (2018) Rehabilitating the nuisance injunction to protect the environment. Wash Lee Law Rev 75:1859
Ristroph EB (2012) Integrating community knowledge into environmental and natural resource decision-making. Wash Lee J Energy Climate Environ 3:81
Roberts JD et al (2022) "I can't breathe": examining the legacy of American racism on determinants of health and the ongoing pursuit of environmental justice. https://doi.org/10.1007/s40572-022-00343-x
Rothstein R (2017) The color of the law: a forgotten history of how our government segregated America
Schwartz VE et al (2010) Torts: cases and materials 1–2, 12th edn.
Spiller E et al (2021) Mortality risk from PM2.5: a comparison of modeling approaches to identify disparities across racial/ethnic groups in policy outcomes. Environ Health Perspect 129:127,004-1

Sprainer H (2022) Air quality equity: why the clean air act failed to protect low-income communities and communities of color from COVID-19. N Y Univ Environ Law J 30:123
Tate E et al (2021) Flood exposure and social vulnerability in the United States. Nat Haz 106:435
Taylor D (2014) Toxic communities: environmental racism, industrial pollution and residential mobility
Todd J (2020) A "sense of equity" in environmental justice litigation. Harv Environ Law Rev 44:169
Villa CJ (2022) No "box to be checked": environmental justice in modern legal practice. N Y Univ Environ Law J 30:157
Wilson B (2022) Making me ill: environmental racism and justice as disability. Univ Pa Law Rev 170:1722
Yang T (2002) Melding civil rights and environmentalism: finding environmental justice's place in environmental regulation. Harv Environ Law Rev 26:1

Sara Colangelo is the Director of the Environmental Law & Justice Clinic and a Visiting Professor of Law at the Georgetown University Law Center.

Abigail E. André is a lecturer at The Ohio State University, Moritz College of Law.

Ripple Effect: A Look at *Sackett v. EPA* and the Non-Water Quality Values That the Clean Water Act Protects, *in* Environmental Law Before the Courts: A US-EU Narrative

Cale Jaffe and Aspen Ono

Abstract This chapter first explains how the federal Clean Water Act is linked with the implementation of other major environmental laws in the United States. That is, while the Clean Water Act serves an overarching goal "to restore and maintain the chemical, physical, and biological integrity of the Nation's waters," it often does much more than that. Since its enactment in 1972, the Clean Water Act has consistently—and positively—helped to vindicate other public interest values that are not obviously connected to water quality, such as the preservation of historic resources or ensuring environmental justice. Critically, the protection of these non-water quality benefits occurs by congressional design. Issuance of a Clean Water Act permit often triggers other necessary reviews under the National Environmental Policy Act, the National Historic Preservation Act, and the Endangered Species Act, among other laws. We refer to this regulatory interplay as the Clean Water Act's 'ripple effect.'

This chapter's second contribution is to document how holistic environmental reviews engendered by the 'ripple effect' are now at risk, as the Supreme Court of the United States reconsiders the jurisdictional reach of the Clean Water Act. With *Sackett v. Environmental Protection Agency*, the Court is reassessing the question of which aquatic resources are covered as "waters of the United States." A new line "at which water ends and land begins" is being drawn. Historically, the Clean Water Act has safeguarded wetlands, marshes, streams, and tributaries that have a "significant nexus" to a navigable waterway. But if wetlands or tributaries are found to be

A ruling from the Supreme Court of the United States in Sackett, et ux., v. U.S. Environmental Protection Agency, et al., No. 21-454, had not been issued at the time this chapter was submitted for publication. Our analysis of Sackett therefore relies on our review of all briefs that had been filed in the case and our assessment of oral argument before the Court, which took place on October 3, 2022.

C. Jaffe (✉) · A. Ono
University of Virginia School of Law, Charlottesville, VA, USA
e-mail: cjaffe@law.virginia.edu; ajo2yf@virginia.edu

G. Antonelli et al. (eds.), *Environmental Law Before the Courts*,
https://doi.org/10.1007/978-3-031-41527-2_4

outside of the Clean Water Act's purview pursuant to a new ruling in *Sackett*, then the 'ripple effect' never occurs. The impact here may be profound. The interconnectivity of the Clean Water Act with other federal statutes, after all, reflects the interconnectedness of the resources these laws seek to protect.

1 Introduction

The Federal Water Pollution Control Act Amendments of 1972, better known as the Clean Water Act, are a core pillar of modern environmental law in the United States. The Act serves an overarching goal "to restore and maintain the chemical, physical, and biological integrity of the Nation's waters" and focuses protection efforts on "navigable waters," which are defined expansively as "the waters of the United States, including the territorial seas."[1] These "navigable waters" have long included aquatic resources that are not navigable-in-fact: wetlands, marshes, bogs, and streams. Given the interinfluence of various waters—*e.g.*, wetlands drain to tributaries which drain to navigable rivers and lakes—this broad interpretation of the Clean Water Act's jurisdictional scope has proven essential to advancing water pollution reduction efforts.

And the law does much more than remedy water pollution. For over half a century, proceedings under the Clean Water Act have consistently—and positively—affected *other* public interest values that might appear unrelated to water quality: preservation of historic resources; providing access to recreation; protecting endangered species; promoting public health and welfare; ensuring environmental justice. Most importantly, defense of these far-reaching, non-water quality benefits occurs by congressional design. Issuance of a Clean Water Act permit often triggers other necessary reviews under the National Environmental Policy Act, the National Historic Preservation Act, the Endangered Species Act, and the Fish and Wildlife Coordination Act, among other laws. Thus, the Clean Water Act's "ripple effect" compels permitting agencies and developers to evaluate a broad range of impacts.

This is the result of interconnected statutes that ensure federally governed activities receive a comprehensive review, providing decision-makers with a clear picture of the real-world impacts of any proposed activity. Indeed, the interconnectivity of the Clean Water Act and other federal statutes reflects the interconnectedness of the things they regulate.

The future of these holistic reviews, however, is now in flux following a ruling from the Supreme Court of the United States in *Sackett v. Environmental Protection Agency*, a case limiting the jurisdictional reach of the Clean Water Act.[2] At the same time, the Biden Administration has finalized its own interpretation of "waters of the

---

[1] 33 U.S.C § 1362(7).

[2] Michael Sackett, et ux., v. Environmental Protection Agency, et al., No. 21-454, (argued October, 3, 2022), https://www.supremecourt.gov/docket/docketfiles/html/public/21-454.html.

United States," which is the key, jurisdictional term in the Act.[3] As lower courts and federal agencies work to interpret a ruling from the Supreme Court in *Sackett*—and apply that ruling in challenges to the Biden Administration's Final Rule—other environmental regimes that depend on Clean Water Act enforcement will be impacted. This is because other environmental reviews are purposefully daisy-chained together with the Clean Water Act.

This Chapter explores the origins, application, and future of this daisy chain. First, Sect. 2 provides a background on the history of the Clean Water Act. Section 3 then documents the nature of holistic analyses under the Clean Water Act by looking at federal regulations used to implement the law and by highlighting a few "ripple effect" cases from the lower U.S. federal courts. With these cases, we show how the requirement to obtain a Clean Water Act permit has led to federal protection of historic resources and other non-water quality related values. Finally, Sect. 4 concludes with a look to the future. It summarizes how the *Sackett* case could alter the "ripple effect" in the coming years.

2 The Origins of the "Ripple Effect"

Section 101 of the Clean Water Act outlines several intuitive goals and policies that are essential to clean water.[4] These include:

1) The elimination of "the discharge of pollutants into the navigable waters;"
2) The protection of "fish, shellfish, and wildlife" and providing for "recreation in an on the water;"
3) Providing for financial assistance "to construct publicly owned waste [water] treatment works;"

These goals might fairly be called the water-quality related values that Congress sought to protect, and much has been written about their history.[5] The enactment of the Clean Water Act in 1972 was seen as an urgent response to an unmitigated crisis of water contamination.[6]

In 1969, the public's imagination was captured by images of the Cuyahoga River in Cleveland catching on fire. A nearly omnipresent oil slick on the water's surface due to unregulated discharges from Cleveland's manufacturing sector provided fuel

[3] U.S. Army Corps of Engineers & U.S. Environmental Protection Agency, Pre-Publication Final Rule Notice: Revised Definition of "Waters of the United States," Docket ID No. EPA-HQ-OW-2021-0602, (Dec. 30, 2022), at https://www.epa.gov/wotus/revising-definition-waters-united-states.

[4] 33 U.S.C. § 1251(a).

[5] See, e.g. Houck and Rolland (1995), p. 1243 (1995); Jaffe (2001), pp. 339–341 (2001); Drelich (2009), pp. 304–307 (2009) Congressional Research Service (2016); Percival et al. (2021), chpt 6.

[6] See 118 Cong. Rec. 36879 (Senate vote of 52 to 12 to override veto of the 1972 bill); id. at 37060-61 (House vote of 247 to 23).

for the fire.[7] National news coverage of the event led singer/songwriter Randy Newman to record "Burn On," with the lyrics: "Now, the Lord can make you tumble/The Lord can make you turn/The Lord can make you overflow/But the Lord can't make you burn."[8] The fire is often credited with inspiring the first Earth Day protests in the United States.

The Cuyahoga River fire was not an isolated incident. Rather, it was emblematic of similar environmental catastrophes in waterways across the country. A *New York Times* article from July 1970 vividly described daily life along the Potomac River in Washington, D.C., prior to enactment of the Clean Water Act:

> The heat of summer is enveloping the nation's capital, and with it has come the annual resurgence of a problem residents have come increasingly to dread: A stomach-turning miasma rising from the Potomac River … Under the bridges that link the capital with Virginia, the 1,500-foot-wide ribbon of water is a repellent, opaque gray brown, so laden with silt, intestinal bacteria and other pollutants that an official of the water quality agency called it 'a severe threat to the health of anyone coming in contact with it.' … 'Floating sludge masses, lifted by gases of decomposition, add to other debris on the water's surface.'[9]

As litigants before the Supreme Court recently observed, "Presidents Johnson and Nixon both made efforts at leveraging the Rivers and Harbors Act of 1899 to address this water contamination crisis, first through the Water Quality Act of 1965 and then via the Refuse Act of 1970. Neither proved successful."[10]

Senator Edmund Muskie (D-Maine), chief sponsor of the Clean Water Act, laid bare the problem:

> The cancer of water pollution was engendered by our abuse of our lakes, streams, rivers, and oceans… We have ignored this cancer for so long that the romance of environmental concern is already fading in the shadow of grim realities of lakes, river, and bays where all forms of life have been smothered by untreated wastes, and oceans which no longer provide us with food.[11]

Senator Muskie's statement captures the heartland of concerns that Congress sought to address in the Clean Water Act—*i.e.*, that the nation's waters would once again be fishable and swimmable.[12] But the Clean Water Act's statutory structure, legislative history, and regulatory framework confirm that the it actually protects much more.

---

[7] Boissoneault (2019).

[8] Randy Newman, Burn On, Sail Away (Reprise Records 1972).

[9] Hill (1970).

[10] Brief of Amicus Curiae Idaho Conservation League in Support of Respondents, Sackett, et ux., v. U.S. Environmental Protection Agency, et al., No. 21-454, (filed June 17, 2022) (Professor Jaffe served as counsel of record to the Idaho Conservation League on that brief).

[11] See Sapp et al. (2006), pp. 10198–10199 (quoting Senator Muskie's statement as reprinted in 1 CRS, Legislative History of the Water Pollution Control Act Amendments of 1972, at 161 (1973)).

[12] Friends of the Earth, Inc. v. Gaston Copper Recycling Corp., 204 F.3d 149, 156 (4th Cir. 2000) ("One of the well-recognized aims of the Act [was] to ensure that the nation's waters are 'fishable and swimmable.'").

Predating the Clean Water Act itself, the U.S. Army Corps of Engineers' ("Corps") 1968 Rivers and Harbors Act regulations included a directive for explicit and far-reaching public interest analyses. The Corps added a requirement that permit decisions must rest on 'an evaluation of all relevant factors, including the effect of the proposed work on navigation, fish and wildlife, conservation, pollution, aesthetics, ecology, and the general public interest.'"[13] That review process was buttressed by the Fish and Wildlife Coordination Act, which compelled the Corps to first consult with the U.S. Fish and Wildlife Service before issuing any permit for an impoundment or water-diversion project, "with a view to the conservation of wildlife resources by preventing loss of and damage to such resources…"[14] All of these holistic reviews were expanded further in 1969 with passage of the National Environmental Policy Act ("NEPA") and its mandate that federal agencies adopt an "interdisciplinary approach" to ensure that "presently unquantified environmental amenities and values may be given appropriate consideration in decision-making along with economic and technical considerations."[15]

NEPA was authored in response to concern that many federal agencies, without a conservation-oriented mission statement, had been racing to complete projects without regard to adverse environmental impacts. NEPA forced these agencies to quantify environmental harms so that destructive projects could be "nipped in the bud."[16] Interpreting the statute, the U.S. Court of Appeals for the D.C. Circuit observed that the "sweep of NEPA is extraordinarily broad, compelling consideration of any and all types of environmental impact of federal action."[17]

The co-development of public interest reviews under 1968 Rivers and Harbors Act regulations and the National Environmental Policy Act of 1969 suggest that federal agencies would be endowed with the authority and the obligation to consider a broad range of factors beyond their narrow areas of expertise.[18] In short, an expansive approach to assessing pollution-causing activities had begun to take hold even before the Clean Water Act of 1972 was enacted.

Today, the real-world impact of these holistic reviews is seen most clearly in Section 404 of the Clean Water Act, which mandates the acquisition of a permit, usually issued by the Corps, "for the discharge of dredged or fill material into the *navigable waters*."[19] Many development projects—whether for a home, a shopping mall, or a road—first require dredging and filling wetlands to create firm ground on which to begin construction. To assist the Corps with its permitting process, the

---

13 Flournoy (2008), p. 544. 33 Fed. Reg. 18,670, 18,671 (Dec. 18, 1968)).

14 16 U.S.C. § 662(a).

15 42 U.S.C. § 4332.

16 Tarlock (2005), pp. 77–108.

17 Calvert Cliffs' Coordinating Committee, Inc. v. U. S. Atomic Energy Commission, 449 F.2d 1109, 1122 (D.C. Cir. 1971).

18 33 Fed. Reg. 18,670 (1968); 42 U.S.C. § 4332; 33 U.S.C. § 403.

19 33 U.S.C § 1344(a).

Environmental Protection Agency ("EPA") is granted the power under the statute to develop "guidelines" that prevent the degradation of waters under § 404(b).[20]

Section 404(b) also mandates a far-reaching review of non-water quality related values. It cross-references 33 U.S.C. § 1343(c), which specifies that "Guidelines for determining degradation of waters" shall account for the "effect of disposal" on "human health or welfare, ... marine life, ... and aesthetic, recreation, and economic values," among other concerns. If the EPA becomes convinced that the Corps has erred in issuing a permit, the Agency can veto the Corps' approval under § 404(c), thereby preventing an "unacceptable adverse effect on municipal water supplies, shellfish beds and fishery areas (including spawning and breeding areas), wildlife, or recreational areas."[21] One clear takeaway from the text of Section 404 is that it encourages the Corps and EPA to work together in service of broad, public-interest goals.

In 1977, Congress amended the Act to further refine the Section 404 program. It added a handful of exemptions to the regulatory process (*e.g.*, for "normal farming" work like "plowing, seeding, cultivating"), and added a general permit process for the Corps to efficiently manage categories of activities that would have only a "minimal cumulative adverse effect on the environment."[22] But it left in place a broad understanding of the Clean Water Act's jurisdictional reach. As the Senate Committee report on the 1977 Amendments explained:

> The objective of the 1972 act is to protect the physical, chemical, and biological integrity of the Nation's waters. Restriction of jurisdiction to those relatively few waterways that are used or are susceptible to use for navigation would render this purpose impossible to achieve. Discharges of dredged or fill material into lakes and tributaries of these waters can physically disrupt the chemical and biological integrity of the Nation's waters and adversely affect their quality. ... To limit the jurisdiction of the Federal Water Pollution Control Act with reference to discharges of the pollutants of dredged or fill material would cripple efforts to achieve the act's objectives.[23]

The 1977 amendments also added text to assert exclusive federal control over some waters used "as a means to transport interstate or foreign commerce ... *including wetlands adjacent thereto*."".[24] This section of the statute, 33 U.S.C. § 1344(g), made it explicit that non-navigable but "adjacent" wetlands would be federally protected.

One motivation for Congress's acceptance of a broad, jurisdictional purview for the Corps and the EPA under the Clean Water Act was a profound understanding of

[20] 33 U.S.C. § 1344(b).

[21] 33 U.S.C. § 1344(c).

[22] 33 U.S.C. § 1344(e)(1).

[23] S. REP. NO. 95-370, at 75. See also Solid Waste Agency of Northern Cook County v. U.S. Army Corps of Engineers, 531 U.S. 159, 183 n.11 (2001) (Stevens, J., dissenting) ("The House Committee on Government Operations noted the disagreement between the EPA and the Corps over the meaning of 'navigable waters' and ultimately expressed its agreement with the EPA's broader reading of the statute. H.R. Rep. No. 93–1396, pp. 23–27 (1974).").

[24] 33 U.S.C. § 1344(g) (emphasis added).

wetlands as interconnected with the health and integrity of the broader environment.[25] William L. Want, author of a treatise on the Law of Wetlands Regulation, has noted that wetlands might "appear to the untrained eye to be simply an upland field," but that view misses the interconnection between wetlands protection and wildlife conservation, water quality, and flood mitigation.[26] "Two-thirds of the commercially important fish and shellfish harvested along the Atlantic and Gulf coasts depend on coastal estuaries and their wetlands for food sources, spawning grounds, or nurseries for their young; almost one half of Pacific coast fish and shellfish are dependent on wetlands."[27] When a hurricane hits, wetlands play a vital role in reducing the storm surge—an impact that can literally mean the difference between the preservation of historic communities and their devastation.[28] Thus, the Clean Water Act's jurisdiction over adjacent wetlands is integral to its aim of protecting the quality and integrity of the Nation's waters.

3 The "Ripple Effect" in Practice

What starts with the Clean Water Act does not necessarily stay with the Clean Water Act. The regulations that the Corps and the EPA have developed (*e.g.*, guidance to permit applicants on how the statutory text of the Clean Water Act will be applied) ensure a holistic review process.

EPA has drafted Section 404(b)(1) Guidelines to provide that "the proposed fill material will not cause any significantly adverse effects on human health or welfare, aquatic life, aquatic ecosystems, or recreational, aesthetic, or economic values."[29] The EPA's Guidelines require that the Corps consider a wide range of impacts that a proposed activity might have and reference impacts to aesthetic concerns, including "the perception of beauty by one or a combination of the senses of sight, hearing, touch, and smell."[30] They compel consideration of impacts to "Parks, national

[25] See Senate Debate August 4, 1977: Clean Water Act of 1977 at 920–923 ("A fundamental element of the Water Act is broad jurisdiction over water for pollution control purposes. Several Federal courts have endorsed the wisdom, and constitutionality, of this committee's observation that: Water moves in hydrologic cycles and it is essential that discharge of pollutants be controlled at the source… The once seemingly separable types of aquatic systems are, we now know, interrelated and interdependent. We cannot expect to preserve the remaining qualities of our water resources without providing appropriate protection for the entire resource.").

[26] William L. Want, L. of Wetlands Reg. para 2:1 (Thomson Reuters 2022).

[27] Id, para 2:3.

[28] See Houck (2006), pp. 27–29.

[29] Altamaha Riverkeeper v. United States Army Corps Of' Eng'rs, 2020 U.S. Dist. LEXIS 180987, *2-3; 2020 WL 5837650 (citing 40 C.F.R §230.10(c)(1)-(4)).

[30] 40 C.F.R. § 230.53(a).

and historical monuments, national seashores, wilderness areas, research sites, and similar preserves."[31] They obligate analysis of threats posed to endangered species.[32]

The EPA's 404(b)(1) Guidelines also reflect the view that a permit should only be granted if the regulator determines, after analyzing various individual and cumulative effects, that a proposed activity will not have an unacceptable adverse impact.[33] This precept produces a generalized instruction that the Corps should consider information regarding a range of impacts and values beyond those affecting water quality. For example, under Section 230.10(b), the EPA's guidance instructs the Corps not to grant a permit if the proposed activity would violate other law, such as the Endangered Species Act. Coordination with the U.S. Fish and Wildlife Service is required. In short, the EPA Guidelines direct the Corps to account for the full ramifications of a proposed project.

To be sure, the Corps initially resisted a broad application of its authority under Section 404 of the Clean Water Act. But in 1975, an environmental group, the Natural Resources Defense Council, challenged that institutional restraint in court and won.[34] A U.S. District Court in Washington, D.C. confirmed that the term "navigable waters" was "not limited to the traditional tests of navigability."[35] Two years later, the Corps amended its regulations to cover "discharges into intermittent waterways and wetlands if they were necessary 'for the protection of water quality,' even if not adjacent to traditionally navigable waterways."[36]

Under its current regulations, the Corps further commits to weighing a range of "public interest" factors: "conservation, economics, aesthetics, general environmental concerns, wetlands, historic properties, fish and wildlife values … energy needs, safety, food and fiber production, mineral needs, considerations of property ownership and, in general, the needs and welfare of the people."[37] Such reviews "reflect the national concern for both protection and utilization of important resources," not merely water-quality related values.[38]

Further underscoring the multifactor nature of its process, the Corps has pledged to consult with relevant state and federal agencies such as the U.S. Fish and Wildlife Service, the National Marine Fisheries Service, the Environmental Protection

[31] 40 C.F.R. § 230.54.

[32] 40 C.F.R. § 230.30.

[33] 40 C.F.R. § 230.10(c)(2007).

[34] See Natural Resources Defense Council v. Callaway, 392 F. Supp. 685, 686 (D.D.C. 1975) (holding that the Corps' 1974 Clean Water Act regulations "unlawfully" failed to protect nonnavigable waters).

[35] Id.

[36] Brief of Former EPA Administrators William K. Reilly and Carol M. Browner as Amici Curiae in Support of Respondents, Sackett v. EPA, No. 21-454, at 8–9 (filed June 17, 2022) (citing 40 Fed. Reg. 31,320, 31,324-25 (July 25, 1975)).

[37] 33 C.F.R. § 320.4(a).

[38] 33 C.F.R. § 320.4(a).

Agency, and the U.S.D.A's Natural Resources Conservation Service (previously the Soil Conservation Service).[39]

All of which is to say that while the statute focuses on water quality, the agencies implementing the Act have long required the evaluation of *non-water quality related values* before issuing any permit. It is a remarkable story in the legacy of the Clean Water Act; the protection of water quality "ripples out" to support other public-interest goals.

And as we explain throughout this chapter, this "ripple effect" is by congressional design. That is, it flows from interconnected, statutory directives to federal agencies. A permit proceeding under Section 404 is deemed a "major Federal action" that initiates review under NEPA, and thus requires federal agencies to comprehensively evaluate project impacts beyond those directly related to water quality.[40] NEPA regulations, developed by the Council on Environmental Quality, have long specified that the "effects" to be evaluated include: "ecological [effects] (such as the effects on natural resources and on the components, structures, and functioning of affected ecosystems), aesthetic, historic, cultural, economic, social, or health [effects], whether direct, indirect, or cumulative." The regulations implementing NEPA also mandate evaluation of "reasonable alternatives" to mitigate harm to these values.[41] All of these impacts may be incorporated into a Clean Water Act permitting proceeding.[42]

Other federal conservation laws have similar triggers. Section 7(a)(2) of the Endangered Species Act ("ESA") requires federal agencies to consult with the U.S. Fish & Wildlife Service to "insure that any action authorized, funded, or carried out by such agency ... is not likely to jeopardize the continued existence of any endangered or threatened species or result in the destruction or adverse modification of habitat of such species which is determined ... to be critical...."[43] The Corps' determination on a Section 404 permit application can thus trigger consultation under the ESA and require the U.S. Fish and Wildlife Service to prepare a Biological Opinion outlining harms to protected flora and fauna.

Similarly, Section 106 of the National Historic Preservation Act ("NHPA") requires that "prior to the approval of the expenditure of any Federal funds" and "prior to the issuance of any license," federal agencies must "take into account the effect of the undertaking on any historic property."[44] The Fish and Wildlife Coordination Act likewise provides that "whenever the waters of any stream or other body of water are proposed or authorized to be impounded, diverted, ... or

---

[39] 33 C.F.R. § 320.4(b)(3).

[40] 42 U.S.C. § 4332(2)(C); 33 C.F.R. § 330.5(b)(3).

[41] 40 C.F.R. § 1502.14(a).

[42] 40 C.F.R. § 1508.8; 43 Fed. Reg. 55978, 56004 (Nov. 29, 1978).

[43] 16 U.S.C. §§ 1531–1544. See § 7(a)(2) of the ESA stating that federal agencies must assure that all actions they conduct, authorize, or fund are "not likely to jeopardize the continued existence of any listed species or threatened species or result in the destruction or adverse modification of designated critical habitat." 16 U.S.C. § 1536(a)(2).

[44] 54 U.S.C. § 306108.

modified … under Federal permit or license, such department or agency [issuing the permit] first shall consult with the United States Fish and Wildlife Service … with a view to the conservation of wildlife resources…."[45]

The foray into these non-water quality related values may seem perplexing, but they are caused by the very structure of the Clean Water Act (requiring a federal permit) and other conservation-minded statutes (requiring further assessments before the issuance of a federal permit). The cases discussed below document how this daisy chain of federal regulation works in practice.

In a 2011 case, *Sierra Club v. United States Army Corps of Engineers*, the U.S. Court of Appeals for the 8th Circuit upheld an injunction halting construction of the John W. Turk, Jr. coal-fired power plant.[46] The Sierra Club's challenge was grounded in NEPA, which had been triggered by a power company's application for a Clean Water Act Section 404 permit to fill wetlands. The environmental group uncovered a wide array of failures in the NEPA analysis—failures that were tied to significant public interest values not directly related to water quality. For example, the Corps' public interest review mistakenly relied on the power company's assertion that the coal plant needed to be located in Arkansas because the electricity would serve Arkansas citizens. As the court noted, "this is simply wrong…. none of the electricity generated by the plant will go to Arkansas ratepayers."[47]

While the Sierra Club has advocated to preserve wetlands and protect water quality in other contexts, those concerns did not seem to be driving the organization's involvement in this particular case. The Sierra Club's core concern here centered on the smog, soot, and global warming pollution that would be generated through the operation of a massive coal-fired power plant.[48] Indeed, the Sierra Club had brought a wide-ranging series of challenges to the coal plant, with litigation on the air permit at the state level, the Clean Water Act permit at the federal level, and cost concerns addressed before the Arkansas Public Service Commission.[49]

The failure to properly assess the alleged need for a coal plant was a dramatic error in the Corps' NEPA analysis. The Corps, after all, is required to select the "least environmentally damaging practicable alternative"[50] under its regulations and the electric utility's own selection study had identified a site in Texas that "would fill fewer wetlands."[51]But, the Corps had rejected the Texas site because it wrongly believed that the coal plant needed to be on the ground in Arkansas. The Section 404 permitting process thus forced the Corps to account for a material factor, not directly tied to water quality, which it had previously overlooked.

---

[45] 16 U.S.C. § 662(a).

[46] Sierra Club v. United States Army Corps of Eng'rs, 645 F.3d 978, 985 (8th Cir. 2011).

[47] Id, p. 990 (internal citation omitted).

[48] Smith (2011) (discussing related challenge to the air permit in state court).

[49] Guter (2011) (noting that "there have been quite a few twists and turns in the fight against Turk, the 600 MW proposed coal-fired plant in Hempstead County.").

[50] Sierra Club, 645 F.3d at 984.

[51] Id, p. 990.

Another example of Section 404's ripple effect comes out of the Fifth Circuit. In *O'Reilly v. United States Army Corps of Engineers,* the court assessed the sufficiency of an Environmental Assessment (also known as an "EA") under NEPA by looking at increases in non-point source pollution and the "total and complete loss of wetland function," all of which were firmly within the ambit of the Clean Water Act. Yet the court also reviewed—as part of the same analysis—harms to "non-aquatic wildlife" and impacts on traffic, transportation, and public safety.[52] Evaluating the totality of these impacts, the court concluded:

> The EA before us lists the potentially significant adverse impacts, and describes, in broad terms, the types of mitigation measures that will be employed. As is evident from our above review of the Corps' treatment of each individual potential impact, however, the EA provides only cursory detail as to what those measures are and how they serve to reduce those impacts to a less-than-significant level. Because the feasibility of the mitigation measures is not self-evident, we agree with the district court that the EA does not provide a rational basis for determining that the Corps has adequately complied with NEPA.[53]

The case provides a great example of the breadth of issues worthy of concern and consideration that are captured under a Section 404 permitting process. Many of these impacts are directly tied to water quality values, but some unquestionably are not.

This daisy chain of federal reviews can also arise in unexpected ways. In *United States v. Land, 62.50 Acres of Land More or Less*, federal authorities sought to condemn coastal marshland in Louisiana as part of an expansion of Jean Lafitte National Historic Park and Preserve. The private landowner, OKC Limited Partnership, believed the National Park Service had undervalued the property and ignored the land's potential development for a commercial shell-harvesting operation.[54] The shells might be resold "for road paving material and seeding oyster beds." The U.S. Court of Appeals sided with the government in rejecting OKC's overly optimistic valuation, which ignored the challenges of obtaining a Section 404 permit. The Court explained, "The shell mining project would not only threaten a wetlands habitat and pose a substantial risk of shoreline erosion but also destroy native American and prehistoric artifacts of interest to archaeologists and historians."[55] All of these factors would be evaluated in the Clean Water Act permitting process, and collectively they would make it far less likely that a shell-harvesting operation would ever see the light of day.

Finally, *Hough v. Marsh*[56] provides an especially strong illustration of how the Clean Water Act and NEPA have worked together for decades to expand the range of impacts that the Corps must consider in administering Section 404. In *Hough*, two individuals applied for a Section 404 permit to fill tidal wetlands sandwiched

---

[52] O'Reilly v. United States Army Corps of Eng'rs, 477 F.3d 225 (5th Cir. 2007).

[53] Id, p. 234.

[54] United States v. Land, 62.50 Acres of Land More or Less, 953 F.2d 886 (5th Cir. 1992).

[55] Id, pp. 892–893.

[56] Hough v. Marsh, 557 F. Supp. 74 (D. Mass 1982).

between a marina and a beachfront on Martha's Vineyard. They proposed to fill this spit of marshland to build a tennis court and two luxury residences.[57] The Corps solicited comments from the EPA, the National Marine Fisheries Service, and the U.S. Fish and Wildlife Service.[58] Two of those agencies, the National Marine Fisheries Service and the U.S. Fish and Wildlife Service, highlighted heartland water-quality concerns about harms to "nutrients and habitat to the surrounding ecosystem" and harm to wetlands that served as a "storage area for storm and flood water." Nevertheless, the Corps granted a permit to fill in the wetlands.[59]

But a district court disagreed with the Corps' decision and ruled for challengers that the permit should never have been issued. The court's holding did not rest solely on the water-quality related impacts from developing a project on the edge of the ocean. Rather, the court found error in a wide array of non-water-quality related factors. Specifically, the court zeroed in on "two additional factors that the Corps failed to address properly in connection with the public interest review .... Those dealing with 'economics' and 'cumulative effects.'" In evaluating these effects, the court went on to discuss impacts of the proposed development on the adjacent, historic Edgartown Lighthouse.

A lighthouse was built on the site in Edgartown in 1828, on an artificial island just off the coast. Over time, a manmade causeway and the natural accretion of sand connected the lighthouse to the main island of Martha's Vineyard.[60] It soon became a beloved destination for "promenade and pleasure."[61] In 1938, the U.S. Lighthouse Service determined that a taller structure was needed to better aid navigation.[62] An 1881-era cast iron lighthouse tower was thus relocated from Ipswich, Massachusetts to the site in Edgartown on Martha's Vineyard.[63]

Flash forward to 1982 and the permitting controversy at issue in *Hough:* the NHPA, along with regulations promulgated by the Advisory Council on Historic Preservation, required review of the "adverse effect on any property listed in or eligible for inclusion in the National Register of Historic Places." The NHPA mandated that the "head of any Federal agency … having authority to license any undertaking, … prior to the issuance of any license, shall take into account the effect of the undertaking on any historic property."[64] The harm to the lighthouse, along with the economic loss of tourism to the lighthouse (which had been "on the itinerary of sightseeing buses") convinced the court to remand the matter to the Corps for additional analyses.

---

[57] Id, p. 76.

[58] Id, p. 77.

[59] Id.

[60] Martha's Vineyard Museum (2022b).

[61] Id.

[62] Martha's Vineyard Museum (2022a).

[63] United States Coast Guard: U.S. Department of Homeland Security (2016).

[64] 54 U.S.C. § 306108 (formerly 16 U.S.C. § 470f (1982), as cited in Hough).

Hough emphasizes the real-world impact of accounting for non-water quality values in Clean Water Act permitting processes. The case demonstrates the protective "ripple effect" of Clean Water Act jurisdiction. The amount of wetlands to be filled was surprisingly small: just one-quarter of an acre on a three-acre parcel. Yet those wetlands were enough to trigger evaluation of the historic lighthouse on Martha's Vineyard. If those aquatic resources had not been protected by the Clean Water Act, impacts to the lighthouse might never have been considered.

4 An Uncertain Future for the "Ripple Effect"

O'Reilly, Sierra Club, Hough, and the Louisiana condemnation case all highlight how real harm to historic and environmental resources can be overlooked. The communities most directly affected in each of those cases would not have had their concerns considered by federal regulators were it not for the Section 404 permitting process. The jurisdictional reach of the Clean Water Act thus ensures that these broad, holistic reviews take place.

The current focus in Clean Water Act litigation, however, is on *Sackett v. EPA* and assessing the jurisdictional question of which aquatic resources are covered as "waters of the United States." That question is distinct from the issue of whether non-water quality related values are protected by the Clean Water Act, but the two queries are inextricably intertwined. If wetlands or tributaries impacted by a future project are found to be outside of the Corps' jurisdiction, then the "ripple effect" never occurs.

To place *Sackett* in context, it is essential to first understand the cases that preceded it. The Supreme Court's seminal precedent on the jurisdictional question is *United States v. Riverside Bayview Homes,* a decision issued in 1985. In that case, the Court largely deferred to the Corps' expertise in defining the scope of waters to be covered. The Court listed several aquatic features that might be protected as navigable waters—"shallows, marshes, mudflats, swamps, bogs"—and remarked, "Where on this continuum to find the limit of 'waters' is far from obvious."[65]

Sixteen years later, the Court began to reign in the deference given to federal regulators. In *Solid Waste Agency of Northern Cook County v. U.S. Army Corps of Engineers*, it ruled that there must be some hydrological connection to navigability for a wetland, pond, or stream to benefit from the protections of the Clean Water Act. Then-Chief Justice Rehnquist famously explained, "It was the significant nexus between the wetlands and 'navigable waters' that informed our reading of the [Clean Water Act] in *Riverside Bayview Homes*."[66]

The phrase "significant nexus" had driven debate over Clean Water Act jurisdiction—and it dominated the Biden Administration's new rule on "waters of the

[65] United States v. Riverside Bayview Homes, 474 U.S. 121, 132 (1985).

[66] SWANCC v. U.S. Army Corps of Engineers, 531 U.S. 159, 167 (2001).

United States," appearing nearly 500 times in the pre-publication text.[67] Yet in the 2006 case of *Rapanos v. United States,* Justice Scalia—writing for himself, Justice Thomas, Justice Alito, and Chief Justice Roberts—would have rejected the significant nexus test and limited Clean Water Act jurisdiction only to wetlands that maintained a continuous, surface water connection to navigable-in-fact waters. Justice Scalia's proposal was not adopted by a majority of the Court in *Rapanos*. Justice Kennedy, while agreeing that the Corps of Engineers had failed to prove its case, would have sent the issue back to the lower courts "for consideration whether the specific wetlands at issue possess a significant nexus with navigable waters."[68] Kennedy's solo concurrence provided the guiding criteria for agencies seeking to apply *Rapanos* until recently. The applicability of Justice Kennedy's opinion, of course, has now largely been ended given the way that the Supreme Court has reconsidered the jurisdictional issue in *Sackett.*

Oral argument was heard on October 3, 2022 in *Sackett v. EPA,* a case that began when Michael and Chantell Sackett dumped sand and gravel into wetlands to shore up a "housing pad" for a water-view home sitting roughly 300 linear feet from the edge of Priest Lake, Idaho—a 23,000-acre water body that serves as an international tourist destination.[69] The Sackett property sits adjacent to Kalispell Creek's inflow to the western shores of Priest Lake and is part of the broader Kalispell Bay Fen, an aquatic wetland resource. Wetlands at the building site drain via groundwater into Priest Lake.[70] What is more, a creek adjacent to the Sackett property supports a population of native westslope cutthroat trout,[71] and the EPA has documented the presence of trout and trout-spawning habitat in the wetlands above the Sackett site.[72]

Notwithstanding the multiple surface and groundwater connections between these aquatic resources and Priest Lake, the Sacketts succeeded in arguing that their wetlands fell outside the reach of the Clean Water Act. That is, they claimed that their wetlands were are not sufficiently adjacent to "navigable waters" to be treated as part of the "waters of the United States."

In addition to the jurisdictional challenges at issue in *Sackett,* the Clean Water Act's ripple effect might also be challenged by Major Questions Doctrine, which posits that federal agencies will often lack the leeway to interpret ambiguous statutory commands.[73] The doctrine, which is a form of "clear statement rule," is in tension with the Supreme Court's longstanding deference to agencies under *Chevron v.*

[67] U.S. Army Corps of Engineers & U.S. Environmental Protection Agency, Pre-Publication Final Rule Notice: Revised Definition of "Waters of the United States," Docket ID No. EPA-HQ-OW-2021-0602, (Dec. 30, 2022), at https://www.epa.gov/system/files/documents/2022-12/Pre-Publication%20Final%20Rule%20Notice.pdf.

[68] Rapanos v. United States, 547 U.S. 715, 787 (2006) (Kennedy, J., concurring in the judgment).

[69] Amicus Brief of Idaho Conservation League, Sackett v. EPA, at 6–7, 11.

[70] Freeman (1995), p. 43.

[71] Idaho Dep't of Fish and Game, Fisheries Bureau (2013).

[72] Amicus Brief of Idaho Conservation League, Sackett v. EPA, at 16–17.

[73] West Virginia v. EPA, 142 S. Ct. 2587, 2616 (2022) (Gorsuch, J., concurring).

Natural Resources Defense Council.[74] That 1984 decision has achieved canonical status. It holds that if a statute is ambiguous—*i.e.*, if "Congress has not directly addressed the precise question at issue"—then "the court does not simply impose its own construction on the statute…." Rather, "the question for the court is whether the agency's answer is based on a permissible construction of the statute."[75]

The Major Questions Doctrine proposes a far different method for resolving ambiguity in the statutory text. "Under that doctrine's terms, administrative agencies must be able to point to 'clear congressional authorization' when they claim the power to make decisions of vast 'economic and political significance.'"[76] One criticism of the Major Questions Doctrine is that it does not eliminate the need to interpret ambiguous statutory text; it simply pre-decides that ambiguous language will prohibit any significant administrative action. That is, it creates "a one-way ratchet favoring deregulation."[77] Yet federal agency inaction is just as much of a choice as action. Instead of deferring to Congress, courts applying the Major Questions Doctrine simply augment their own authority at the expense of a democratically elected President and the executive branch.[78]

The salience of the Major Questions Doctrine came to the fore in another 2022 case, *West Virginia v. EPA,* which marked the first time that the doctrine was invoked in a majority opinion of the Supreme Court.[79] The interpretive theory resurfaced briefly during the October 2022 oral argument in *Sackett v. EPA*, when Justice Alito asked the government's lawyer if EPA's "understanding of 'waters of the United States' take[s] into account any of the clear statement rules."[80] Justice Kavanaugh similarly wondered, "[W]hy not let Congress figure out where the line is?"[81]

The justification for Congress's delegation of authority to agencies is often rooted in assumptions of agency expertise. We thus accept ambiguously worded delegations to protect water quality because we understand that EPA's hydrologists, aquatic biologists, and engineers have the requisite knowledge needed to implement congressional directives—and that Congress lacks the expertise to be "in the weeds" on regulation. EPA's consideration of historical, social, economic, and other non-water quality related values, however, might be viewed as only tangentially related to their areas of competency and thus subject to attack under the Major Questions Doctrine.

[74] Chevron v. Natural Resources Defense Council, 467 U.S. 837 (1984).

[75] Id, pp. 842–843 (internal footnotes omitted).

[76] West Virginia v. EPA, 142 S.Ct. 2587, 2616 (2022) (Gorsuch, J., concurring).

[77] Super (2022).

[78] Support for deregulation—i.e., a limited role for the federal government—is often associated with political conservatives, but that is always the case. See, e.g., Jaffe (2020), pp. 692–698 (discussing several instances of Republican support for broad, federal, regulatory authority).

[79] West Virginia v. EPA, 142 S. Ct. 2587, 2609 (2022); Congressional Research Service (2022).

[80] Transcript of Oral Argument at 106, Sackett v. Environmental Protection Agency, No. 21-454, (Oct. 3, 2022).

[81] Id, p. 77.

Even so, it is hard to predict whether the Major Questions Doctrine will play much of a role in the implementation of the *Sackett* decision in the coming years. Noting that directly adjacent but non-navigable wetlands are unquestionably regulated as "navigable" waters, Justice Kavanaugh cautioned the Sacketts' lawyer that he was "calling [for] a textual limit on something that's divorced from the text to begin with."[82] Even more pointedly, Kavanaugh followed by asking, "Why did seven straight [Presidential] administrations not agree with you" on a narrow interpretation of the Clean Water Act?[83] (Kavanaugh would develop this observation more thoroughly in a concurring opinion filed in *Sackett.*) And Justices Sotomayor, Kagan, Kavanaugh, and Jackson all seemed to agree that adjacent wetlands—including those separated from navigable waters by manmade berms or dikes—would be protected.[84]

At the same time, the Justices made it clear that the significant nexus test presented line-drawing problems. It can be challenging for a developer to determine where buildable property ends and jurisdictional waters begin. Appearing to summarize a majority of the Court's thinking at the time of oral argument, Justice Sotomayor asked:

> So is there another test? Not the *Rapanos* test, not the adjacency test, not the significant nexus test. But is there another test that could be more precise and less open-ended than the adjacency test or the significant nexus test that you use? Is there some sort of connection that could be articulated?[85]

To be sure, congressional leaders have affirmed the function and importance of the Clean Water Act's public interest reviews, thus granting the EPA and the Corps additional legitimacy in pursuing these goals. Such support was demonstrated in a 1982 House Subcommittee of the Committee on Government Operations hearing,[86] during which Chairman Toby Moffit admonished the Corps to consider broad public interest concerns: "When a private party wants to use a national resource, such as a wetland, for its private gain, there are legitimate public interest criteria that must be met. The Corps obviously needs to be reminded of whose resources are at stake."[87]

And, as stated throughout this chapter, the Clean Water Act functions in collaboration with other federal statutes to create a holistic process that is set in motion with the filing of an application for a Section 404 permit. Regarding the incorporation of non-water quality values into project review, NEPA remains the most influential of these linked statutes. NEPA applies to all "major Federal actions significantly

[82] Id, p. 52.

[83] Id, p. 53.

[84] Justice Barrett cautioned the Sacketts' lawyer that Section "1344(g) is the biggest problem for you, clearly." Id. at 29.

[85] Id, p. 92.

[86] Hearing before the Subcommittee of the Committee on Government Operations House of Representatives ninety-seventh Congress Second Session, Army Corps of Engineers: The North Haven, Conn., Mall, September 9, 1982.

[87] Id, p. 3.

affecting the quality of the human environment."[88] The issuance of a Section 404 permit by the Corps is a "major Federal Action" and thus requires federal agencies to comprehensively evaluate a project's broader impacts.[89]

The connection between Section 404 of the Clean Water Act and NEPA was acknowledged in a 1992 Senate Report of the Committee on Environment and Public Works,[90] which noted, "Before the Corps can grant such a section 404 authorization, the Corps is required by the Clean Water Act and /or the National Environmental Policy Act (NEPA) to: Solicit comments … regarding any Corps proposal to authorize water disposal of dredged material; Apply EPA's section 404(b)(1) Guidelines…"[91] Evaluation of a Section 404 permit application is inherently linked to NEPA review.

Section 101(b) of NEPA further provides that "it is the continuing responsibility of the federal government to use all practicable means, consistent with other essential considerations of national policy to avoid environmental degradation, preserve historic, cultural, and natural resources, and promote the widest range of beneficial uses of the environment without undesirable and unintentional consequences."[92] Section 404's triggering of NEPA thus authorizes and demands the consideration of non-water quality values in permitting decisions. To put it in the context of a possible Major Questions Doctrine challenge, the statutory linkage between NEPA and Section 404 is clear and unambiguous.

Indeed, this linkage was highlighted in a 2007 Congressional Research Service Report, which characterized NEPA as an umbrella statute:

> The appropriate NEPA documentation must also indicate any federal permits, licenses, and other entitlements required to implement the proposed project. This means that compliance requirements of any additional environmental laws, regulations, or executive orders must be determined (but not necessarily completed) during the NEPA process.[93]

The report acknowledged that if a project impacted a historic site, then the Advisory Council on Historic Preservation should be engaged as a cooperating agency.[94] Similarly, Section 106 of the National Historic Preservation Act of 1966 instructs agencies to consider an activity's impact on any district, building, structure, site, or object that does or could fall under the purview of the National Register that before issuing any permit or license.[95] In a contemporaneous report published with

[88] 42 U.S.C. § 4332 (C).

[89] 33 C.F.R. § 330.5(b)(3).

[90] Report of the Committee on Environment and Public Works, United States Senate together with additional views to accompany, S. 2734. Water Resources Development Act of 1992 (May 15, 1992).

[91] Id, p. 46.

[92] 42 USC 4331(b).

[93] CRS Report for Congress: The National Environmental Policy Act: Streamlining NEPA, December 6, 2007 at (CRS-7).

[94] Id; 40 C.F.R. §§ 1501.6 and 1508.5.

[95] 16 U.S.C. § 470f.

enactment of the NHPA, the Senate noted that "[Section 106] is intended to insure that the Federal agencies will not work at cross purposes with the goals of historic preservation and provides meaningful review of Federal or federally assisted projects which affect historic properties identifies on the national register."[96]

A Clean Water Act permitting decision can also instigate review under Section 7 of the Endangered Species Act ("ESA"), which requires agencies to consult with the U.S. Fish and Wildlife Service or the National Marine Fisheries Service to evaluate the impacts of any permitting or authorization decisions on threatened or endangered species or critical habitats.[97] The Corps' granting of a Clean Water Act permit plainly constitutes an agency authorization under section 7 of the ESA, indicating that Congress expects ESA review to be part of any impactful Clean Water Act permitting decision.

Simply put, the requirement of NEPA, NHPA, and ESA review along with a Clean Water Act permit—the ripple effect we describe—should continue to prompt holistic executive agency consideration of non-water quality values. This should remain the case regardless of how the *Sackett* Court adjusts the jurisdictional reach of the Clean Water Act.

5 Conclusion

Issuance of a Section 404 permit under the Clean Water Act often triggers a NEPA review and may also trigger analyses under the Endangered Species Act, National Historic Preservation Act, and the Fish and Wildlife Coordination Act, among others—all by congressional design. That is, Congress drafted NEPA, the ESA, and the NHPA to piggyback on other federal permitting proceedings. In doing so, Congress ensured that non-water quality related values would be evaluated in Clean Water Act permit proceedings. Looking back on 50 years of Clean Water Act implementation, our conclusion is that this is a good thing.

The Biden Administration in January 2023 issued a Final Rule asserting its interpretation of the statutory phrase "waters of the United States." The Corps and the EPA claim that theirs is "a durable rule that retains the protections of the longstanding regulatory framework and avoids harms to important aquatic resources, informed by the best available science and consistent with the agencies' determination of the statutory limits on the scope of the 'waters of the United States,' informed by relevant Supreme Court case law."[98]

[96] Preservation of Historic Properties, Report To accompany S. 3035, 89th Congress 2d Session, Senate, July 7, 1966 at 8.

[97] 16 U.S.C. 1536(a)(2). It should be noted that such consultation is unnecessary if the managing agency determines that a particular action will not impact a listed species or designated critical habitat.

[98] U.S. Army Corps of Engineers & U.S. Environmental Protection Agency, Pre-Publication Final Rule Notice: Revised Definition of "Waters of the United States," Docket ID No. EPA-

Yet it remains to be seen how extensively the U.S. Supreme Court's re-evaluation of the Clean Water Act's jurisdictional reach in *Sackett* will threaten to undermine this "durable" definition. As we have shown, the interplay between a broad swath of federal environmental laws is at stake. Stated differently, *Sackett* is obviously a very important Clean Water Act case, but it is likely to be much more than that.

References

Boissoneault L (2019) The Cuyahoga River caught fire at least a dozen times, but no one cared until 1969, Smithsonian Magazine. https://www.smithsonianmag.com/history/cuyahoga-river-caught-fire-least-dozen-times-no-one-cared-until-1969-180972444/

Congressional Research Service (2016) Clean Water Act: a summary of the law. https://crsreports.congress.gov/product/pdf/RL/RL30030

Congressional Research Service (2022) The Major Questions Doctrine. https://crsreports.congress.gov/product/pdf/IF/IF12077#:~:text=Under%20the%20Court's%20formulation%20of,require%20it%20to%20do%20so

Drelich D (2009) Restoring the cornerstone of the Clean Water Act. Columbia J Environ Law 34:267–330

Flournoy AC (2008) Supply, demand, and consequences: the impact of information flow on individual permitting decisions under Section 404 of the Clean Water Act. Ind Law J 83:537–582

Freeman KM (1995) An evaluation of ground water nutrient loading to Priest Lake, Bonner County, Idaho. M.S. Thesis, University of Idaho

Guter L (2011) Sierra Club – Arkansas Chapter. Update on the Turk coal-fired power plant in southwest Arkansas. Spring 2011 Newsletter of the Sierra Club. https://www.sierraclub.org/sites/www.sierraclub.org/files/sce/arkansaschapter/newsletters/ARSC_newsletter_spring2011.pdf

Hill G (1970) The polluted Potomac: sewage and politics create acute capital problem, New York Times. https://www.nytimes.com/1970/07/12/archives/the-polluted-potomac-sewage-and-politics-create-acute-capital.html

Houck O (2006) Katrina to Louisiana: you can't have your wetlands and eat 'em too. In: Arnold G (ed) After the storm: restoring America's Gulf Coast wetlands. Environmental Law Institute, Washington DC, pp 27–29

Houck O, Rolland M (1995) Federalism in wetlands regulation: a consideration of delegation of Clean Water Act Section 404 and related programs to the states. Md Law Rev 54:1242–1314

Idaho Dep't of Fish and Game, Fisheries Bureau (2013) Management Plan for the Conservation of Westslope Cutthroat Trout in Idaho. https://idfg.idaho.gov/old-web/docs/fish/planWestslopeCutthroat.pdf

Jaffe C (2001) Tragedy of the wetlands commons. Va Environ Law J 20:329–364

Jaffe C (2020) Environmental federalism as forum shopping. Wm Mary Environ Law Policy Rev 44:669–699

Martha's Vineyard Museum (2022a) Lighthouses. https://mvmuseum.org/lighthouses/

Martha's Vineyard Museum (2022b) Treasured Beacon: The Edgartown Harbor Light, Martha's Vineyard Museum. https://express.adobe.com/page/kTRoPF0kPI0Wl/

Percival RV et al (2021) Environmental regulation: law, science, and policy, 9th edn. Aspen Publishing, Boston

HQ-OW-2021-0602, at page 185 of 514 (Dec. 30, 2022), at https://www.epa.gov/system/files/documents/2022-12/Pre-Publication%20Final%20Rule%20Notice.pdf.

Sapp WW et al (2006) From the Fields of Runnymede to the Waters of the United States: a historical review of the Clean Water Act and the term "navigable waters". Environ Law Rep 36:10190, 10213

Smith D (2011) 2 Groups Appeal SWEPCO Ruling, Arkansas Democrat-Gazette. https://www.arkansasonline.com/news/2011/jan/15/2-groups-appeal-swepco-ruling-20110115/

Super D (2022) The court reads free-market economics into the Constitution(again), Washington Post. https://www.washingtonpost.com/outlook/2022/07/05/epa-supreme-court-environmental-regulations/

Tarlock AD (2005) The story of Calvert Cliffs: a court construes the National Environmental Policy Act to create a powerful cause of action. In: Lazarus R, Houck O (eds) Environmental law stories. Foundation Press, Santa Barbara, pp 77–108

United States Coast Guard: U.S. Department of Homeland Security (2016) Historic light station information & photography: Massachusetts. https://web.archive.org/web/20170501202446/http://www.uscg.mil/history/weblighthouses/LHMA.asp

Cale Jaffe is Professor of Law at the General Faculty, University of Virginia School of Law.

Aspen Ono Class of 2023, University of Virginia School of Law.

Specialized State Environmental Courts in the U.S.: The Experience of Vermont and Hawai'i

Merideth Wright

Abstract Vermont and Hawai'i are the only two U.S. States that so far have adopted specialized statewide environmental judicial branch courts, Vermont's beginning its work in 1990 and Hawai'i's in 2015. These two states are similar in that they are both small states dependent on tourism and agriculture, with strong environmental laws and awareness, and historically having experienced problems with environmental enforcement and permitting in practice. Hawai'i and Vermont also obviously differ from one another in their climate, biological and cultural diversity, isolation, history and the nature of their vulnerability to different aspects of climate change. This chapter examines the development and structure of the environmental courts in both states, explores their differences and similarities, their innovative practices and judicial training programs, and suggests some areas for further comparative study.

1 Introduction

It is important at the outset to bear in mind that the United States of America is a federal country, and that in the areas of environmental protection, land use, agriculture, and forestry, the states have developed programs, policies, and institutions that may differ greatly from one another. The Tenth Amendment to the U.S. Constitution provides that "[t]he powers not delegated to the United States by the Constitution, nor prohibited by it to the States, are reserved to the States respectively, or to the people."

The author particularly wishes to thank Judge Jeffrey P. Crabtree, Senior Environmental Court judge in the Hawai'i Circuit Court, based in Honolulu, and Professor Denise Antolini of the Environmental Law Program Faculty of the William S. Richardson School of Law, University of Hawai'i at Mānoa, for their contributions to the Hawai'i portions of this Chapter. Any errors in this chapter are, of course, my own.

M. Wright (✉)
Environmental Law Institute, Washington, DC, USA

G. Antonelli et al. (eds.), *Environmental Law Before the Courts*,
https://doi.org/10.1007/978-3-031-41527-2_5

It is also important to bear in mind that the U.S. federal judicial system, which handles matters arising under the federal laws and Constitution, as well as disputes between the states and involving citizens of different states, is independent of and different from each state's own judicial system. In some states, judges are elected; in others, including both Vermont and Hawai'i, they are appointed by the governor after a merit selection process,[1] and undergo a retention process to seek each additional term.

Finally, it is necessary to note that civil courts in the U.S. handle both matters that would be classed as private or civil law in most European legal systems, and matters that would be classed as public law or disputes between citizens and governmental entities considered to be the province of administrative courts in most European systems. Both are distinct from the courts that handle criminal matters.

The state courts have jurisdiction over many environmental matters arising under state law and regulation, and state constitutional provisions, including pollution control, land use and development, agriculture, and forestry, as well as civil lawsuits for compensation due to environmental and health harm such as toxic torts, and major or minor criminal environmental violations under state law.

Vermont and Hawai'i are the only two U.S. States that so far have adopted specialized environmental judicial branch courts. Several other states have some specialized judicial branch courts for other topics, such as the Delaware Court of Chancery specialized in corporate law,[2] and three of the Western U.S. states (Colorado,[3] Montana,[4] and Idaho[5]) for water allocation.

2 Vermont's Experience with an Environmental Court Since Its Founding in 1990

2.1 The U.S. State of Vermont

Vermont is a small, relatively rural state in the northeast of the United States, with a land area of just under 25,000 square kilometers[6] and a resident population of approximately 647,000. Its economy is characterized by agriculture and tourism, receiving approximately 13 million visitors a year,[7] who come to enjoy its forests, streams, lakes and landscape in the summer months and its skiing and other winter

[1] https://ncsctableauserver.org/t/Communications/views/JudicialSelection/JudicialSelection?%3Aembed=y&%3AisGuestRedirectFromVizportal=y.

[2] https://courts.delaware.gov/chancery/.

[3] https://www.courts.state.co.us/Courts/Water/Index.cfm.

[4] https://courts.mt.gov/courts/water/.

[5] http://www.srba.state.id.us.

[6] Approximately the land area of North Macedonia.

[7] accd.vermont.gov/tourism.

sports in the winter months. Vermont is particularly vulnerable to climate change with respect to the snow to support winter tourism, as well as the habitat and conditions for its sugar maple trees that are tapped in the early spring to produce its famous maple syrup.[8]

Vermont is divided into 14 counties, each of which contains at least one courthouse. However, to understand land use and development regulation within the state, discussed further below, it is necessary to understand that the state is divided into 251 towns and municipalities, so that no land within the state does not also lie within a town (or city or gore).[9]

2.2 *Vermont's Judiciary and a Brief History of Its Environmental Court*

Vermont now has a unified court system.[10] The trial courts consist of the general jurisdiction Superior Court, divided into Civil, Criminal, Family, Environmental, and Probate Divisions, and there is a limited jurisdiction Judicial Bureau, which handles traffic violations, and hunting and fishing violations. The Civil, Criminal, Environmental, and Family Divisions are courts of record. The Civil, Criminal, Family and Probate courts are divided into 14 units following the county lines; the Environmental Division is a court of statewide jurisdiction.[11] The Vermont Supreme Court is the sole appellate court; Vermont does not have an intermediate appellate court.

The Vermont Superior Court Environmental Division, formerly called the Vermont Environmental Court[12] (the Court), is a trial-level judicial branch court of record, with state-wide jurisdiction,[13] consisting now of two judges, each of whom sits alone. Its judges have all the powers of the Superior Court judges and may be assigned to hear other matters by the Chief Justice. The organization and jurisdiction of the Court did not change when it was renamed as a division of the Superior Court, and it continues to operate on a statewide basis, with two environmental judges.[14] The environmental judges do not rotate into the other divisions of the Superior Court, although they may be assigned to handle cases or dockets in the

---

[8] https://vt.audubon.org/news/end-maple-maple-sugaring-amid-changing-climate.

[9] http://vtransmaps.vermont.gov/staticMaps/TownCounty.pdf.

[10] 4 V.S.A. § 1. The current version of all the Vermont statutes cited in this chapter may be found at https://legislature.vermont.gov/statutes/.

[11] 4 V.S.A. § 30(b).

[12] The statute initially named it the Environmental Law Division (of the Judiciary as a whole), which created unnecessary confusion as to whether it was within an executive branch agency. The legislature renamed it the Environmental Court in 1995, to clarify that it was a statewide judicial branch court.

[13] 4 V.S.A. §§ 30(a)(1)(D), 30(b).

[14] 4 V.S.A. § 1001(a), (b).

other courts as available and to round out their judicial experience. They decide cases according to the statutory standards applicable to the particular case, and through judicial education and experience, develop expertise in considering the laws and scientific evidence presented in these often complex cases.

The Court was originally created as part of the 1989 Uniform Environmental Enforcement Act.[15] The statute delayed its effective date until the Agency of Natural Resources adopted regulations governing the administrative assessment of monetary penalties (which occurred in July of 1990) and the single environmental judge contemplated by that statute was appointed (which occurred on November 2, 1990).

When the court was created, the legislature did not set it up to handle any cases that might be entitled to a trial by jury, therefore, it was not allocated jurisdiction over criminal environmental cases; these remain within the jurisdiction of the general criminal court. Similarly, due to the right to a jury trial in civil damages cases for private compensation for environmental harm, such cases remain within the jurisdiction of the general civil court, although in both of these types of cases, the Chief Judge of the Superior Court has authority to specially assign an environmental judge to hear the case.[16] Hunting and fishing license violations, on the other hand, are handled by the Judicial Bureau, which is also responsible for traffic tickets.

Chronologically, after the Court's initial creation in 1990 to handle a then-new type of environmental enforcement case brought by the state environmental agency, landfill closure cases[17] and heavy cutting forestry cases were added to its jurisdiction. In 1995, already-existing local land use and development jurisdiction was transferred to it from the general civil courts, and these cases made up the bulk of the Court's caseload. In 2005, the Court was expanded to two judges, to handle further additional jurisdiction of state-level land use and state pollution control permit appeals.

2.3 *Vermont Superior Court Environmental Division Jurisdiction*

The legislation applicable to the Vermont Superior Court Environmental Division covers more than 40 statutory chapters and environmental programs, setting out the court's jurisdiction over state environmental enforcement and state environmental permit appeals,[18] as well as municipal land use permit appeals and enforcement.

[15] 1989, No. 98, § 2.

[16] 4 V.S.A. § 1001(b).

[17] For a period of time in the early 1990s, it had jurisdiction of landfill closure extension orders, during the phasing out of unlined landfills. See 10 V.S.A. § 8008a.

[18] 10 V.S.A. §§ 8003, 8503. This jurisdiction includes air pollution, flood hazard areas, dams and stream flow regulation, stream alteration, water pollution, public water supply and wastewater systems, aquatic nuisance control, wetlands protection, management of lakes and ponds and land

Jurisdiction has been added to the Vermont Superior Court Environmental Division over time, so that the Court's jurisdiction now covers essentially the following types of cases: (1) enforcement of Vermont's state environmental laws and regulations; (2) appeals from decisions of the state environmental agency (Agency of Natural Resources) ruling on a myriad of state environmental water discharge, air emissions, waste disposal, stormwater, heavy logging and other environmental permits; (3) appeals from decisions of the regional district environmental commissions and district coordinators under Vermont's state land-use law, informally known as Act 250 (10 V.S.A. ch. 151), (4) appeals from decisions of the Agency of Agriculture, Food and Markets, regulating the management of animal waste produced by certain farm operations, pursuant to 6 V.S.A. §§ 4855, 4861; and (5) appeals and enforcement of all the municipal planning and zoning (land-use) decisions and ordinances state-wide.

2.3.1 Vermont State Environmental Enforcement Jurisdiction

The Court was initially created to improve the enforcement of Vermont's state environmental laws, including its state land use law. Vermont has had strong environmental and state land use laws since the late 1960s and early 1970s, but their enforcement was uneven.

Each of the different environmental statutes had different enforcement provisions. Some provided for criminal prosecution but not for civil injunctive relief (remedies ordered by the court to take action or refrain from action). Some allowed the state environmental agency to issue administrative orders, but provided no mechanism to enforce those orders. Some allowed the state Attorney General to apply to the civil court for court orders, but did not provide for money penalties to be imposed in those proceedings. Moreover, even when a monetary fine or penalty was available, the statute made no explicit linkage between the economic gain from the violation and the appropriate amount of the penalty, or even any specific linkage between the magnitude of the environmental or public health harm (or risk of harm) and the appropriate penalty.

Moreover, the inspection and prosecution of cases differed, due to the uneven workload of the environmental inspectors. The lack of clear and certain enforcement from program to program led to differences in treatment between one environmental violator and another that were perceived as unfair. Those who spent money to bring their operations into compliance with the laws, or to seek a permit prior to

---

lying under public waters, groundwater protection and withdrawal, beverage containers, under- and aboveground liquid storage tanks, solid waste, hazardous waste, hazardous materials, collection and recycling of electronic waste, comprehensive mercury management, batteries, salvage yards, abandoned vessels, lake shoreland protection, trade in covered animal parts or products, threatened and endangered species, heavy cutting of timber, importation of firewood, regional coordination to reduce greenhouse gases, and the state's land use permitting program for major development (Act 250).

beginning operation, felt at an economic disadvantage if others were able to violate the law without being penalized. Because of this, there was support among the regulated community, as well as from governmental agencies and citizen groups, for a more uniform and predictable approach to environmental enforcement. The Uniform Environmental Enforcement Act[19] was enacted in 1989 to create an environmental enforcement system that is meant to foster both the existence of and the public awareness of even-handedness, consistency, and predictability in the system.[20]

Under this statute, the Agency of Natural Resources was given new authority to issue unilateral administrative enforcement orders[21] that could contain monetary penalties as well as remedial provisions. The statute also provided new inspection authority,[22] and provided for the issuance of emergency orders.[23] The statute allows for consent orders to be filed with the court so that they will later be enforceable,[24] and provides for enforcement of final administrative orders.[25]

The statute also created the Environmental Court and established it in the judicial branch of government, distinct from the executive branch agency responsible for issuing the initial orders. It provided for an unusually prompt[26] hearing to be held on the merits[27] of the order in the independent judicial branch court. By establishing a specialized court within the judicial system for these hearings, the legislature wanted to ensure fair treatment for the respondent in court, and also to ensure consistency from one part of the state to another.

In these types of cases, the hearing is de novo, and the Court can assess a monetary penalty anew, applying the penalty factors provided for in the statute. The Court also has authority to affirm, modify, or reverse some remedial orders, but for other, more technical remedial orders, the Court has authority only either to affirm the order or to vacate and remand it to the agency after making a finding that the order is not reasonably likely to achieve the intended result. This provision

[19] 10 V.S.A., ch. 201.

[20] The stated purposes of the statute are to enhance the protection of environmental and human health, to prevent the unfair economic advantage obtained by persons who operate in violation of environmental laws, to provide for more even-handed enforcement of those environmental laws, to foster greater compliance with, and deter repeated violation of those laws, and to establish a fair and consistent system for assessing penalties. 10 V.S.A. § 8001.

[21] 10 V.S.A. § 8008.

[22] 10 V.S.A. § 8005.

[23] 10 V.S.A. § 8009.

[24] 10 V.S.A. § 8007.

[25] 10 V.S.A. § 8014.

[26] A respondent who receives an administrative order and wishes to contest either the remedial provisions of the order or the amount of a monetary penalty files a "notice of request for hearing." The statute requires the hearing on the merits to be held within 30 days. Respondents rarely are prepared to have the hearing scheduled this rapidly.; in an early pretrial conference, the judge determines whether the respondent wishes to postpone this hearing.

[27] It is important to understand that this is not an appeal of the administrative enforcement order; rather, the statute provides a right to an evidentiary trial on the merits of the order, in which the environmental agency must present evidence and prove its case in the first instance.

recognizes the technical expertise of the state's environmental agency in formulating the remedial requirements of an enforcement order.

Because an important purpose of the statute is to prevent the unfair economic advantage obtained by those who violate the state's environmental laws, the Court has authority to set a penalty amount to remove or recapture the economic benefit resulting from a violation, in addition to and calculated separately from the penalty amount assessed under the other statutory factors.[28] The environmental enforcement statute thus recognizes that effective environmental enforcement depends on accounting for the economics of the violation from the point of view of the violator. The principle is to create an economic incentive for compliance, that is, to make it more expensive to commit a violation of the laws and regulations than to comply with them.

The other statutory factors[29] the Court must consider in assessing a penalty include not only the actual harm to the environment or to public health, safety or welfare resulting from the violation, but also the potential for such harm even if it did not occur or has not yet occurred. Other factors in setting a penalty include the length of time the violation has existed, the respondent's record of compliance, and whether the respondent knew or should have known that the violation existed. The Court is directed to consider the state's actual costs of enforcement, but also to consider any mitigating factors, including whether the state delayed unreasonably between the violation and seeking enforcement.

2.3.2 State Environmental and Land Use Permits: Judicial Review/Merits Appeals Jurisdiction

As well as the many permits issued by the state Agency of Natural Resources that may be necessary for a particular project, Vermont's state land use development review statute[30] ("Act 250") requires review under specific environmental criteria for larger projects that meet the statutory threshold.[31] The Consolidated Environmental Appeals statute[32] effective as of 2005, sought to increase the participation of all affected parties at the earliest stages of a proposed project by making such participation a prerequisite for bringing an appeal, and to consolidate and streamline the various permit application appeals so that they would all go to the Vermont Superior Court Environmental Division.[33] Almost all of these appeals are

[28] 10 V.S.A. § 8010(b), (c).

[29] 10 V.S.A. § 8010(b).

[30] 10 V.S.A. ch. 151.

[31] Some other states, including Hawai'i, require a state-level environmental impact assessment process for larger projects. See, e.g., HI Rev. Stat. ch 343.

[32] 10 V.S.A. ch. 220.

[33] 10 V.S.A. § 8501.

heard de novo on their merits,[34] limited to the issues raised in the Statement of Questions. The Court applies the substantive standards that were applicable before the tribunal appealed from.[35]

With this jurisdiction, the Court is able to coordinate or consolidate proceedings involving all the municipal and state permits required for any particular project, sometimes referred to as one-stop shopping. Although the Act 250 and state environmental appeals are not as numerous as the municipal appeals discussed below, these cases tend to be far more complex and time-consuming.

2.3.3 Municipal Land Use Permits: Appeals and Enforcement Jurisdiction

Approximately 80% of Vermont's 251 municipalities have chosen to regulate land use and development.[36] Most of the local development review boards are not composed of experts and most of them conduct relatively informal proceedings. Jurisdiction over all municipal land use and planning appeals and enforcement cases was transferred in 1995 to the Environmental Court from the general civil courts.[37] Since that time, the highest volume of the work of the Vermont Superior Court Environmental Division has been to handle appeals from municipal zoning and planning decisions statewide, as well as enforcement of municipal land use ordinances and permit decisions.

Almost all the cases are heard de novo, meaning that the Court does not review what the administrative or permit-issuing body has done, but instead hears the evidence in a trial and decides the matter itself based on the local ordinance and the state enabling statute.[38] However, the statute allows municipalities to adopt more formal procedures at the local level,[39] which then allows appeals to the Court to be reviewed on the record.

In order to bring an appeal, appellants are required to have participated in the proceeding at the municipal level. The scope of any appeal is governed by the statement of questions filed by the appellant at the outset of the case, so that a de novo trial, if necessary, is limited to the issues in the appeal.

Cases to enforce municipalities' zoning ordinances directly may be brought by the municipality under 24 V.S.A. §§ 4451, 4452. Enforcement of decisions of the municipal zoning boards, planning commissions, and development review boards,

[34] Vt. Env. Ct. Rule 5(g).

[35] Unlike judicial review in the U.K., but similarly to the Swedish Land and Environment Courts, the court conducts merits or de novo review in most of its cases, rather than only sending cases back to the administrative agency to be done over.

[36] http://vpic.info/Publications/Reports/Implementation/Zoning.pdf (2007).

[37] See 24 V.S.A. §§ 4471, 4472.

[38] 24 V.S.A. ch. 117.

[39] 24 V.S.A. ch. 36.

however, may be brought either by the municipality or by any interested person under 24 V.S.A. § 4470(b).

2.4 Practices and Procedures in the Vermont Superior Court Environmental Division

The Court's procedures are governed by rules promulgated by the Vermont Supreme Court, referred to in the current statute (4 V.S.A. § 30(a)(1)(D)) and in the Reporter's Notes as the "Vermont Rules for Environmental Proceedings."[40] The work of the Court would not be possible without a strong case management system, tailored to the needs of the individual cases, in which the judges, the case manager, the court manager, and the administrative staff of the Court all play important roles.[41]

After a case is filed and docketed at the Court, it is assigned to one of the two judges. To avoid the possibility of predicting which judge will be assigned to any particular case, each judge is assigned alternating cases arising in each of the state's counties. If related cases are filed they will be assigned to the judge who was assigned the first case appealed on a particular project.

In appeals, the appellant must file a Statement of Questions defining the issues on appeal, and must give other potential parties the opportunity to come into the case. Most of the cases involve several different parties: for example, the developer of a project, the neighbors or people who may be affected by the project, possibly other people who support or oppose the project for various economic or environmental reasons, and the municipality or state agency responsible for regulating the project. It is therefore not necessarily easy to determine the way in which litigants' interests are aligned with one another.

Litigants may be represented by an attorney, but there is no requirement for attorney representation. People may and do represent themselves; in fact, most of the cases in the Vermont Superior Court Environmental Division involve at least some self-represented parties, appearing without a lawyer. The Court developed materials to explain procedures and the process of mediation to self-represented litigants; these are now available on the Vermont Judiciary website.[42] The challenge for the judges in handling cases with self-represented litigants is both to accommodate their need for procedural information and, at the same time, to treat them the same as represented parties with respect to the merits or substance of the case.

[40] The rules are available are available through the Vermont Judiciary website (www.vermontjudiciary.org), by clicking on "Attorneys" and then on "Proposed and Promulgated Rules", which page (https://www.vermontjudiciary.org/attorneys/rules) provides a free link to "Lexis-Nexis - Vermont Statutes, Court Rules and Administrative Code", landing on the page for the court rules, which refers to them as the Vermont Rules for Environmental Court Proceedings as provided in the text of Rule 7, but cites them as Vt. Env. Ct. Rule [#].

[41] See generally Vt. Env. Ct. Rule 2.

[42] https://www.vermontjudiciary.org/environmental.

In this regard, it is worth noting that disputes involving litigants' homes, property, and surroundings may be extremely emotional in the same way as disputes that reach family court. In some respects, these disputes can be more difficult for the judge to manage than the level of emotion in family court, because after a divorce or custody dispute is concluded, the litigants have the possibility of moving away and putting the dispute behind them. But the environmental court litigants, unless they move away, will have to continue to live next to each other or next to the project on into the future, whether they have succeeded in the litigation or not. For this reason, it is very important to maintain a level of civility and respect in the entire process, especially because, for many people, their experience in the courtroom with these cases will be their only experience with the judicial system.

After all the parties have come into the case, an initial pretrial conference is held with the judge by telephone, and recorded. The purpose of these conferences, governed by Vt. Env. Ct. Rule 2, is to give each case its appropriate scheduling, to require mediation[43] in appropriate cases, and to establish an appropriate sequence for related cases, including whether they should be scheduled together for a single hearing, or whether some cases should be placed on inactive status pending resolution of other related cases.

Ensuring that each case gets its appropriate and timely consideration may require an expedited schedule, but some cases instead require a delayed schedule to achieve efficiency. For instance, if an applicant is going back to submit a revised application to the local authority, or has other related permit applications pending before other state agencies, it may make sense to put the initial appeal on hold, so that any challenges to any of the permits required for the project could be heard together, rather than holding successive trials on similar evidence. In many cases, it is necessary to address pre-trial motions which may resolve or narrow the issues in the case or resolve issues of party status, jurisdiction, and timeliness.

The pretrial conference also develops an estimate of the time required for trial and when it should be scheduled, and results in a written scheduling order setting deadlines for all the steps discussed at the conference. A follow-up conference may be set with the judge or the case manager if anything in the order needs to be changed as the case moves forward. A few weeks before trial, the case manager holds a final

[43] Although the Environmental Court had some success with mediation on a voluntary and occasional basis prior to 2005, the revised rules as of 2005 gave the Environmental Court the authority to require the parties to mediate. Mediators are not provided by the Court and the parties may use any mediator, not only the ones on the roster of mediators who have taken the Court's training about jurisdiction and procedures. A study conducted after about five years of operation of the mediation rules found that the judges were ordering mediation in more than a third of the active disputes, that is, in cases that were not filed as consent orders or settled before the judges' initial conference with the parties. Of the cases in which mediation was ordered, more than three-quarters of them resolved through mediation. Mediation is not only an important case management tool, but also provides an opportunity for the litigants to air and possibly resolve important issues that are beyond the scope of the case before the Court. Once the litigants understand that mediation may provide an opportunity to resolve these underlying issues, it can be successful in the most contentious and surprising cases.

pretrial conference with the parties, covering issues such as the marking of exhibits, whether any prefiled evidence will be submitted,[44] and whether and when a site visit is needed.

The cases that go to trial are heard by the judge sitting alone, without a jury; all trials and other hearings are recorded. Unless the parties prefer to have the trial scheduled for another place, it will be scheduled in a courthouse near the area where the case arises.[45] Trials are conducted like any other civil non-jury trial. Under the so-called American rule as to litigation costs, each party bears its own costs of litigation.

Because no record can be made at a site visit, a site visit can only be illustrative of evidence presented at a hearing. However, in the Vermont Superior Court Environmental Division the judges conduct site visits in almost every case that goes to trial, because they are so useful in fully understanding the parties' testimony, plans, and photographs.[46] Depending on the available time, the season of the year, and the nature of the case, the site visit may be conducted on the day of trial, or may be conducted in advance of or after the trial. For example, in cases in which the nature of vegetative screening of a project is at issue in a temperate climate like Vermont, it may be necessary to take two site visits, one at a time at which leaves are present on the deciduous trees, and another when the trees are bare.

At or before the conclusion of the trial, a schedule is set for the filing of any post-trial memoranda, usually within a short time of trial. However, in complex cases in which a great deal of evidence has been presented, the parties may request a more extended schedule to file these documents. When self-represented parties are involved, the judge explains that this is an opportunity to make any arguments in writing about "what you want the court to decide in this case, and why."

Most of the Vermont Superior Court Environmental Division decisions after trial are issued in writing. As a practical matter, the parties, their contractors and anyone later searching titles of the involved properties, need to be able to determine what permit constraints and conditions affect a particular property.

---

[44] If prefiled testimony is submitted, the witness must appear in court at the trial to answer questions on cross-examination. Vt. Env. Ct. Rule 2(e)(2)(B).

[45] 4 V.S.A. § 1001(e).

[46] The practice of conducting site visits in environmental cases is specifically authorized by 4 V.S.A. §1001(g)(4) and Vt. Env. Ct. Rule 2(e)(3).

3 Hawai'i's Experience with an Environmental Court Since Its Founding in 2015[47]

3.1 The U.S. State of Hawai'i

Hawai'i is a small and isolated state in the Pacific Ocean; the archipelago of islands that makes up the state of Hawai'i is the most remote and isolated land mass in the world. The land area of Hawai'i amounts to a little over 28,000 square kilometers,[48] but extends over a far larger area of ocean. It has a resident population of 1.41 million people. Hawai'i's biodiversity is extraordinary; although it makes up less than 0.2% of U.S. land, it is home to 25% of the species on the U.S. endangered species list. The traditional cultural, religious, and environmental rights of the Native Hawai'ian population are specifically recognized and protected in Hawai'i's constitution,[49] as well as in specific statutes such as those protecting caves and ancient burials.[50]

Hawai'i's economy is characterized by tourism, agriculture and the military. Hawai'i receives approximately 10 million visitors a year, which creates unique challenges for both its infrastructure and its natural resources. Hawai'i is at risk from the effects of climate change; both the tourist hotel area of Waikiki on the island of Oahu and the military base are extremely vulnerable to sea level rise.

Unlike most of the states in the U.S.,[51] the Hawai'i Constitution contains specific and powerful provisions relating to the environment. Article XI, Section 1 lays out the state's obligations to protect Hawai'i's natural resources for the benefit of future as well as present generations, and recognizes the public trust in natural resources

> For the benefit of present and future generations, the State and its political subdivisions shall conserve and protect Hawaii's natural beauty and all natural resources, including land, water, air, minerals and energy sources, and shall promote the development and utilization of these resources in a manner consistent with their conservation and in furtherance of the self-sufficiency of the State. All public natural resources are held in trust by the State for the benefit of the people.

Article XI, Section 9 recognizes that "[e]ach person has the right to a clean and healthful environment, as defined by laws relating to environmental quality,

---

[47] Judge Jeffrey Crabtree presentation to U.S. Conference of Chief Justices, 2 February 2020; Prof. Denise Antolini lecture "New State Environmental Courts in Hawaii - Challenges and Opportunities", available at leap.unep.org/countries/us/case-studies/video-lecturelnew-state-environmental-courts-hawaii-challenges-and, 2021.

[48] Similar to that of Albania, or a little smaller than Belgium.

[49] Article XII, Section 7. https://lrb.hawaii.gov/constitution/.

[50] See, generally, MacKenzie et al. (2015), Forman and Serrano (2012).

[51] Six U.S. states have environmental protections in their constitutions: Hawai'i, Illinois, Massachusetts, Montana, New York, and Pennsylvania.. See Dernbach, J. chapter in this publication; also https://www.bdlaw.com/publications/new-york-becomes-the-third-state-to-adopt-a-constitutional-green-amendment/.

including control of pollution and conservation, protection and enhancement of natural resources," and recognizes a cause of action, stating that "[a]ny person may enforce this right against any party, public or private, through appropriate legal proceedings … ."

3.2 *Hawai'i's Judiciary and a Brief History of Its Environmental Court*

Hawai'i's court system consists of two levels of trial court: the District Court, which handles primarily criminal cases and more minor violations, and the Circuit Court, which deals with major civil cases, jury trials, and felonies (more major criminal cases). Courts and judges are located in each of four counties (comprising an island or group of islands).

Hawai'i also has two levels of appellate court: an intermediate appellate court and the Hawai'i Supreme Court. The appellate courts handle appeals from the Environmental Court but do not have specially designated judges for those cases.

Recognizing that enforcement of Hawai'i's environmental law had fallen behind, and that enforcement programs had been historically underfunded and understaffed, in about 2010, citizen groups and NGOs on the island of Maui and others began to suggest to the state legislature that Hawai'i could benefit from the creation of an environmental court.[52]

The Hawai'i Environmental Court was established by legislation[53] in 2014, and took full effect in July of 2015 after a year's work by a study commission required by the statute. The stated purpose of the statute is "to promote and protect Hawaii's natural environment through consistent and uniform application of environmental laws by establishing environmental courts." The legislature made the following findings in support of that purpose:

(1) that environmental disputes are currently dealt with in a variety of courts, [and that t]his organizational structure inadvertently promotes inconsistent application of the wide variety of environmental laws.
(2) that the continued maintenance and improvement of Hawaii's environment requires constant vigilance and continued stewardship to ensure its lasting beauty, cleanliness, uniqueness, and the stability of its natural systems, all of which enhance the mental and physical well-being of Hawaii's people.
(3) that Hawaii's natural resources are compromised every day resulting in numerous violations of the law [and that a]n environmental court will better ensure

[52] Prof. Denise Antolini 2021 lecture, "New State Environmental Courts in Hawaii - Challenges and Opportunities", available at leap.unep.org/countries/us/case-studies/video-lecturelnew-state-environmental-courts-hawaii-challenges-and.

[53] https://www.capitol.hawaii.gov/sessions/session2014/bills/SB632_CD1_.pdf.

that the State upholds its constitutional obligation to protect the public trust for the benefit of all beneficiaries.

3.3 *Hawai'i Environmental Court Jurisdiction*

Hawai'i created its Environmental Courts as divisions within its District and Circuit trial-level courts rather than by establishing a separate state-wide court.[54] Under the statute, the chief justice of the Hawai'i Supreme Court designates an environmental judge or judges for each of the four judicial circuits, and designates multiple District Court judges in each circuit.[55] These are judges who have gone through the same nominating and appointment process as any of the other Circuit or District judges; and they exercise the full power and authority of the Circuit or District court judges.[56]

The designated Environmental Court judges are given judicial education and training specific to the environmental laws and science that they will face in carrying out their responsibilities. This training is especially valuable, given that many of the cases are complex cases of first impression, and trigger broad and sometimes immediate public impact.

These judges continue to handle their court's full regular calendars, as well as their environmental case responsibilities, which can be challenging. The environmental cases, particularly in circuit court, can be complex and time consuming. The environmental cases are generally subject to the same procedural rules as all other state-wide litigation.

The legislation governing the Hawai'i Environmental Court covers over fifty chapters and sub-chapters, setting out the court's original and exclusive jurisdiction for both civil and criminal cases.[57] Any given case is either in or out of Environmental Court depending on whether it meets the statutory subject matter jurisdiction. For example, the Environmental Courts have adjudicated cases involving prohibited self-help for shoreline erosion of homes; the size of fishing nets; fishing in certain seasons and locations; crimes against wildlife; declaratory relief regarding gas water heaters; cases with a cultural component involving iwi (ancestral bones); and injunctive relief regarding an Army Corp of Engineers project for Waikiki and its upland valleys, a project involving storage tanks containing 180 million gallons of

---

[54] HI Rev Stat § 604A-1(a).

[55] HI Rev Stat § 604A-1(b).

[56] HI Rev Stat §604A-2(b) :"shall exercise general equity powers as authorized by law".

[57] HI Rev Stat § 604A-2(a). This jurisdiction includes protection of caves, historic preservation including ancient burial sites, the environmental response law, litter control, electronic and other special waste and recycling, solid waste management and pollution, safe drinking water, air pollution, ozone layer protection, water pollution, non-point-source (runoff) pollution, noise pollution, hazardous waste, underground storage tanks, asbestos and lead, uniform environmental covenants, environmental impact statements, and public lands.

aviation fuel located immediately above an underground natural aquifer providing most of Honolulu's water.

3.3.1 Hawai'i State Environmental Enforcement Jurisdiction

The District Courts handle criminal enforcement cases for the smaller criminal offenses or civil infractions, such as taking too many fish or fishing in a protected area. The Circuit Courts handle felonies, any cases requiring a jury trial, and "direct actions" including cases between private sector and government agencies.

On the islands[58] of Maui, Hawai'i (sometimes referred to as The Big Island), and Kauai, the designated circuit court judges hear both criminal and civil Environmental Court cases. On Oahu, the most populous island, where Honolulu is located, the Circuit Court has criminal and civil divisions, within each of which judges are designated to hear the Environmental Court criminal or civil circuit court cases.

All eligible cases go to those judges by virtue of the original and exclusive jurisdiction provisions in the enabling legislation. The Chief Justice also has discretion to assign any case to the Environmental Court to promote uniform application of environmental laws.[59]

3.3.2 Hawai'i State Environmental Permits and Environmental Impact Statements: Judicial Review Jurisdiction

The Circuit Court designated environmental judges handle those cases requiring a jury trial, cases requesting declaratory relief, appeals from administrative agency decisions, and "direct filing" civil and criminal cases.

When hearing an agency appeal, conclusions of law are reviewed *de novo*, while an agency's factual findings are reviewed for clear error. Mixed questions of fact and law are reviewed under the clearly erroneous standard, that is, if the administrative record lacks substantial evidence to support the finding or determination. Regarding remedies, by statute the Environmental Court is authorized to exercise general equity powers.[60]

---

[58] Under Hawaii's venue statute (HI Rev Stat § 603-36), many Environmental Court cases can have a choice of venue on the island where they arose, and also on Oahu. For example, a case arising from streams on Maui could be filed in either Maui Environmental Court or Oahu Environmental Court. Oahu is an alternative venue primarily because that is often where defendants are domiciled.

[59] HI Rev Stat § 604A-2(a)(2).

[60] HI Rev Stat § 604A-2(b).

4 Reflections from the Experience of Vermont's and Hawaii's Environmental Courts

4.1 The Impartiality and Integrity of Judges

It is always important to clarify to the general public, and to the lawyers and other users of the courts, that our courts and judges with specialized jurisdiction and who receive continuing judicial education about environmental law and science, remain impartial judges, ready to consider each case on its merits, based on the law, the facts, and the rules of procedure.

4.2 Judges' Authority to Ask Questions of Witnesses and to Appoint a Witness

It is not generally recognized by people familiar with so-called adversarial legal systems that judges in U.S. jurisdictions also have authority to question witnesses and to call witnesses on behalf of the court.[61] In both Vermont and Hawai'i, this authority is embodied in Rule 614 of the states' respective Rules of Evidence, entitled "Calling and interrogation of witness by court", which provides in subsection (a) that "[t]he court may, on its own motion or at the suggestion of a party, call witnesses, and all parties are entitled to cross-examine witnesses thus called"; and in subsection (b) that "[t]he court may interrogate witnesses, whether called by itself or by a party."

Nevertheless, even with this authority, the tradition for judges in the common-law jurisdictions is generally to refrain from questioning witnesses. This tradition is most understandable in jury trials, as the judge's involvement, from the jury's perspective, can give undue importance or emphasis to that evidence. But at least in non-jury cases, and even in cases with a jury if enough care is taken, judicial questioning may be helpful to elicit relevant evidence that has not emerged under questioning by the parties and their lawyers. In environmental cases the judge has the responsibility to decide the case according to the applicable constitutional, statutory, and regulatory standards, and therefore may elicit evidence necessary to carry out those responsibilities.[62]

---

[61] See, e.g., Vt. R. Evid. 614; Hawai'i R. Evid. 614; Fed. R. Evid. 614. In many U.S. jurisdictions, the court also has specific authority to appoint expert witnesses, either on its own motion or by agreement of the parties. See, e.g., Vt. R. Evid. 706; Fed. R. Evid. 706; and Kamahalo v. Coelho, 24 H. 689, 694 (1919), in which the Hawai'i Supreme Court upheld the trial court's calling of a handwriting expert, ruling that: "[C]ourts have from the earliest period exercised the right to call in experts to aid them in their deliberations and this right we concede."

[62] In *United States v. Ostendorff*, 371 F.2d 729, 732 (4th Cir. 1967), the appellate court noted that "[w]e think counsel misconceives the proper function of a … [trial] judge. [The judge] is not a

The author's practice has been to wait until after the parties or their lawyers have presented their evidence and have asked all the questions they wish to ask of a particular witness, and then to ask any additional question only if it is necessary in order to carry out the court's task under the particular statute, regulation, or ordinance. After a judicial question, the parties should then be allowed to ask any follow-up question, not limited by the usual rules of direct- and cross-examination questioning.

With respect to the court's appointment of a witness, especially an expert, issues of funding to pay for such a witness, or how to allocate the cost among the parties, and whether it should only be done if the parties have requested it by motion, may make it difficult for the judge to take such a step.

4.3 Judicial Education and Accumulation of Experience

The First Global Report on the Environmental Rule of Law, published by United Nations Environment in 2019, emphasized the importance of judicial education in Section 5.3.2 headed "Capable Judges Acting with Integrity." The report stated that "[d]ue to the complexity and technical nature of many environmental matters, it is particularly important that judges be knowledgeable and competent regarding environmental law."

Judges designated to serve on an environmental court will benefit from continuing judicial education courses to maintain and develop competency in understanding the specialized environmental laws and in being able to assess[63] scientific and technical evidence.

Under the direction of law professor Denise Antolini, the Environmental Law Program of the University of Hawaii at Manoa William S. Richardson School of Law has partnered with the Hawai'i Judiciary to hold trainings for the Hawai'i Environmental Court every other year since the court started in 2015. Topics include the operation of the court, legal developments, natural resources management, and cutting edge science. The Environmental Law Program has also produced a Benchbook for the Environmental Court judges and a parallel Hawai'i Natural Resources Law Enforcement Manual for the legal community.

This kind of judicial education allows law-trained judges to acquire a certain facility or level of comfort with scientific and technical evidence, so as to be able to

---

bump on a log, nor even a referee at a prizefight. [The judge] has not only the right, but ... has the duty to participate in the examination of witnesses when necessary to bring out matters that have been insufficiently developed by counsel. [The judge] is in charge of the trial and may exercise ... control to assure that the jury is not [misled] by unfair phrasing of questions by counsel."

[63] The judges in the Vermont and Hawai'i courts are law-trained judges; they are not expected to have a background in the underlying scientific or engineering fields. There are successful environmental courts, notably in Sweden and in New South Wales, that use technically trained judges as well as law-trained judges. Darpö (2019), Schultz (2019), and Preston (2012).

consider it independently. It is extremely important that the judge be able to hear the expert testimony with a critical ear, not just to weigh the testimony according to the credentials of the particular expert. In addition to gaining technical expertise, training helps the trial judge analyze and research issues faster and in more depth, which is helpful to any judge, especially those with a heavy caseload.

In addition, the designated trial judge who handles more environmental cases gains experience, context, and a deeper understanding of the subject matter. These cases can be extremely complex, with major public impact based on trial court rulings which will often remain in place through lengthy appeals. This is particularly true since many environmental cases involve unsettled areas of law with a serious potential for long term consequences. Moreover, it is critical for the trial judge to understand the subject matter and procedures as thoroughly as possible to help establish a good record for the appellate courts.

4.4 Transparency of Judicial Reasoning and Publication of Decisions

When the Vermont Court began its work in 1990, it was not the custom, at least in Vermont, for any trial court (first-instance) decisions to be published. That is, although decisions on motions for summary judgment and on the merits of non-jury trials were issued in writing by the trial judge, and were available to the participants in the case and their lawyers, these decisions were not readily available to the public.

Especially for a newly-established court, it is important to reveal to all observers what the reasoning is for any given decision—a feature that is sometimes called the system's "transparency." The transparency of the rationale for each decision, and the clarity of language in which it is written, is particularly important so that the decision can be understood by the litigants themselves, not only by their lawyers, and can be understood by members of the community who may not have been following the ongoing litigation.

From the beginning, therefore, the Vermont Superior Court Environmental Division issued its decisions in writing and published them electronically, and they are available on the Vermont Judiciary website[64] These published decisions include dispositive pretrial motions as well as decisions on the merits of cases; the fact that they are freely available in electronic form has greatly assisted the development of consistency and predictability in the areas of the law within the Court's jurisdiction.

It is important to understand that the usefulness of a body of published decisions is not restricted to the common-law legal systems.[65] Access to these decisions is important not because the decisions are precedent, but because their reasoning is

[64] https://www.vermontjudiciary.org/environmental.

[65] In any event, the jurisdiction of both courts discussed in this chapter is primarily in the realm of statutory and regulatory public law, rather than judge-made doctrine.

persuasive or useful in future cases. That is, to the extent that the decisions as a whole present the rationale of particular recurring topics, the body of decisions functions like a persuasive treatise on those areas of the law.

5 Conclusion and Topics for Future Research

As these two courts continue their work, it may be interesting and useful to conduct comparative research on some of their features as compared with each other, as compared with these cases in the general jurisdiction courts of other U.S. States, and as compared with the practices of similar specialized environmental courts and green benches around the world.[66] In particular, topics could include:

- the designation of general jurisdiction judges who retain their other case docket responsibilities, compared with environmental judges who are primarily responsible only for environmental court cases, and how to determine when and whether they should also be assigned to other cases in the general judicial system.
- having major criminal environmental cases heard in the environmental court, with all the special procedures for jury trials.
- having major civil environmental tort and nuisance cases heard in the environmental court, with all the special procedures for jury trials.
- evaluating such courts' experience with different standards of judicial review and merits appeals from administrative action.
- having minor (misdemeanor and civil) infractions going to some level of environmental court, as compared with a judicial bureau set up to handle a high volume of small non-jury cases.
- providing designated environmental judges with continuing judicial education provided by an in-state law school.
- evaluating whether the application of environmental law becomes more uniform and predictable when fewer but more experienced judges are administering it.
- evaluating whether an environmental court created by legislation, such as those in both these states, represents support for and acceptance of the new court by all three branches of government.
- evaluating the practice of mediation in environmental court cases and safeguards against imbalance of power among the mediating parties.
- evaluating the practice and usefulness of site visits in environmental cases.

[66] See, generally, Pring and Pring (2009, 2016). http://wedocs.unep.org/bitstream/handle/20.500.11822/10001/environmental-courts-tribunals.pdf?sequence=1; Journal of Court Innovation (3)(1) (Winter 2010) Issue on The Role of the Environmental Judiciary; Pace Environmental Law Review, 29(2) (Winter 2012) Special Edition, Environmental Courts and Tribunals: Improving Access to Justice and Protection of the Environment Around the World. Asian Development Bank (2011, 2013, 2015), Greiber (2006), Lavrysen (2009, 2014), Macrory (2014), May and Daly (2015), Mergen (2016), Preston (2014), United Nations Environment Programme (2019, 2022), Wright (2010, 2019).

- evaluating the remedies available to be imposed and actually imposed in environmental cases, and whether they are effective in having the result actually carried out in the real world.

In conclusion, this chapter provides a window into the work of two specialized environmental courts at the state level in the United States, and reflects on some of the most important features critical for the overall success of these two statewide environmental courts. We recognize that there is always room for improvement, and continue to work to maintain courts characterized by the fairness and respect with which all litigants are treated, and adapted to the needs of often-complex environmental litigation. This is important not only for the actual procedural fairness in any particular case, but for the fundamental respect for the rule of law that develops with the consistent experience of fairness in the environmental court system, especially as the state courts take on more cases related to the challenges of climate change and sustainability, and must consider the needs of the present world and of future generations.

References

Antolini D (2021) Lecture, "New State Environmental Courts in Hawaii - Challenges and Opportunities". Available at leap.unep.org/countries/us/case-studies/video-lecturelnew-state-environmental-courts-hawaii-challenges-and

Asian Development Bank (2011) Asian judges symposium on environmental decision making, the rule of law, and environmental justice: the proceedings of the symposium. https://www.adb.org/sites/default/files/publication/29631/symposium-environmental-decisions-law-justice.pdf

Asian Development Bank (2013) Proceedings of the Second South Asia Judicial Roundtable on Environmental Justice. Mandaluyong City, Philippines

Asian Development Bank (2015) Proceedings of the Fourth ASEAN Chief Justices Roundtable on Environment: Role of the Judiciary in Environmental Protection. Mandaluyong City, Philippines

Darpö J (2019) Understanding the nuts and bolts: scientific and technical knowledge in environmental litigation – national solutions, EU requirements and current challenges. In: Squintani L, Darpö J, Lavrysen L, Stoll PT (eds) Managing facts and feelings in environmental governance. Edward Elgar Publishing, Cheltenham, pp 82–102

Forman DM, Serrano SK (2012) Ho'ohana Aku, a Ho'ōla Aku: a legal primer for traditional and customary rights in Hawai'i

Greiber T (ed) (2006) Judges and the rule of law. Creating the links: environment, human rights and poverty. IUCN, Gland and Cambridge

Journal of Court Innovation (3)(1) (Winter 2010) Issue on The Role of the Environmental Judiciary

Lavrysen L (2009) Chapter 2: Belgium. In: Kotzé LJ, Paterson AR (eds) The role of the judiciary in environmental governance: comparative perspectives. Kluwer Law International, Alphen aan den Rijn, pp 85–122

Lavrysen L (2014) Access to justice in environmental matters: perspective from the European Union forum of judges for the environment. Global Symposium on the Environmental Rule of Law, Proceedings. https://biblio.ugent.be/publication/4426506/file/4426507.pdf

MacKenzie MK, Serrano SK, Sproat DK (eds) (2015) Native Hawaiian law - a treatise. Kamehameha Publishing, Honolulu

Macrory R (2014) Regulation, enforcement and governance in environmental law, 2nd edn. Oxford, UK

May J, Daly E (2015) Global environmental constitutionalism. CUP, Cambridge

Mergen A (2016) An environmental court for Hawai'i-will other states follow? (ABA Section on Environment, Energy and Resources). Available at https://www.americanbar.org/groups/environment_energy_resources/publications/trends/2015-2016/january-february-2016/an_environmental_court_for_hawaii-will_other_states_follow/

Pace Environmental Law Review, 29(2) (Winter 2012) Special Edition, Environmental Courts and Tribunals: Improving Access to Justice and Protection of the Environment Around the World

Preston B (2012) Benefits of judicial specialization in environmental law: the land and environment court of New South Wales as a case study. Pace Environ Law Rev 29(2):396–440

Preston B (2014) Characteristics of successful environmental courts and tribunals. J Environ Law 26(3):281–282. https://doi.org/10.1093/jel/equ019

Pring G, Pring C (2009) Greening justice: creating and improving environmental courts and tribunals. The Access Initiative. https://www.eufje.org/images/DocDivers/Rapport%20Pring.pdf

Pring G, Pring C (2016) Environmental courts & tribunals. A guide for policy makers. UNEP, Nairobi

Schultz M (2019) Scientific evidence in Swedish courts: the use of technical judges for better integration of scientific data in environmental decision-making. In: Squintani L, Darpö J, Lavrysen L, Stoll PT (eds) Managing facts and feelings in environmental governance. Edward Elgar Publishing, Cheltenham, pp 118–135

United Nations Environment Programme (2019) First Global Report on the Environmental Rule of Law, Chapter 5: Justice. Nairobi. https://www.unep.org/resources/assessment/environmental-rule-law-first-global-report

United Nations Environment Programme (2022) Environmental Courts and Tribunals - 2021: A Guide for Policymakers, Nairobi. https://wedocs.unep.org/handle/20.500.11822/40309

Wright M (2010) The Vermont Environmental Court. J Court Innov 3(Winter):201–214. Available at https://www.nycourts.gov/court-innovation/Winter-2010/index.shtml

Wright M (2019) Scientific facts and litigants' feelings: practical innovations from the Vermont Environmental Court and other jurisdictions. In: Squintani L, Darpö J, Lavrysen L, Stoll PT (eds) Managing facts and feelings in environmental governance. Edward Elgar Publishing, Cheltenham, pp 103–117

Merideth Wright is associated with the Environmental Law Institute as its Distinguished Judicial Scholar, involved with judicial education projects, research, and capacity-building to support the environmental work of the judiciary globally. She served as an environmental judge in Vermont from the founding of its Environmental Court in 1990 to 2011. She can be reached at envj.wright@gmail.com.

Protection of the Environment and the Court of Justice of the European Union

Jan M. Passer and Věra Pazderová

Abstract Using the example of air protection, this Chapter presents the role of the Court of Justice of the European Union (CJEU) in shaping the field of environmental law. Its role is twofold. First, as will be demonstrated by a series of infringement proceedings in which various Member States have been condemned for failing to fulfil their obligations under EU law, the CJEU contributes to the enforcement of EU law, specifically to the enforcement of the Ambient Air Quality Directives. Second, by way of the preliminary ruling procedure, the CJEU provides guidance to national courts on the interpretation and application of EU law. As will be shown, preliminary references sometimes raise challenging and politically sensitive issues, such as whether EU law setting the limit values for certain air pollutants confers rights on individuals, the breach of which could render a Member State liable for loss and damage caused to them. In addition, it will be highlighted that the CJEU is not the sole player in this field. National courts have also proved to be effective guardians of compliance with the obligations imposed on Member States by the Ambient Air Quality Directives.

1 Introduction

The ever increasing amount of EU legislation on protection of the environment confirms that this topic has been at the heart of the EU's attention. As a consequence, there is certainly no shortage of the case-law of the Court of Justice of the EU (CJEU) illustrating various facets of the protection of the environment. The ambition to present this case-law in its entirety would be destined to fail and would significantly exceed the scope and purpose of the Chapter. We have, therefore, opted

J. M. Passer (✉)
Court of Justice of the EU, Luxembourg City, Luxembourg
e-mail: jan.passer@curia.europa.eu

V. Pazderová
Regional Court in Prague, Prague, Czech Republic

G. Antonelli et al. (eds.), *Environmental Law Before the Courts*,
https://doi.org/10.1007/978-3-031-41527-2_6

for a narrower approach. The role of the CJEU in the field of environmental protection can be well illustrated on the example of air protection. This is one of the fundamental components of the environment and achieving adequate limits on air pollution or reducing air pollution is a challenge in the global context and within the European Union (EU).

The core of this Chapter is divided in three main parts. Section 2, providing the reader with the necessary background information on the legislative framework in the field of air protection, will serve as a catalyst for the subsequent presentation of the case-law of the CJEU in this field in Sect. 3. The case-law includes a series of direct actions showing that Member States are still struggling to comply with the requirements of the relevant directives, as well as several preliminary rulings dealing, inter alia, with the question of compensation for adverse effects to health resulting from an exceedance of the limit values for different air pollutants. Additionally, Sect. 4 demonstrates that national courts have a role to play in this debate as well. Finally, Sect. 5 will highlight the main arguments and conclusions submitted throughout this Chapter.

2 The Legislative Framework in the Field of Air Protection

2.1 Air Pollution, a Global and European Challenge

On September 2021, the World Health Organization (WHO) released new global air quality guidelines for six pollutants, namely particulate matter $PM_{2.5}$ and PM_{10}, ozone (O_3), nitrogen dioxide (NO_2), sulfur dioxide (SO_2), and carbon monoxide (CO).[1]

These guidelines "present clear evidence of adverse health effects from air pollution at even lower concentrations than previously thought". According to the WHO, "[e]xposure to air pollution causes an estimated seven million premature deaths and the loss of millions of years of healthy life each year. In children, air pollution may even impair lung development, limit lung function, cause respiratory infections and aggravate asthma. In adults, ischemic heart disease and stroke are the most common causes of premature death due to outdoor air pollution, and new evidence also shows that it can cause diabetes and neurodegenerative diseases. In terms of disease burden, air pollution is therefore comparable to other major health risk factors worldwide, such as poor diet and smoking." Along with climate change, air pollution is said to be "one of the major environmental threats to health".[2]

---

[1] WHO global air quality guidelines: particulate matter ($PM_{2.5}$ and PM_{10}), ozone, nitrogen dioxide, sulfur dioxide and carbon monoxide (https://apps.who.int/iris/handle/10665/345329; accessed 21 February 2023).

[2] See further WHO, *New WHO Global Air Quality Guidelines aim to save millions of lives from air pollution*, News release, 22 September 2021 (https://www.who.int/news/item/22-09-2021-new-

According to European Environment Agency (EEA) "Air pollution is the largest environmental health risk in Europe and significantly impacts the health of the European population, particularly in urban areas. While emissions of key air pollutants and their concentrations in ambient air have fallen significantly over the past two decades in Europe, air quality remains poor in many areas."[3] "In 2020 in the European Union, 96% of the urban population was exposed to levels of fine particulate matter above the health-based guideline level set by the World Health Organization."[4]

The sources of air pollution are many, and solving the problem of air pollution is a complex task. Some of the main sources of air pollution are anthropogenic and inherent to the way we live our life today. One can mention the burning of fossil fuels to generate electricity, for transport, industry and households, and the releasing of chemicals and solvents in chemical and agricultural industries. Data of the EEA indicate that more than 40% of emissions of nitrogen oxides come from road transport, and almost 40% of $PM_{2,5}$ emissions come from transport. Businesses, public buildings and households contribute to around half of $PM_{2,5}$ and carbon monoxide emissions.[5]

Because of the adverse effects of air pollution not only on human health, but also on ecosystems and on the environment in general, and because of the complexity of the problem, the EU has adopted a large number of policies to reduce air pollution, which has substantially decreased the emissions of many air pollutants over the past decades.

Nevertheless, the EU action to protect human health from air pollution has not delivered its expected impact, according to the 2018 report from the European Court of Auditors (ECA). The ECA warns that every year, air pollution causes hundreds of billions of euros in health-related external costs. It also suggests that some of the EU air quality standards are much weaker than the WHO guidelines and what the latest scientific evidence suggests. Moreover, "[while] emissions of air pollutants have been decreasing, most Member States still do not comply with the EU's air quality standards and are not taking enough effective action to improve air quality. [...] Air Quality Plans – a key requirement under the Ambient Air Quality Directive – have often not delivered their expected results. There are limitations in the European Commission's monitoring of Member States' performance in meeting air quality targets. Its enforcement procedures so far have not ensured that Member States comply with the air quality limits set by the Directive. Despite the Commission taking legal action against many Member States and achieving favourable rulings, Member States continue to breach air quality limits frequently [...] Air pollution is

who-global-air-quality-guidelines-aim-to-save-millions-of-lives-from-air-pollution; accessed 21 February 2023).

[3] EEA, Air pollution (https://www.eea.europa.eu/themes/air; accessed 21 February 2023).

[4] EEA, Air quality in Europe 2022 (https://www.eea.europa.eu/publications/air-quality-in-europe-2022; accessed 21 February 2023).

[5] EEA, Air pollution sources (https://www.eea.europa.eu/themes/air/air-pollution-sources-1; accessed 21 February 2023).

a pressing public health issue, and public awareness and information has a critical role to play in addressing it. Recently, citizens have been getting more involved in air quality issues and have brought cases before national courts, which in several Member States have ruled in favour of their right to clean air."[6]

2.2 *The Ambient Air Quality Directive*

The legislative instrument used by the EU in the context of air regulation has been mainly directives. Directives are addressed to Member States of the EU and they are binding, as to the result to be achieved, upon each Member State, but they leave to the national authorities the choice of form and method.[7]

The CJEU has established a number of principles in its case-law that allow individuals to invoke rights arising from incorrectly or not at all implemented directives.[8] However, there are limits to this possibility for individuals, stemming from the nature of the directives—which is reflected, inter alia, in the considerable differences in the ability of various actors to enforce Member States' obligations. A direct action against a Member State that has failed to fulfil its obligations is in principle open only to the Commission or another Member State.[9] Individuals are dependent on the cooperation of the CJEU and the courts of the Member States in preliminary ruling procedures,[10] which, combined with the limitations imposed on individuals in claiming rights arising from directives that are not sufficiently or not at all implemented, are less effective than direct actions.

These differences are also very clearly illustrated in the case-law of the CJEU on air pollution limits described below.

Our focus is on the Directive 2008/50/EC of the European Parliament and of the Council on ambient air quality and cleaner air for Europe.[11] We will outline the challenges faced by the CJEU in the field of environmental protection through the example of its interpretation and application.

The Directive 2008/50 is part of a legislative package, collectively known as the Ambient Air Quality Directives. The package aims to define common methods for monitoring, assessing and reporting on ambient air quality in the EU and to set targets for ambient air quality in order to prevent, avoid or reduce harmful effects on

---

[6]ECA, Press release, 11 September 2018 (https://www.eca.europa.eu/Lists/ECADocuments/INSR18_23/INSR_AIR_QUALITY_EN.pdf; accessed 21 February 2023).

[7]Article 288 of the Treaty on the Functioning of the European Union (hereinafter "TFEU").

[8]The most important one being the direct effect, for the first time introduced in the judgment of the CJEU of 4 December 1974, 41/74, *Van Duyn v Home Office*, followed by a principle of state liability for damages for non-implementation of a directive, for the first time introduced in the judgment of 19 November 1991, C-6, 9/90, *Francovich and Bonifaci v Italy*.

[9]Articles 258 and 259 TFEU.

[10]Article 267 TFEU.

[11]Hereinafter "Directive 2008/50".

human health and the environment. The Ambient Air Quality Directives set EU air quality standards for 12 air pollutants: sulphur dioxide, nitrogen dioxide/nitrogen oxides, particulate matter (PM_{10}, $PM_{2.5}$), ozone, benzene, lead, carbon monoxide, arsenic, cadmium, nickel and benzo(a)pyrene.[12]

The Directive 2008/50 itself provides for measures aimed at defining and setting ambient air quality objectives to prevent, avoid or reduce harmful effects on human health and the environment as a whole; assessing ambient air quality in the Member States on the basis of common methods and criteria; obtaining information on ambient air quality to help combat air pollution and nuisance and to monitor long-term trends and improvements resulting from national and Community measures; ensuring that this information on ambient air quality is made available to the public; maintaining air quality where it is good and improving it otherwise; and promoting greater cooperation between Member States in reducing air pollution.[13]

Member States are obliged to ensure that, throughout their zones and agglomerations, levels of sulphur dioxide, PM_{10}, lead, and carbon monoxide and of nitrogen dioxide and benzene in ambient air do not exceed the limit values laid down in the Annexes of that directive.[14]

This provision thus creates an obligation of result not subject to any exceptions according to the Directive 2008/50. According to the 18th recital in the preamble of that directive, air quality plans should be developed for zones and agglomerations within which concentrations of pollutants in ambient air exceed the relevant air target values or limit values.

More specifically, Article 23(1) of the Directive 2008/50 stipulates that where, in given zones or agglomerations, the levels of pollutants in ambient air exceed any limit value or target value, plus any relevant margin of tolerance in each case, Member States are obliged to ensure that air quality plans are established for those zones and agglomerations in order to achieve the related limit value or target value. In the event of exceedances of those limit values for which the attainment deadline is already expired, the air quality plans should set out appropriate measures, so that the exceedance period can be kept as short as possible. The air quality plans may also additionally include specific measures aiming at the protection of sensitive population groups, including children.

In 2022, the Commission presented a proposal to revise the Ambient Air Quality Directives.[15] These directives are part of a comprehensive clean air policy framework built on three main pillars. The first pillar is represented by the Ambient Air Quality Directives themselves, which set quality standards for concentration levels of 12 ambient air pollutants. The second pillar includes the Directive on the

---

[12] Air Quality (https://environment.ec.europa.eu/topics/air/air-quality_en; accessed 21 February 2023).

[13] Article 1 of the Directive 2008/50.

[14] Article 13(1) of the Directive 2008/50.

[15] Proposal for a directive of the European Parliament and of the Council on ambient air quality and cleaner air for Europe (recast), COM(2022) 542 final/2 (https://eur-lex.europa.eu/legal-content/EN/TXT/HTML/?uri=CELEX:52022PC0542&from=EN; accessed 21 February 2023).

reduction of national emissions of certain air pollutants (the NEC Directive),[16] which sets out national commitments to reduce emissions of the main air pollutants and their precursors and works within the EU to achieve collective reductions in pollution across national borders. This is complemented by international efforts, notably through the UNECE Air Convention, to reduce emissions across national borders outside the EU14. And finally, the third pillar consists of legislation setting emission standards for key sources of air pollution such as road vehicles, domestic heating and industrial installations. As noted in the proposal, the amount of pollution from these sources is also influenced by other policies that affect key activities and sectors in areas such as transport, industry, energy and climate and agriculture. The proposal also emphasises that a number of these policies are part of recent initiatives adopted under the European Green Deal, such as the Zero Pollution Action Plan, the European Climate Act and the Fit for 55 package with initiatives on energy efficiency and renewable energy, the Methane Strategy, the Sustainable and Smart Mobility Strategy, the related new Urban Mobility Framework 2021, the Biodiversity Strategy and the Farm to Fork Initiative. In addition, the adoption and implementation of the forthcoming Euro 7 proposal[17] is expected to result in significant reductions in pollutant emissions from cars, vans, trucks and buses. The revision of the Ambient Air Quality Directives would merge these directives into one with the aim to align EU air quality standards with WHO recommendations, further improve the legislative framework (e.g. in relation to penalties and public information), and better support local authorities in achieving cleaner air by strengthening monitoring, modelling and air quality plans.

It follows from the above that the Commission's approach to air quality is very ambitious and has led to sustained pressure on Member States to deal with this component of the environment. Moreover, the drive to further revise and tighten the Ambient Air Quality Directives comes at a time when even the directives currently in force face difficulties in implementation by (some) Member States.

It is a fact, that even if Member States take their responsibilities under the Ambient Air Quality Directives seriously, many of them have failed to reach the limit values or to establish air quality plans, which has resulted in a considerable amount of infringement procedures under Article 258 TFEU having been brought to the CJEU by the Commission.

Moreover, one important aspect that cannot be overlooked in this context is the considerable political sensitivity of the measures taken by the Member State. As mentioned above, although in certain areas the source (or one of the significant sources) of pollution may be identifiable, this is not the case in particular in urban areas. Sources of pollution defined in a general way (e.g. transport, emissions from buildings, etc.) are related to the lives and lifestyles of a large number of people (and thus potential voters). Thus, on the one hand, the political representation is confronted with a demand from the general public to ensure the air quality standards of

[16] Directive 2016/2284/EU.

[17] See PLAN/2020/6308.

the Ambient Air Quality Directives, while, on the other hand, the same public is not necessarily prepared to bear the costs of a dramatic change in their lifestyle. The adoption of an air quality plan in the sense of the Directive 2008/50 can thus be a real Gordian knot in terms of balancing different interests, which defies the simple application of certain principles of environmental law, such as the "polluter pays" principle.

3 Air Protection in the Case-Law of the Court of Justice

3.1 Infringement Procedures

A typical type of an action in response to a Member State's failure to comply with its obligations under the Directive 2008/50 is an action for breach of obligations under EU law.

Within these actions, the CJEU decides whether a Member State has failed to fulfil its obligations under EU law.[18] This type of proceedings reflects to some extent the early roots of the EU as an international law organisation *sui generis*, whereby organs of the community of states or those states directly can sue others for breach of obligations related to the membership in the community/organisation (some might certainly argue that this is rather a parallel to the federal model, but such an argument has no basis in the foundations from which the EU has developed).

If the Commission considers that a Member State has not fulfilled an obligation under the Treaties, it must first give the State the opportunity to make a statement and then the Commission will issue a reasoned opinion on the matter. Only if the State fails to comply with the opinion within the time limit set by the Commission, may the Commission bring an action for infringement of EU law before the CJEU.[19]

Although the action is typically brought by the Commission, it can also be brought by another Member State, which is obliged to refer the case to the Commission first. However, the Commission's opinion does not affect the ability of that Member State to bring a subsequent action against another Member State.[20]

If the CJEU finds that a Member State has failed to comply with an obligation under EU law, it must immediately put an end to the failure. If, after further action by the Commission, the CJEU finds that the Member State concerned has not complied with its judgment, it may impose a fixed or periodic financial penalty. However, if an action is based on the grounds that the Member State concerned has failed to fulfil its obligation to notify measures transposing a directive adopted under a legislative procedure, the Commission may propose to the CJEU to impose a financial

[18] CURIA Presentation - Court of Justice of the European Union (https://curia.europa.eu/jcms/jcms/Jo2_7024/en/; accessed 21 February 2023).

[19] Article 258 TFEU.

[20] Article 259 TFEU.

penalty on the Member State concerned as soon as the initial judgment finding non-compliance has been delivered.[21]

In the context of proceedings for breach of the Treaty/obligations under EU law and in relation to non-compliance with air pollution limits, we can find a number of judgments, some of which relate to legislation predating the current Ambient Air Quality Directives.

As indicated before, it is clear that many Member States find it difficult to comply with their obligations in this area. As stated by Advocate General J. Kokott in the case *JP*[22] discussed below, the CJEU has found in twelve infringement proceedings that ten Member States did not meet the ambient air quality standards. The nine most recent judgments even found a systematic and persistent breach of the standards. At the time of writing of her opinion, there were seven pending proceedings involving, among others, three other Member States and at least in Belgium, Germany, France and the United Kingdom, the national courts have heard cases concerning air quality standards.

For the exceedance of limit values of particular matter 10 (PM_{10}) under the Directive 1999/30/EC,[23] there are judgments against Slovenia,[24] Sweden,[25] Portugal[26] and Italy.[27] With the exception of the case *Commission v Sweden*, which was decided by a three-judge panel, the basic five-judge panel has always ruled. It can therefore be concluded that, in terms of relevance to EU law, the CJEU considered these cases to be standard ones. At the same time, in all of these cases, the CJEU has ruled without an Advocate General's opinion, and it can therefore be deduced that it has not considered the cases to be of any legal difficulty or novelty.

As examples of the exceedance of limit values of particular matter 10 (PM_{10}) under the Directive 2008/50, we find judgments against Bulgaria,[28] Poland,[29] Hungary,[30] Romania,[31] Italy[32] and Slovakia[33] and additionally, there is a Greek case still pending.[34] Out of all these cases, only the Bulgarian one was decided by a

---

[21] Article 260 TFEU.

[22] Opinion of AG J. Kokott, 5 May 2022, C-61/21, *JP v Ministre de la Transition écologique, Premier ministre*, para 96.

[23] Council Directive 1999/30/EC of 22 April 1999 relating to limit values for sulphur dioxide, nitrogen dioxide and oxides of nitrogen, particulate matter and lead in ambient air, effective until 10 June 2010.

[24] Judgment of 24 March 2011, C-365/10, *Commission v Slovenia.*

[25] Judgment of 10 May 2011, C-479/10, *Commission v Sweden.*

[26] Judgment of 15 November 2012, C-34/11, *Commission v Portugal.*

[27] Judgment of 19 December 2012, C-68/11, *Commission v Italy.*

[28] Judgment of 5 April 2017, C-485/15, *Commission v Bulgaria.*

[29] Judgment of 22 February 2018, C-336/16, *Commission v Poland.*

[30] Judgment of 3 February 2021, C-637/18, *Commission v Hungary.*

[31] Judgment of 30 April 2020, C-638/18, *Commission v Romania.*

[32] Judgment of 10 November 2020, C-644/18, *Commission v Italy.*

[33] Judgment of 9 February 2023, C-342/21, *Commission v Slovakia.*

[34] Case C-70/21, *Commission v Greece*, pending.

five-judge panel with an opinion of the Advocate General. Thus, it should have apparently serve as a pilot case. The following case, *Commission v Poland* was still judged by a five-judge panel, but already without an opinion of an Advocate General. The other cases (against Romania, Hungary and Slovakia) were decided by a three-judge panel without an opinion of an Advocate General which indicates that they were considered to be a more or less automatic application of well-established case-law in clear factual circumstances. The exception represented by the case *Commission v Italy*, that was decided by the Grand Chamber, does not signal anything about the importance or particularities of that case since the composition of the chamber was the result of a qualified request of Italy[35] under Article 16 of the Statute of the Court of Justice.[36] Even this case was decided without an opinion of an Advocate General. With the exception of the Polish case, in all other of these cases the CJEU found systematic and persistent exceedance of limit values/violation of the EU law.

Furthermore, there are also five judgments of the CJEU concerning the exceedance of limit values of nitrogen dioxide (NO_2) under the Directive 2008/50. These were delivered in proceedings of the Commission against Germany,[37] the United Kingdom,[38] Spain,[39] Italy[40] and Greece.[41] All of these cases were decided by a three-judge panel without an Advocate General's opinion and were therefore apparently considered to be an automatic application of well-established case-law in clear factual circumstances. Their reasoning follows judgments in cases dealing with the exceedance of limit values of PM_{10}. The fact that different pollutant was involved did not play any role in considerations of the CJEU. Moreover, like in the previously mentioned line of the case-law, in all of these cases the CJEU found systematic and persistent exceedance of limit values/violation of the EU law.

So far, there is only one example of the procedure for the exceedance of limit values of sulphur dioxide (SO_2), represented by the judgment *Commission v Bulgaria*.[42] Identically with the previous cases, the CJEU in a three-judge panel without an opinion of an Advocate General concluded that Bulgaria exceeded limit values/violated the EU law in systematic and persistent manner. This case, as we shall see below, has had wider implications. Exceedances of air pollution limits in a given area, specifically SO_2, together with this Court's judgment played a role in one of the preliminary ruling procedures discussed further below (Sect. 3.3).

In addition, there are two other decisions of the CJEU which fit into the series of infringement proceedings described above. In the case against France, the CJEU

---

[35] Judgment of 10 November 2020, C-644/18, *Commission v Italy*, para 25.

[36] tra-doc-en-div-c-0000-2016-201606984-05_00.pdf; accessed 21 February 2023.

[37] Judgment of 3 June 2021, C-635/18, *Commission v Germany*.

[38] Judgment of 3 March 2021, C-664/18, *Commission v the United Kingdom*.

[39] Judgment of 22 December 2021, C-125/20, *Commission v Spain*.

[40] Judgment of 12 May 2022, C-573/19, *Commission v Italy*.

[41] Judgment of 16 February 2023, C-633/21, *Commission v Greece*.

[42] Judgment of 12 May 2022, C-730/19, *Commission v Bulgaria*.

found systematic and persistent exceedances of limit values for NO_2 in 12 French agglomerations and air quality zones, including Paris.[43] France has thus failed to fulfil its obligations under Article 13(1) and Article 23(1) of the Directive 2008/50, in particular the obligation to ensure that the period of exceedance is as short as possible.

According to the CJEU, the exceedance of the limit values had already been recorded at the time when the obligation to draw up an air quality plan under the Directive 2008/50 arose. The French Government was therefore obliged to adopt and implement appropriate measures within the meaning of the Article 23(1) of that directive, which it manifestly failed to do, since the exceedances of the limit values persisted for seven consecutive years, even though France was obliged to take all appropriate and effective measures to comply with the requirement that the period of exceedances be as short as possible.

In the second infringement proceedings, also against France, concerning PM_{10}, the CJEU concluded that France had failed to ensure that the limit value for the concentration of the pollutant in the Paris area was complied with.[44] France has thus failed to fulfil its obligations under Article 13(1) of the Directive 2008/50 and, also in this case, has failed to ensure that the period during which the limit value is exceeded is as short as possible.

These two cases are no different in substance or result from the cases cited above. Once again, the CJEU ruled only in a three-judge panel without an opinion of an Advocate General. The reason why these two cases are mentioned separately is the context of the national proceedings, which will be described in more detail below.[45] It confirms the view that the CJEU cannot be regarded as the sole player even in the field of rights guaranteed by the EU's legal order, since the full protection of those rights is afforded to individuals only through the CJEU's cooperation with the courts of the Member States. The readiness of the national courts to deal with the EU aspects of the cases they decide is crucial in this respect.

3.2 *Infringement Procedures: Penalties?*

All of the above infringement proceedings against Member States have been brought under Article 258 TFEU. No Member State has yet been fined or ordered to pay a lump sum under Article 260 TFEU for failure to comply with its obligations under Articles 13 and 23 of the Directive 2008/50. This could theoretically change as the Commission has brought an action before the CJEU against Bulgaria under the latter Article of the TFEU for Bulgaria's failure to comply to date with the judgment

[43] Judgment of 24 October 2019, C-636/18, *Commission v France*.

[44] Judgment of 28 April 2022, C-286/21, *Commission v France*.

[45] See below Sect. 4.

of the CJEU in the case C-488/15[46] (by which the CJEU found that Bulgaria had breached its obligations under Articles 13 and 23 of the Directive 2008/50).[47] The Commission's action seeks an order requiring Bulgaria to pay a daily lump sum as well as a penalty payment until Bulgaria complies with the above-mentioned judgment and its obligations under the Directive 2008/50.

The Court has not yet ruled and the case is still pending. However, the opinion of Advocate General J. Kokott has already been delivered.[48] The Advocate General proposes that the CJEU should partially grant the Commission's application and conclude that, as at 9 February 2019, Bulgaria had failed fully to comply with the judgment in the case C-488/15 with regard to its obligation under the second subparagraph of Article 23(1) of the Directive 2008/50. At the same time, however, the Advocate General advises the CJEU that on the basis of the findings to date, there is no reason to impose a penalty payment. Nor does a lump sum on account of the insufficient compliance with that judgment as at 9 February 2019 seems logical to her, as Bulgaria should instead invest the scarce resources in air quality improvement.

3.3 *Preliminary References*

Much of EU law is applied at national level. In such cases, the CJEU cooperates with the courts of the Member States in preliminary ruling proceedings.[49] The division of competences can be simplistically characterised as the CJEU having the power to interpret EU law and the national courts having the power to apply EU law to a specific case. In the absence of any recourse from national courts to the CJEU on matters of EU law, the preliminary ruling procedure is in principle the only way of effectively and consistently interpreting and applying EU law. Thus, national courts generally can, and in some cases even must, refer to the CJEU a preliminary question concerning the interpretation of EU law. In addition, a reference for a preliminary ruling may also seek a review of the validity of an act of EU law. Although the CJEU is not empowered to interpret national law, an interpretation of a related provision of EU law may be necessary in order for a national court to ascertain whether a national law is compatible with EU law.

The Court's response takes the form of a judgment or a reasoned order. The national court to which the preliminary ruling is addressed is bound by the CJEU's interpretation in deciding the dispute before it. Likewise, the judgment of the CJEU is binding on the other national courts before which the same issue has arisen.

It is therefore by means of references for a preliminary ruling that any European citizen can seek clarification of the EU rules that concern him. Although only a

[46] Judgment of 5 April 2017, C-488/15, *Commission v Bulgaria*.

[47] Case C-174/21, *Commission v Bulgaria*, pending.

[48] Opinion of AG J. Kokott, 17 November 2022, C-174/21, *Commission v Bulgaria*.

[49] Article 267 TFEU.

national court may make such a reference, all parties to the proceedings before the CJEU, the Member States and the institutions of the EU may participate in the proceedings before the CJEU.[50] It can be added that as a consequence of the relatively limited standing in direct actions, the typical route for an individual to the CJEU is through the preliminary ruling procedure.

3.3.1 *Sdruzhenie "Za Zemjata – dostap do pravosadie"*[51]

The first of the cases referred to in the section on preliminary questions is indirectly related to the infringement proceedings against Bulgaria in relation to SO_2 air pollution mentioned above.[52]

"TETS Maritsa iztok 2 EAD" is the largest of the four thermal power plants forming part of the energy complex for the production of electricity in the "Maritsa iztok" mining basin, located in Bulgaria. The power plant is considered to be one of the main sources of industrial pollution in the territory of the municipality of Galabovo as was proved also in an infringement procedure pending against Bulgaria for systematic and persistent infringement of the rules on sulphur dioxide in the municipality of Galabovo.[53]

The main proceedings (before a national court) concern a decision taken by the Bulgarian authorities to grant the thermal power plant "TETS Maritsa-iztok 2 EAD" a derogation, accepting emission limit values for sulfur dioxide (SO_2) and mercury (Hg) different from those set in accordance with the Commission's implementing decision 2017/1442.

In principle, the competent authority shall, pursuant to Article 15(3) of the Directive 2010/75,[54] set emission limit values ensuring that emissions under normal operating conditions do not exceed the emission levels associated with the best available techniques (BAT) as described in the decisions on BAT conclusions referred to in Article 13(5) of that directive. However, the first subparagraph of Article 15(4) of the Directive 2010/75 provides that, by way of derogation from paragraph 3 and without prejudice to Article 18, the competent authority may, in specific cases, set less stringent emission limit values. In accordance with Article 18, if an environmental quality standard requires more stringent conditions than those achievable by the use of the BAT, additional measures shall be included in the permit, without prejudice to other measures that may be taken to meet the environmental quality standards.

---

[50] CURIA Presentation - Court of Justice of the European Union (https://curia.europa.eu/jcms/jcms/Jo2_7024/en/; accessed 21 February 2023).

[51] Case C-375/21, *Sdruzhenie 'Za Zemyata – dostap do pravosadie' and Others*, pending.

[52] Judgment of 12 May 2022, C-730/19, *Commission v Bulgaria.*

[53] Judgment of 12 May 2022, C-730/19, *Commission v Bulgaria.*

[54] Directive 2010/75/EU of the European Parliament and of the Council of 24 November 2010 on industrial emissions (integrated pollution prevention and control).

The reference in Article 15(4) of the Directive 2010/75 to Article 18 of that directive is the reason which led the national court to ask whether the permissible limit values for sulphur dioxide laid down in the Directive 2008/50 and/or the measures provided for in the air quality plan drawn up in accordance with Article 23 of the Directive 2008/50 constitute such environmental quality standards within the meaning of Article 18 of the Directive 2010/75.

At the heart of this case is therefore the novel question of the delineation of the relationship between the Directive 2010/75 and the Directive 2008/50. These two directives share common objectives, namely to achieve a high level of environmental protection (recital 29 and Article 1 of the Directive 2010/75 as well as recitals 2 and 30 and Article 1 of the Directive 2008/50). These two directives adopt complementary approaches. The Directive 2010/75 sets out the rules for the operation of industrial installations on the basis of BAT, the "polluter pays" principle and the principle of preventing pollution by intervening at source (recital 2 of the Directive 2010/75). The Directive 2008/50, in Article 13, determines the air quality limit values that may not be exceeded as an hourly or daily average and, in Article 23(1), states that if the air quality limit values are exceeded after the deadline for their application, the air quality plans must provide for measures to ensure that the air quality standards are met as soon as possible.

The judgment of the CJEU has not been delivered yet. However, the opinion of the Advocate General J. Kokott is already available.[55] She has proposed to the CJEU to rule that according to the fourth subparagraph of Article 15(4) of the Directive 2010/75, the granting of a derogation under Article 15(4) of that directive for the release of an air pollutant whose air quality limit value under Article 13 of and Annex XI to the Directive 2008/50 is exceeded in the area affected by that emission has to fulfil two conditions: first, one or more air quality plans under Article 23 of the Directive 2008/50 must ensure beyond reasonable scientific doubt that the period of non-compliance can be kept as short as possible throughout the affected area. Second, the derogation must comply with the requirements of those air quality plans. She further proposes that the grant of a permit for an industrial installation for which, under Articles 11, 14, 15 and 18 of the Directive 2010/75, limit values are to be set for the release of an air pollutant whose air quality limit value under Article 13 of and Annex XI to the Directive 2008/50 is being exceeded in the area affected by that release is also subject to two conditions: first, one or more air quality plans under Article 23 of the Directive 2008/50 must ensure that the period of non-compliance can be kept as short as possible. Second, the permit must comply with the requirements of those air quality plans.

The Advocate General's opinion illustrates the importance and critical role of air quality plans. Moreover, it suggests that the failure of a Member State to comply with its obligations is not necessarily only an exceedance of air-pollution limits.

[55] Opinion of AG J. Kokott, 22 September 2022, C-375/21, *Sdruzhenie 'Za Zemyata – dostap do pravosadie' and Others*.

Rather, what is important is the trend and strategy (adoption of the air quality plan), which shows goodwill and a solution-oriented approach.

3.3.2 Deutsche Umwelthilfe

In the *Deutsche Umwelthilfe judgment*,[56] the CJEU, sitting in the Grand Chamber, expressed its views on whether national courts are entitled, or even obliged, to order the compulsory detention of persons representing national authorities who persistently refuse to comply with a court order requiring them to comply with obligations under EU law.

The CJEU was asked to give a ruling in a dispute between Deutsche Umwelthilfe (a German environmental organisation) and the Land of Bavaria concerning the latter's persistent refusal to take the measures necessary to comply with a limit value for NO_2 in the city of Munich in the context of the implementation of the Directive 2008/50. Despite of an order (in 2012), requiring it to amend its air quality action plan applicable in that city, and another order (in 2016) requiring it to comply with its obligations under threat of a financial penalty, inter alia, by imposing bans on the operation of certain diesel vehicles in various urban areas, the Land of Bavaria nevertheless refused to comply with those orders. It was therefore ordered to pay a financial penalty of EUR 4000 in 2017, which it did. As it continued to refuse to comply with the above mentioned orders and publicly stated that it would not comply with its obligations, Deutsche Umwelthilfe brought a new action seeking both the payment of a new financial penalty of EUR 4000, which was granted, and the compulsory detention of the persons at the head of the Land of Bavaria, which was rejected.[57]

In the action brought by the Land of Bavaria, the referring court upheld the payment of the fine and decided to refer a preliminary question to the CJEU concerning the possible order for coercive detention. The referring court concluded that the order to pay the fines could not lead to a change in the behaviour of the Land of Bavaria, since those fines are counted as income of the Land and therefore do not lead to any economic loss, and that the use of coercive detention was precluded on national constitutional grounds. In the preliminary question referred to the CJEU, the national court essentially sought to determine whether EU law, in particular the right to an effective remedy guaranteed by Article 47 of the Charter of Fundamental Rights of the European Union (hereinafter "the Charter"), must be interpreted as authorising or even obliging the national courts to adopt such a measure.[58]

The CJEU has held that in circumstances where a national authority persistently refuses to comply with a court order requiring it to comply with a clear, precise and unconditional obligation under EU law, in particular the Directive 2008/50, the

[56] Judgment of 19 December 2018, C-752/18, *Deutsche Umwelthilfe*.

[57] Ibid, paras 15–20.

[58] Ibid, paras 21–29.

competent national court is obliged to order the compulsory detention of persons at the head of the Land if two conditions are met. First, national law must contain a legal basis for the adoption of such a measure which is sufficiently accessible, precise and predictable in its application. Secondly, the principle of proportionality must be respected.[59]

In that regard, the CJEU recalled that the Member States are required, when implementing EU law, to ensure that the right to effective judicial protection, which is guaranteed both by Article 47 of the Charter and, in the field of the environment, by Article 9(4) of the Aarhus Convention,[60] is respected. That right is all the more important since failure to adopt the measures required by the Directive 2008/50 would endanger human health. National legislation which renders a court judgment ineffective does not correspond to the essential content of that right and deprives it of any useful effect. The CJEU pointed out that, in such a situation, it is for the national court to interpret its national law in a manner which is as consistent as possible with the objectives pursued by those provisions or, failing that, to disapply any provision of national law which is contrary to EU law having direct effect.[61]

However, the CJEU has also explained that compliance with the latter obligation cannot lead to a violation of another fundamental right, namely the right to liberty, which is guaranteed by Article 6 of the Charter and which is limited by compulsory detention. Given that the right to effective judicial protection is not absolute and may be limited in accordance with Article 52(1) of the Charter, the fundamental rights concerned must be weighed against each other. In order to meet the requirements of Article 52(1) of the Charter, the law empowering the court to deprive a person of his liberty must, first of all, be sufficiently accessible, precise and predictable in its application to avoid any risk of arbitrariness, which is a matter for the referring court to assess. Moreover, since the order for detention in custody entails deprivation of liberty, such an order may be made only if, in compliance with the requirements of the principle of proportionality, there are no less restrictive measures (such as, in particular, heavy fines which are repeated over a short period of time and the payment of which does not ultimately benefit the budget from which they are financed), which was also a matter for the referring court to assess. Only if it were to be concluded that the restriction of the right to liberty resulting from a detention order complied with those conditions would EU law not only permit but also require the application of such a measure.[62]

However, the CJEU added that a breach of the Directive 2008/50 may be found by the CJEU in an action for failure to fulfil obligations under EU law, as indeed

---

[59] Ibid, paras 46, 50 and 56. See also https://curia.europa.eu/juris/document/document.jsf?text=&docid=221831&pageIndex=0&doclang=EN&mode=req&dir=&occ=first&part=1&cid=3345855; accessed 21 February 2023.

[60] Convention on access to information, public participation in decision-making and access to justice in environmental matters, signed in Aarhus on 25 June 1998 and approved on behalf of the European Community by Council Decision 2005/370/EC of 17 February 2005.

[61] Ibid, para 34–42.

[62] Ibid, paras 43–52.

happened in the German case,[63] or may give rise to liability of the State for the resulting damage, within the meaning of the Francovich principle.[64]

In particular, the latter reference, which was made by the CJEU more or less as an obiter dicta, without being preceded by specific reasoning dealing with the fulfilment of the criteria of that principle (although it may be assumed that the Court had a more detailed debate on the subject in the deliberations than the obiter dicta itself suggests), became subsequently the focal point of another case in which the CJEU dealt with the protection of the individual in the context of exceedances of air pollution limits and related infringements of EU law.

3.3.3 Ministre de la Transition écologique and Premier ministre

This case has its roots in an action brought before a French Administrative Court by JP, a resident in part of the agglomeration of Paris, who sought, inter alia, compensation from France for damage related to the deterioration of his health allegedly caused by the deterioration of the ambient air quality in that agglomeration. That deterioration resulted from exceedances of the NO_2 and PM_{10} concentration limit values, fixed by the Directive 2008/50, due to the failure of the French authorities to comply with their obligations under Articles 13 and 23 of that directive.

JP's action was dismissed with the argument, in essence, that the provisions relied on by him of the Directive 2008/50 do not confer any right on individuals to obtain compensation for any damage suffered as a result of the deterioration of air quality. Subsequently, JP brought an appeal against that judgment before the Administrative Court of Appeal in Versailles, which decided to ask the CJEU for preliminary ruling.

The CJEU, sitting in the Grand Chamber, clarified the conditions under which a Member State incurs liability for damage caused to an individual by the deterioration of air quality as a result of the exceedance of the limit values for pollutants in the ambient air.[65]

The CJEU recalled that the engagement of State liability by individuals requires three cumulative conditions to be satisfied, namely that: the rule of EU law infringed must be intended to confer rights on them; the infringement of that rule must be sufficiently serious; and there must be a direct causal link between that infringement and the loss or damage sustained by those individuals.[66]

Relying on its well-established case-law the CJEU emphasised that as regards the first condition those rights conferred on individuals arise not only where they are expressly granted by provisions of EU law, but also by reason of positive or negative

[63] Judgment of 3 June 2021, C-635/18, *Commission v Germany*.

[64] *Deutsche Umwelthilfe*, paras 53–55.

[65] Judgment of 22 December 2022, C-61/21, *JP v Ministre de la Transition écologique, Premier ministre*.

[66] Ibid, para 44.

obligations which those provisions impose in a clearly defined manner, whether on individuals, on the Member States or on the EU institutions. The breach of such positive or negative obligations by a Member State is liable to hinder the exercise of rights implicitly conferred on individuals by the provisions in question and thus to alter the legal situation which those provisions seek to establish for them. That is the reason why the full effectiveness of those rules and the protection of the rights that they confer require that individuals have the possibility of obtaining redress, irrespective of whether the provisions in question have direct effect, the quality of direct effect being neither necessary nor sufficient in itself for that first condition to be satisfied.[67]

The CJEU noted that Articles 13(1) and 23(1) of Directive 2008/50, like the analogous provisions of the preceding directives, lay down fairly clear and precise obligations as to the result to be achieved by Member States. Therefore, they oblige Member States, in essence, first, to ensure that the levels of, inter alia, PM_{10} and NO_2 do not exceed, in their respective territories and with effect from certain dates, the limit values set by those directives and, second, where those limit values are nonetheless exceeded, an obligation to provide for appropriate measures to remedy those exceedances, inter alia by means of air quality plans. However, those obligations pursue a general objective of protecting human health and the environment as a whole and it cannot be inferred that they implicitly confer rights on individuals, the breach of which would be capable of giving rise to a Member State's liability for loss and damage caused to them. Therefore, the CJEU concluded that the first of the three conditions, which are cumulative, for State liability to be incurred is not satisfied.[68]

The CJEU added that its conclusion could not be altered as a result of the right that individuals are recognised as having, under it's case-law, to require the national authorities, if necessary by bringing an action before the courts having jurisdiction, to adopt an air quality plan in the event that the limit values referred to in the Directive 2008/50 are exceeded. That right, which stems in particular from the principle of effectiveness of EU law, effectiveness to which affected individuals are entitled to contribute by bringing administrative or judicial proceedings based on their own particular situation, does not mean that the obligations resulting from Article 13(1) and Article 23(1) of the Directive 2008/50 and the analogous provisions of the earlier directives were intended to confer individual rights on interested persons, for the purpose of the first of the three conditions referred to above.[69]

In the light of the foregoing, the CJEU concluded that Articles 13(1) and 23(1) of the Directive 2008/50 must be interpreted as meaning that they are not intended to confer rights on individuals capable of entitling them to compensation from a

[67] Ibid, paras 46 and 47.

[68] Ibid, paras 54–57.

[69] Ibid, paras 58–62.

Member State under the principle of State liability for loss and damage caused to individuals as a result of breaches of EU law attributable to that Member State.[70]

In this context, it is worth noting the opinion of Advocate General J. Kokott,[71] although the CJEU departed from it. According to her, Article 13(1) of the Directive 2008/50 imposes a precise, directly effective obligation on the Member States to prevent exceedances of the limit values laid down for the air pollutants to which it applies. Moreover, Article 23(1) of that directive implies a clear separate obligation to draw up air quality plans arising from breaches of limit values. The Advocate General further maintained that these obligations correspond to individual rights. And therefore she concluded that the limit values set for ambient air pollutants and the obligations to improve ambient air quality laid down, inter alia, in Articles 13 and 23 of the Directive 2008/50 are intended to confer rights on individuals. A claim for compensation for damage resulting from health problems caused by exceeding the limit values set for PM_{10} and nitrogen dioxide in ambient air requires according to her that the injured party proves the existence of a direct link between that injury and his residence in the place where the relevant limit values were exceeded, without a plan for improving air quality having been drawn up which complied with the requirements set out in Section A of Annex XV to the Directive 2008/50 and without any other manifest deficiencies. Nonetheless, as already noted, the CJEU did not share the Advocate General's views.

4 Limits of the EU Competences and Role of National Courts

However, it should be noted that actions against non-compliance with the obligations imposed by the Ambient Air Quality Directives are not only brought to the CJEU. In France, the Conseil d'État has imposed sanctions on the French government for not complying with its obligations in three relatively recent decisions.

Following the referral of the matter by an environmental protection association, in 2017 the Conseil d'État had ordered the Government to implement plans to reduce concentrations of NO_2 and PM_{10} in 13 areas in France in the shortest possible time, in order to comply with the requirements of the Directive 2008/50, which sets limit values not to be exceeded for these concentrations.

Noting in 2020 that limit values were still exceeded in 8 zones and that the Government had not taken all the measures to ensure the execution of the decision

[70] Ibid, para 65; see also https://curia.europa.eu/juris/document/document.jsf?text=&docid=268801&pageIndex=0&doclang=EN&mode=lst&dir=&occ=first&part=1&cid=3933116; accessed 21 February 2023.

[71] Opinion of AG J. Kokott, 5 May 2022, C-61/21, *JP v Ministre de la Transition écologique, Premier ministre*.

of 2017, the Conseil d'État ordered the latter to take the necessary measures within six months, under penalty of EUR 10 million per half-year of delay.[72]

Subsequently, although some measures have been taken, the Conseil d'État in 2021 considered that they would not improve the situation in the shortest possible time, as their implementation has remained uncertain in some cases and their effects have not been evaluated. Furthermore, no new air protection plan has been adopted for the areas concerned, although these plans are now a known and appropriate tool for specifying the actions to be taken and assessing the timeframe in which they will enable a return to below the limit values. For these reasons, the Conseil d'État considered that, despite the measures taken and notwithstanding a real improvement of the situation in several regions, the measures taken by the French Government were not sufficient to conclude that the 2017 decision has been fully implemented.

Accordingly, the Conseil d'État has ordered the French Government to pay a fine of EUR 10 million for the first half of 2021 to the association Les Amis de la Terre, the referring body, as well as to several organizations and associations involved in the fight against air pollution. Following this decision, it was reported that the Conseil d'État would evaluate the actions of the Government for the second half of the year in early 2022 and decide whether the State must pay an additional fine.[73]

And indeed, in 2022, the Conseil d'État ordered the French Government to pay two new penalty payments of EUR 10 million each for the two periods from July 2021 to January 2022 and from January to July 2022. Although there have been improvements over time, the thresholds for NO_2 pollution—which must be respected since 2010—were still exceeded in several areas in France, notably in the conurbations of Paris, Lyon and Marseille. To date, the measures taken by the French State have not guaranteed that air quality would improve so that the pollution limits would be respected in the shortest possible time.[74]

---

[72] Le Conseil d'État ordonne au Gouvernement de prendre des mesures pour réduire la pollution de l'air, sous astreinte de 10 M€ par semestre de retard (https://www.conseil-etat.fr/actualites/le-conseil-d-etat-ordonne-au-gouvernement-de-prendre-des-mesures-pour-reduire-la-pollution-de-l--air-sous-astreinte-de-10-m-par-semestre-de-retard; accessed 21 February 2023), see also Conseil d'État, decision of 10 July 2020, N° 428409 (https://www.conseil-etat.fr/fr/arianeweb/CE/decision/2020-07-10/428409; accessed 21 February 2023).

[73] Pollution de l'air : le Conseil d'État condamne l'État à payer 10 millions d'euros (https://www.conseil-etat.fr/actualites/pollution-de-l-air-le-conseil-d-etat-condamne-l-etat-a-payer-10-millions-d-euros; accessed 21 February 2023), see also Conseil d'État, decision of 4 August 2021, N° 428409 (https://www.conseil-etat.fr/fr/arianeweb/CE/decision/2021-08-04/428409; accessed 21 February 2023).

[74] Pollution de l'air : le Conseil d'État condamne l'État à payer deux astreintes de 10 millions d'euros (https://www.conseil-etat.fr/actualites/pollution-de-l-air-le-conseil-d-etat-condamne-l-etat-a-payer-deux-astreintes-de-10-millions-d-euros; accessed 21 February), see also Conseil d'État, decision of 17 October 2022, N° 428409 (https://www.conseil-etat.fr/fr/arianeweb/CE/decision/2022-10-17/428409; accessed 21 February 2023).

5 Conclusion

This Chapter has demonstrated, through the example of air protection and the setting of air pollution limits, the strong role played by the CJEU on the one hand and the limits of its competence on the other.

Thanks to the Commission's actions, we have a long line of convictions against Member States in direct actions. The composition of the chambers of the CJEU that have ruled on these cases confirms that their nature is legally and factually uncomplicated. At the same time, the number of convictions reasonably raises the question of whether non-compliance with the limits is the result of a political reluctance on the part of individual Member States to comply with the limits set, or whether the legislation does not rather run up against the limits of practical political feasibility. It will be all the more interesting to follow the legislative changes being prepared by the Commission leading to a further tightening of the limits.

The situation is different in the case of individual actions. Here the CJEU is dependent on cooperation with the courts of the Member States. Environmental associations and NGOs have an important role to play. Given the lack of specific individual rights under EU law, actions can only lead to limited pressure on Member States to comply with air pollution limits. However, there is more scope in the process of permitting of new projects, where exceedances of the limits already at the time of the assessment of a new project can play a significant role. The nature of environmental rights also leads to a limited possibility to apply some of the principles that follow from the lack of implementation of directives, such as compensation/damages.

In addition, the possibility of Member States' courts to act independently of the CJEU should not be underestimated. The example of the series of decisions of the French Conseil d'État confirms that individuals are not solely dependent on the activity of the CJEU, and that they can effectively use the scope provided by the national legal and institutional order (in combination with EU law).

References

I. Judgments of the CJEU

Judgment of 4 December 1974, 41/74, Van Duyn v Home Office
Judgment of 19 November 1991, C-6, 9/90, Francovich and Bonifaci v Italy
Judgment of 24 March 2011, C-365/10, Commission v Slovenia
Judgment of 10 May 2011, C-479/10, Commission v Sweden
Judgment of 15 November 2012, C-34/11, Commission v Portugal
Judgment of 19 December 2012, C-68/11, Commission v Italy
Judgment of 5 April 2017, C-485/15, Commission v Bulgaria
Judgment of 22 February 2018, C-336/16, Commission v Poland
Judgment of 19 December 2018, C-752/18, Deutsche Umwelthilfe eV v Freistaat Bayern

Judgment of 24 October 2019, C-636/18, Commission v France
Judgment of 30 April 2020, C-638/18, Commission v Romania
Judgment of 10 November 2020, C-644/18, Commission v Italy
Judgment of 3 February 2021, C-637/18, Commission v Hungary
Judgment of 3 March 2021, C-664/18, Commission v the United Kingdom
Judgment of 3 June 2021, C-635/18, Commission v Germany
Judgment of 22 December 2021, C-125/20, Commission v Spain
Judgment of 28 April 2022, C-286/21, Commission v France
Judgment of 12 May 2022, C-573/19, Commission v Italy
Judgment of 12 May 2022, C-730/19, Commission v Bulgaria
Judgment of 22 December 2022, C-61/21, JP v Ministre de la Transition écologique, Premier ministre
Judgment of 9 February 2023, C-342/21, Commission v Slovakia
Judgment of 16 February 2023, C-633/21, Commission v Greece

II. Opinions of Advocate General

Opinion of AG J. Kokott, 5 May 2022, C-61/21, *JP v Ministre de la Transition écologique, Premier ministre*
Opinion of AG J. Kokott, 22 September 2022, C-375/21, *Sdruzhenie 'Za Zemyata – dostap do pravosadie' and Others*
Opinion of AG J. Kokott, 17 November 2022, C-174/21, Commission v Bulgaria

III. Judgments of National Courts

Conseil d'État, decision of 10 July 2020, N° 428409. https://www.conseil-etat.fr/fr/arianeweb/CE/decision/2020-07-10/428409
Conseil d'État, decision of 4 August 2021, N° 428409. https://www.conseil-etat.fr/fr/arianeweb/CE/decision/2021-08-04/428409
Conseil d'État, decision of 17 October 2022, N° 428409. https://www.conseil-etat.fr/fr/arianeweb/CE/decision/2022-10-17/428409

IV. On-line Sources

Conseil d'État, Press Release, *Le Conseil d'État ordonne au Gouvernement de prendre des mesures pour réduire la pollution de l'air, sous astreinte de 10 M€ par semestre de retard*, 10 July 2020. https://www.conseil-etat.fr/actualites/le-conseil-d-etat-ordonne-au-gouvernement-de-prendre-des-mesures-pour-reduire-la-pollution-de-l-air-sous-astreinte-de-10-m-par-semestre-de-retard; accessed 21 February 2023
Conseil d'État, Press Release, *Pollution de l'air : le Conseil d'État condamne l'État à payer 10 millions d'euros*, 4 August 2021. https://www.conseil-etat.fr/actualites/pollution-de-l-air-le-conseil-d-etat-condamne-l-etat-a-payer-10-millions-d-euros; accessed 21 February 2023
Conseil d'État, Press Release, *Pollution de l'air : le Conseil d'État condamne l'État à payer deux astreintes de 10 millions d'euros*, 17 October 2022. https://www.conseil-etat.fr/actualites/pollution-de-l-air-le-conseil-d-etat-condamne-l-etat-a-payer-deux-astreintes-de-10-millions-d-euros ; accessed 21 February 2023
CURIA, Presentation - Court of Justice of the European Union. https://curia.europa.eu/jcms/jcms/Jo2_7024/en/ ; accessed 21 February 2023
CURIA, Résumé of the judgment of 19 December 2018, C-752/18, Deutsche Umwelthilfe. https://curia.europa.eu/juris/document/document.jsf?text=&docid=221831&pageIndex=0&doclang=EN&mode=req&dir=&occ=first&part=1&cid=3345855; accessed 21 February 2023

CURIA, Résumé of the judgment of 22 December 2022, C-61/21, JP v Ministre de la Transition écologique, Premier ministre. https://curia.europa.eu/juris/document/document.jsf?text=&docid=268801&pageIndex=0&doclang=EN&mode=lst&dir=&occ=first&part=1&cid=3933116; accessed 21 February 2023

European Commission, Air Quality. https://environment.ec.europa.eu/topics/air/air-quality_en; accessed 21 February 2023

European Court of Auditors, Press release, 11 September 2018. https://www.eca.europa.eu/Lists/ECADocuments/INSR18_23/INSR_AIR_QUALITY_EN.pdf; accessed 21 February 2023

European Environment Agency, Air pollution. https://www.eea.europa.eu/themes/air; accessed 21 February 2023

European Environment Agency, Air pollution sources. https://www.eea.europa.eu/themes/air/air-pollution-sources-1; accessed 21 February 2023

European Environment Agency, Air quality in Europe 2022. https://www.eea.europa.eu/publications/air-quality-in-europe-2022; accessed 21 February 2023

Statute of the Court of Justice. tra-doc-en-div-c-0000-2016-201606984-05_00.pdf ; accessed 21 February 2023

World Health Organization, *WHO global air quality guidelines: particulate matter (PM2.5 and PM10), ozone, nitrogen dioxide, sulfur dioxide and carbon monoxide*, 2021. https://apps.who.int/iris/handle/10665/345329; accessed 21 February 2023

World Health Organization, *New WHO Global Air Quality Guidelines aim to save millions of lives from air pollution*, News release, 22 September 2021. https://www.who.int/news/item/22-09-2021-new-who-global-air-quality-guidelines-aim-to-save-millions-of-lives-from-air-pollution; accessed 21 February 2023

Jan M. Passer is Judge at the Court of Justice of the EU.

Věra Pazderová is Judge at a Regional Court in Prague, the Czech Republic.

Anchoring the Right to a Healthy Environment in the European Convention on Human Rights: What Concretized Normative Consequences Can Be Anticipated for the Strasbourg Court?

Natalia Kobylarz

Abstract This chapter explores, in a concretized and systematized manner, how the projected additional protocol to the ECHR on the right to a healthy environment would likely affect the existing normative framework regarding the ECtHR's territorial, subject-matter, and personal jurisdiction. It is concluded that institutionalizing a separate specialized regime for the right to a healthy environment would release tensions generated by environmental claims based on the expansive interpretations of the current doctrine of indirect rights-based protection of the environment. The proposed right would further fill in several normative gaps and bring uniformity to standards concerning participatory rights. The chapter intends to contribute to the ongoing legislative deliberation within the Council of Europe and, indirectly, to the similar process within the United Nations.

The views expressed in this paper are those of the author and do not represent the official position of the ECtHR or the CoE. The Registry gives support to the Court and cannot in any way influence the Court's decisions on the admissibility and/or the merits of any case. The author wishes to thank Jan Darpö for his valuable comments on the earlier version of this paper.

N. Kobylarz (✉)
Registry of the European Court of Human Rights, Strasbourg, France
e-mail: natalia.kobylarz@echr.coe.int

G. Antonelli et al. (eds.), *Environmental Law Before the Courts*, https://doi.org/10.1007/978-3-031-41527-2_7

1 Introduction

The right to a healthy environment ("the right") has tenaciously returned to the international political agenda by means of two votes: one in 2021, at the Parliamentary Assembly of the Council of Europe ("CoE"),[1] and the other one in 2022, at the General Assembly of the United Nations.[2] The result of the former vote is the recommendation for the preparation, among others, of an additional protocol to the European Convention on Human Rights ("ECHR") on the right to a safe, clean, healthy and sustainable environment.[3] The ensuing political process has, at this stage, included preliminary drafting and ongoing work on necessity and feasibility studies.[4]

Academia and international experts have made extensive arguments about the right's philosophical sense,[5] and about its added value, essentially, in terms of improved environmental governance, as well as increased levels of environmental quality and protection.[6] Conversely, commentary about the right's anticipated normative consequences for the European Court of Human Rights ("ECtHR" or "Court") has mostly been limited to abstract effects, such as giving the environment legal protection equal to that of other fundamental rights; providing a basis for integrated and progressive development of the Court's jurisprudence; and closing gaps in the current legal framework.[7] Several specific effects have also been signaled.[8]

[1] CoE Parliamentary Assembly Resolution 2396 (2021) Anchoring the right to a healthy environment: need for enhanced action by the Council of Europe, 29 September 2021 and Recommendation no. 2211 (2021) Anchoring the right to a healthy environment: need for enhanced action by the Council of Europe, 29 September 2021.

[2] United Nations General Assembly Resolution The human right to a clean, healthy and sustainable environment, 26 July 2022, A/76/300.

[3] Recommendation (n 1), § 3.1; Resolution (n 1), § 14.3.

[4] Reply to Recommendation 2211 (2021) from the Committee of Ministers of the CoE, Doc. 15623, 4 October 2022, §§ 3 and 4; Recommendation CM/Rec(2022)20 of the Committee of Ministers to member States on human rights and the protection of the environment, 27 September 2022 (CM/Rec(2022)20); and website of the CoE's Drafting Group on Human Rights and Environment (CDDH-ENV). https://www.coe.int/en/web/human-rights-intergovernmental-cooperation/environment-and-human-rights#{%22113149991%22:[5].

[5] Lambert (2020), pp. 3 and 4.

[6] Knox (2018), paras 39–45; Boyd (2012); Boyd (2018), pp. 25–26; and The Right to a Healthy Environment in Ireland: Supporting the proposal for a new Protocol to the ECHR, Environmental Rights Recognition Project, ERRP and Kings College London (May 2022), pp. 6 and 8.

[7] Separate opinion of Judge Serghides in the case of *Pavlov and Others v. Russia*, § 21; JH Knox, Extended summary of the exchange of views with external independent experts and representatives of the Parliamentary Assembly and the European Committee on Social Rights, 13–15 September 2022, pp. 24–27; Knox (2018), para 39; Lambert (2020), pp. 13–15; Baumann (2018), p. 38.

[8] Knox, Extended summary (n 7), pp. 27 and 29; Elisabeth Lambert, Extended summary (n 7), pp. 43 and 44; Léa Raible, Extended summary (n 7), pp. 33–36; and Helen Keller, Extended summary (n 7), pp. 2 and 3.

Those, however, would benefit from a more consolidated, comprehensive and practical study.

This study departs from and goes beyond the existing research in that it explores, in a concretized and systematized manner, how the right to a healthy environment—if indeed adopted by means of a new protocol to the ECHR—would likely affect the existing normative framework concerning admissibility requirements. The chapter provides answers to the following underlying questions: To what extent would the right broaden the scope of the Court's territorial, subject-matter, and personal jurisdiction? Could similar changes emerge within the current doctrine of indirect rights-based protection in the absence of the additional protocol? For reasons of space, the work does not extend to normative consequences for the subsequent stages of case examination by the ECtHR. Reserved for another publication is the analysis of issues such as, how the right would affect the scope of States' human rights obligations or the Court's assessment of "fair balance" where conflicting interests are at stake, or whether the right would engender novel remedies.

The author employs the concept of normative consequences that denotes, as proposed by Michał Araszkiewicz and Krzysztof Płeszka,[9] outcomes accepted by courts "as those that satisfy the hypotheses of the projected legal norms and thus generate legal effects".[10] Such outcomes may result from various models of legal reasoning.[11] As opposed to abstract consequences, where the relevant outcome is determined by the text of the statute only, concretized consequences emerge from the application of the projected rules to cases.[12] Here, inferences may be made from the terminology that goes beyond the statute. This is opportune for the present study, given that the content of the right proposed is, at this stage, only tentative. The chapter thus examines the "transition" from the right to a healthy environment that is under preliminary elaboration (input), to normative consequences that would likely be "assigned to the regulated states of affairs" by the ECtHR (outcome).[13] To make the present analysis valid and practically functional, the concretized normative consequences of the right are foreseen, insofar as possible, with reference to cases that have in the past been examined by the ECtHR.

By anticipating concretized normative consequences of the right to a healthy environment framed in an additional protocol to the ECHR, the article intends to contribute to the ongoing necessity and feasibility studies, as well as to the possibly ensuing legislative deliberation within the CoE. As observed by Araszkiewicz and Płeszka, considering the consequences of planned regulations is imperative for each

---

9 Araszkiewicz and Płeszka (2015), pp. 253–297.

10 Ibid, pp. 254, 261–267.

11 Ibid, pp. 254, 272, and 273–280.

12 Ibid, p. 267.

13 Ibid, p. 266.

"rational legislator"[14] if the regulation is to stand a chance of solving the problems caused by the existing legal order.[15]

The chapter is divided into two main parts. The first one describes the content of the right to a healthy environment, as currently projected (Sect. 2). The second one examines the concretized normative consequences of the right in the sphere of: extraterritorial jurisdiction of States (Sect. 3.1), the Court's subject-matter jurisdiction (Sect. 3.2), and its personal jurisdiction (Sect. 3.3).

2 The Right's Content

The input that serves the present analysis of normative consequences is the right to a healthy environment that draws on the following text retained at this stage within the CoE:[16] "Everyone has the right to a safe, clean, healthy and sustainable environment."[17] The four above-mentioned adjectives describe the substance of the right and are commonly understood to mean "a safe climate, clean air, clean water and adequate sanitation, healthy and sustainably produced food, non-toxic environments …, and healthy biodiversity and ecosystems."[18] As it happens, the adjective "safe" has recently been dropped within the United Nations process.[19] It is therefore possible that the CoE's formulation may be changed accordingly.[20] The proposed right is further described as being "of present and future generations", and defined as "the right to live in a non-degraded, viable and decent environment that is conducive to their health, development and well-being."[21]

Based on the above-mentioned wording, it is fair to posit that the proposed right breaks with the traditional individualistic and largely extractivist approach to environmental human rights, in terms of which the natural environment is only protected for its utilitarian value, insofar as it guarantees conditions or resources that are immediately necessary for human life and well-being (extractive anthropocentrism).[22] Instead, by referring to future generations and to the environment's sustainability and viability, the text reflects immersive anthropocentrism. That paradigm fosters the recognition that humans and the environment are interconnected and interdependent. Human dependence on environmental benefits is not viewed here, solely in

---

[14] Ibid, pp. 253 and 268.

[15] Ibid, p. 254.

[16] Recommendation (n 1), § 3.1.

[17] Ibid, Appendix, Article 5.

[18] Boyd (2019), para 43.

[19] United Nations General Assembly Resolution The human right to a clean, healthy and sustainable environment, 26 July 2022, A/76/L.75, § 1.

[20] Recommendation (n 1), § 3.1.

[21] Ibid, Appendix, Article 1.

[22] Kobylarz (2022), pp. 33–37.

terms of direct or short-term utility for individuals in a geographical vicinity. The use of natural resources and services is predominantly shaped by the principles of sustainability, intergenerational equity, precaution, social solidarity, international cooperation and overall responsibility for environmental quality.[23] Articles 2–4 and the Preamble of the suggested text indeed confirm the interpretation of the right as immersively anthropocentric.[24]

The proposal, however, goes further in that the right to a healthy environment is to be interpreted not only in its anthropocentric (subjective) dimension, recognizing that nature undeniably has a utility for humans, but also in its ecocentric (objective) dimension, recognizing the intrinsic value of nature and ecosystems.[25] The ecocentric paradigm is clearly identified in the 2021 Resolution and in the Preamble of the 2021 Recommendation.[26] It is also confirmed by Article 2 of the proposed draft protocol that imposes a duty to protect "the environment and biodiversity" and to prevent damage "to life on Earth",[27] as well as by its Article 4 insofar as this provision establishes the principle *in dubio pro natura*, and expressly determines that the obligations of prevention and precaution apply to threats of severe damage not only to human, but also, to animal or plant health.[28] As suggested by Helen Keller, a former judge at the ECtHR, the ultimate right's formulation should be more specific as to whether the legislator's intention is indeed to maintain the ecocentric approach.[29] Notwithstanding, it must be stressed that nothing in the proposed text or in the preparatory works underway goes as far as to suggest the rights of nature approach or biocentrism,[30] pursuant to which nature would be recognised as a legal person[31] and a holder of rights such as the right to exist, thrive, and evolve,[32] or the right to protection, conservation, maintenance and restoration.[33] Although the ecocentric paradigm suggested within the CoE does not consider the biosphere or its ecosystems as

[23] Ibid, pp. 38–40.

[24] Recommendation (n 1), Appendix, Articles 2–4 and Preamble.

[25] Resolution (n 1), § 6 *in fine*. In international law, ecocentrism was first introduced by the 1979 CoE Bern Convention on the Conservation of European Wildlife and Natural Habitats.

[26] Resolution (n 1), § 6 *in fine*; and Recommendation (n 1), Preamble.

[27] Recommendation (n 1), Article 2.

[28] Ibid, Article 4.

[29] Keller, Extended summary (n 7), p. 3.

[30] Kobylarz (2022), p. 45.

[31] For example, in Spain, Ley 19/2022 para el reconocimiento de personalidad jurídica a la laguna del Mar Menor y su cuenca, 30 September 2022; Article 71 of the 2008 Constitution of Ecuador; in Bolivia, Ley no. 071, 21 December 2010 and Ley no. 300, 15 October 2012; in New Zealand, Te Urewera Act 2014, 27 July 2014; in India, High Court of Uttarakhand, judgment of 20 March 2017, No.126 of 2014; in Colombia, Centro de Estudios para la Justicia Social 'Tierra Digna' and Others v President of the Republic and Others, no. T-622, Corte Constitucional [Constitutional Court, Colombia] 10 November 2016; and Andrea Lozano Barragán, Victoria Alexandra Arenas Sánchez, Jose Daniel y Felix Jeffry Rodríguez Peña y otros v Presidente de la República y otros, STC4360-2018, Corte Suprema de Justicia [Supreme Court, Colombia] 5 April 2018.

[32] 2016, IUCN World Declaration on the Environmental Rule of Law, Principle 2.

[33] Centro de Estudios (para 9.27, 9.32 and 10.2.(1).

subjects of rights, it certainly raises the level of their legal protection by treating them as legal interests in themselves, and by imposing responsibilities and duties on humans to protect nature for its intrinsic worth.[34] As explained in the 2017 Advisory Opinion of the Inter-American Court of Human Rights ("IACtHR"), the right to a healthy environment "protects nature and the environment, not only because of the benefits they provide to humanity or the effects that their degradation may have on other human rights, such as health, life or personal integrity, but because of their importance to the other living organisms with which we share the planet that also merit protection in their own right."[35]

The proposed right is therefore viewed as dualistic in that it aims at ensuring the environment that is "healthful", meaning, conducive to a healthy living for people, and that is "healthy in itself", meaning, "free of "diseases" that hinder its ecological balance and sustainability."[36] Deep interdependence between people and a healthy biosphere[37] makes the anthropocentric and ecocentric dimensions of the right to a healthy environment inextricable.[38]

Furthermore, the proposed right to a healthy environment contains a series of participatory (procedural) rights reflecting the three pillars of the 1998 Convention on Access to Information, Public Participation in Decision-Making and Access to Justice in Environmental Matters ("Aarhus Convention").[39] It is explicitly stated, in the project, that the right to information can be relied on by anyone, without having to prove an individual interest.[40] An unqualified "everyone" is the holder of the right to prior consultation and participation in a decision-making process, and the right to access to justice and to an effective remedy.[41]

Even though this goes beyond the context of admissibility requirements, it must be noted that the right to a healthy environment is not advanced as absolute in that its exercise may be subject to restrictions that are prescribed by law and are necessary in a democratic society to protect a series of general interests, including those of national security, public safety, the protection of health or of the rights and freedoms of others.[42] On the other hand, it is stated that no derogation and no reservation can be made, with the exception of the right to participate in a decision-making

---

[34] Recommendation (n 1), Preamble.

[35] Inter-American Court of Human Rights, Advisory Opinion OC-23/17 of 15 November 2017, para 62; and Case of the Indigenous Communities Members of the Lhaka Honhat (Our Land) Association v. Argentina, Inter-American Court of Human Rights, judgment of 6 February 2020, para 203.

[36] Ksentini (1994), para 180.

[37] Boyd (2020); and IUCN World Declaration on the Environmental Rule of Law, 2016.

[38] World Charter for Nature, United Nations General Assembly Resolution 37/7, 18 October 1982 (A/RES/37/7); and Kunming-Montreal Global Biodiversity Framework, Section C (9), 22 December 2022 (CBD/COP/15/L25).

[39] Recommendation (n 1), Article 6.

[40] Ibid, Article 6a.

[41] Ibid, Article 6 b–d.

[42] Ibid, Article 7.

process (public consultation) about projects, programmes or policies that have an environmental impact.[43] By comparison, Article 8 (right to respect for private and family life, and for home)—which is the provision most often engaged in cases regarding environmental pollution—can be derogated from, pursuant to Article 15 of the ECHR, in time of war or other public emergency threatening the life of the nation.[44]

The intended goals of the protocol containing the right to a healthy environment have been phrased as: (1) establishing "the clear responsibility of member States to maintain a good state of the environment that is compatible with life in dignity and in good health and the full enjoyment of other fundamental rights"; and (2) supporting "much more effective protection of a safe, clean, healthy and sustainable environment at national level."[45] These aims arise from a concern about "the speed and extent of environmental degradation, the loss of biodiversity and the climate crisis that directly affect human health, dignity and life"; and from the need to "show ambition and strategic vision for the future by facing up to this major transformative challenge to human rights and securing their enhanced protection in the era of systemic environmental threats to present and future generations."

3 Concretized Normative Consequences

The proposed right to a healthy environment would necessarily engender a series of normative consequences for the admissibility (and the merits) of future environment-related applications before the ECtHR. What follows is a systematized compilation of the existing legal norms that have been crucial for the Court's assessment of the admissibility of such cases under the current regime of indirect rights-based protection. It is discussed how these norms are susceptible to be affected by the projected additional protocol and if they can incrementally change without the right being adopted. In fact, some of these norms are already being challenged by a series of climate change and pollution cases pending before the ECtHR.[46] Subject to ongoing judicial deliberation, these norms will inevitably either be confirmed or reshaped. The former would indicate that the current doctrine of "green human rights" has reached its interpretative limits and, consequently, that any further expansion could only be triggered through the political process underway. The latter would mean

[43] Ibid, Appendix, Articles 8 and 9.

[44] ECtHR Guide on Article 15. In the current system, the only non-derogable rights are Article 2 (right to life), except in respect of deaths resulting from lawful acts of war, Article 3 (prohibition of ill-treatment), Article 4 § 1 (prohibition of slavery or servitude), and Article 7 (no punishment without law).

[45] Recommendation (n 1), § 3.1.

[46] *Duarte Agostinho and Others v. Portugal and 32 other Member States*, no. 39371/20; *Verein KlimaSeniorinnen Schweiz v. Switzerland*, no. 53600/20; *Carême v. France*, no. 7189/21; and *Di Caprio and Other v. Italy*, no. 39742/14.

that a more evolutive interpretation of the present legal framework can be prompted, even without the right, by the international *corpus iuris* comprising environmental and climate change law. Given the advanced stage of the proceedings in question, it appears that the Court's rulings may intervene before the end of the first stages of the political process in the Committee of Ministers. Without speculating on the content of these rulings, the norms in question are mentioned in the chapter for the sake of exhaustiveness. Ultimately, to demonstrate concretized normative consequences, the hypotheses of the projected legal norms are tested against several applications that have in the past been examined by the ECtHR.

3.1 Extraterritorial Jurisdiction of Member States

The draft text of the proposed protocol to the ECHR does not set out any specific rule on jurisdiction within which the member States of the CoE would have to secure to everyone the right to a healthy environment. The applicable general principles would therefore have to come from the Court's case-law on Article 1 of the ECHR that has always been developed with close reference to the specific facts of cases examined.[47]

While almost none of these cases is related to the environment, the record of past and ongoing international environmental litigation or of historic environmental incidents indicates that the issues of extraterritorial jurisdiction in the context of State actions or omissions allegedly causing environmental degradation or climate change could hypothetically be raised in connection with impacts on the enjoyment of human rights of people in another member State of the CoE[48] or in a third country.[49] Such impacts on human rights may allegedly be caused by emissions of toxic substances[50] or greenhouse gasses,[51] or by other activities, such as exploitation or

[47] ECtHR Guide on Article 1, para 42; *M.N. and Others v. Belgium* (dec.) [GC], no. 3599/18, § 102, 5 May 2020.

[48] *Zeynep Ahunbay and Others v. Turkey, Austria and Germany* (dec.), no. 6080/06, 21 June 2016; *Duarte Agostinho* (n 46); and *Soubeste v. Austria and 11 Other States*, no. 31925/22, and 4 other applications.

[49] Sacchi and Others v. Argentina, Brazil, France, Germany and Turkey, no. 104/2019, 8/10/2021, Committee on the Rights of the Child (CRC/C/88/D/104/2019); and Luciano Lliuya v. RWE AG, Case No. 2 O 285/15, pending before the Essen Regional Court.

[50] Pulp Mills on the River Uruguay (Argentina v. Uruguay), Judgment, I.C.J. Reports 201, p. 14.

[51] Armando Carvalho and Others v European Parliament and Council of the EU, CJEU, C-565/19 P, 25/03/2021; Sacchi (n 49); and *Duarte Agostinho* (n 46).

management of natural resources,[52] land development,[53] or war.[54] Theoretically, such triggering events may either take place in the member State's territory or in a foreign territory. In the latter scenario, the responsibility of a member State may be engaged either directly—in its capacity of an investor,[55] or indirectly—in its capacity of a corporate domicile of a non-State entity (company or parent company).[56] The impugned impacts on human rights may stem either from an instantaneous act, such an industrial accident,[57] planned isolated action,[58] or an act of war,[59] or from continuous and uninterrupted operations.[60] Therefore, depending on the scenario, the question of Convention-compliance may concern member States' obligation to refrain from causing harm (negative obligation) or obligation to regulate industry and/or business, and/or to take necessary operational measures (positive obligations).

The existing general principles on extraterritorial jurisdiction are not directly transposable to the above-mentioned scenarios. They are nevertheless summarized below for two reasons. Firstly, they may inspire or inform the necessary elements of any new special norm on transboundary violations of the right to a healthy environment. Secondly, the existing standards constitute a normative framework into which any such new norm would have to fit harmoniously.

As regards substantive obligations arising from the ECHR and its Protocols, the first of the two traditional and main criteria governing the exercise of extraterritorial jurisdiction is that of "effective control" by the State over an area outside its territory.[61] In this context, the Court has looked at elements such as State's military

[52] Gabčikovo-Nagymaros Project (Hungary/Slovakia), Judgment, I.C.J. Reports 1997, p. 7; Czech Republic v Poland (Mine de Turów), C-121/21, CJEU, closed on 20 May 2022.

[53] Certain Activities Carried Out by Nicaragua in the Border Area (Costa Rica v. Nicaragua) and Construction of a Road in Costa Rica along the San Juan River (Nicaragua v. Costa Rica), Judgment, I.C.J. Reports 2015, p. 665.

[54] *Russia v. Ukraine*, no. 36958/21; Report Environmental impact of armed conflicts, CoE Parliamentary Assembly Committee on Social Affairs, Health and Sustainable Development, 5 January 2023 (Doc. 15674); see also Young (2009).

[55] *Zeynep Ahunbay* (n 48).

[56] Social and Economic Rights Action Center (SERAC) and Center for Economic and Social Rights (CESR) v. Nigeria, Decision, Comm. 155/96 (ACmHPR, Oct. 27, 2001); Rechtbank Den Haag, 1 May 2019, C/09/540872/HA ZA 17-1048, ECLI:NL:RBDHA:2019:4233 (case ultimately rejected); Lungowe v Vedanta Resources Plc [2019] UKSC 20; [2020] AC 1045; UKSC 2017/0185, 10 Apr 2019; and Okpabi and others v. Royal Dutch Shell Plc and another, UK Supreme Court [2021] UKSC 3, 12 February 2021.

[57] Environmental Consequences of the Chernobyl Accident and their Remediation: Twenty Years of Experience, Report of the Chernobyl Forum Expert Group 'Environment', IAEA, 2006.

[58] Zheng (2021).

[59] E.g. Kuwaiti oil fires caused by the Iraqi military in 1991. Husain and Bakr Amin (1995).

[60] E.g., Gabčikovo (n 52), p. 7; Czech Republic v Poland (Mine de Turów), C-121/21, CJEU, closed on 20 May 2022; and *Duarte Agostinho* (n 46).

[61] *Ukraine and the Netherlands v. Russia* (dec.) [GC], nos. 8019/16 et al., §§ 560–64, 30 November 2022, with further references; and *Al-Skeini and Others v. the United Kingdom* [GC], no. 55721/07, §§ 133–40, ECHR 2011.

presence in the area; or its influence and control over the area based on its military, economic and political support for the local subordinate administration.[62] Where such "spatial jurisdiction" is established, the controlling State has the responsibility to secure, within the area under its control, the entire range of substantive rights set out in the ECHR and will be liable for any violations of those rights in exactly the same way as it would be in a purely territorial context.[63] Although never expressly stated by the Court, the case-law appears to indicate that where spatial control is established, extraterritorial obligations of the controlling State can only concern the area within the sovereign territory of the CoE member States.[64] The second main criterion is that of "State agent authority and control" over individuals.[65] A State Party's jurisdiction may arise, among others, where, with the consent or at the invitation of the government of another country, it exercises *via* its agents or others under their command and direct supervision, public powers normally to be exercised by that government.[66] Extraterritorial jurisdiction may also cover the exercise by State agents of physical power and control over a person or property.[67] Where such "personal jurisdiction" is established, the extent of the controlling State's obligations is to secure to the individuals concerned only the rights and freedoms that are relevant to their situation.[68] Unlike jurisdiction based on control over an area, the Court has on numerous occasions found personal jurisdiction to exist "outside the Convention legal space", that is to say in places outside the sovereign territory of the CoE member States.[69]

Although responsibility cannot, in principle, stem from an instantaneous extraterritorial act,[70] extraterritorial jurisdiction has been extended to situations arising from full and exclusive control that was continuous and uninterrupted.[71] The responsibility of member States can also cover isolated and specific acts committed by State agents who exercise physical power and control over the victim in a situation of proximity.[72] Conversely, the Court has declined to extend extraterritorial responsibility to actions that occurred "in a context of chaos" during a military operation

---

[62] *Ukraine and the Netherlands* (n 61), § 560.

[63] *Ukraine and the Netherlands* (n 61), § 561.

[64] *Ukraine and the Netherlands* (n 61), §§ 562–63.

[65] *Ukraine and the Netherlands* (n 61), §§ 565–72, with further references; and *Öcalan v. Turkey* [GC], no. 46221/99, § 91, ECHR 2005-IV.

[66] *Ukraine and the Netherlands* (n 61), § 567, with further references.

[67] *Ukraine and the Netherlands* (n 61), §§ 568–69; and *Öcalan* (n 65), § 91; and *Medvedyev and Others v. France*, [GC], no. 3394/03, § 67, ECHR 2010.

[68] *Ukraine and the Netherlands* (n 61), § 571.

[69] *Ukraine and the Netherlands* (n 61), § 572.

[70] E.g., bombardement, see *Banković and Others v. Belgium and Others* (dec.) [GC], no. 52207/99, § 75, 12 December 2001; and *Medvedyev* (n 67), § 64.

[71] E.g., intercepting the crew of a ship, followed by their detention and trial, see *Medvedyev* (n 67), § 66.

[72] E.g., beating, shooting or targeted killing, see *Carter v. Russia*, no. 20914/07, §§ 129–30 and 161, 21 September 2021; *Georgia v. Russia (II)* (dec.), no. 38263/08, §§ 130–32, 21 January 2021;

"in the active phase of hostilities".[73] The Court considered that there could be no effective control over an area or any form of State agent authority or control over individuals.[74] More recently, however, the Court has found it possible to pierce "the fog of war" in relation to particular incidents committed in the active phase of hostilities, provided that there is "clarity" as to the circumstances surrounding these incidents.[75] Additionally, with more direct relevance to hypothetical environment-related applications, the Court has declined to impute extraterritorial responsibility to member States co-financing a project with a foreign State, that allegedly caused, among others, environmental degradation.[76] This conclusion was based on the following elements: (1) the disputed actions and omissions linked to the project (construction of the hydroelectric dam) were attributed to the authorities of the State on which territory it was developed and which was its national investor; (2) the territorial State had exclusive jurisdiction to settle the issues raised in the case; and (3) all the related legal proceedings have taken place under the jurisdiction of that State.[77] Lastly, instituting court proceedings in a State Party, with which the applicant has no connections, is, in principle, insufficient to attribute to this State jurisdiction within the meaning of Article 1 of the ECHR, in respect of the substantive act which is the subject of the procedure.[78] Such a jurisdictional link, however, can establish jurisdiction in relation to procedural obligations under Article 6 (right to fair trial) or Article 2 (right to life) of the ECHR.

As regards procedural obligations, guarantees of fair trial under Article 6 of the ECHR may indeed cover proceedings concerning extraterritorial events, provided that the domestic law of the member State recognizes a right to bring such an action in the determination of the applicants' civil rights and obligations.[79] The Court's examination would, however, be limited to the procedural rights guaranteed under that provision. This does not open up to the Court the competence to re-decide the merits of the issues examined by the domestic courts.[80] The Court has also established that a specific obligation to carry out an effective investigation under Article 2 or Article 3 (prohibition of ill-treatment) of the ECHR can bind the State even

Isaak v. Turkey (dec.), no. 44587/98, 28 September 2006; and *Andreou v. Turkey* (dec.), no. 45653/99, 3 June 2008.

[73] For example, bombing, shelling and artillery fire, see *Georgia v. Russia (II)* (n 72), §§ 126–37; and *Ukraine and the Netherlands* (n 61), § 558.

[74] *Georgia v. Russia (II)* (n 72), §§ 126–37.

[75] E.g., downing of a flight by a surface-to-air missile, see *Ukraine and the Netherlands* (n 61), §§ 703–706.

[76] *Zeynep Ahunbay* (n 48), § 94.

[77] Ibid, § 94.

[78] *Hanan v. Germany* [GC], no. 4871/16, § 143, 16 February 2021; and *Abdul Wahab Khan v. the United Kingdom* (dec.), no. 11987/11, § 28, 28 January 2014.

[79] *Marković and Others v. Italy* [GC], no. 1398/03, §§ 53–54, 14 December 2006.

[80] *Chagos Islanders v. the United Kingdom* (dec.), no. 35622/04, § 66, 11 December 2012.

when the death or ill-treatment occurred outside its jurisdiction.[81] In such cases, the Court looks for a "jurisdictional link", for example, that the investigative or judicial authorities of a Contracting State institute, by virtue of their domestic law, their own criminal investigation or proceedings concerning the incident in question.[82] Such a jurisdictional link may also be established, where no investigation or proceedings have been instituted, based on some other special features of a case.[83]

While several authors find all the above-mentioned tests confounding and controversial,[84] the very fact that they have gradually evolved shows that jurisdiction within the meaning of Article 1 of the ECHR is not an inflexible concept. It is therefore fair to conclude that there is no normative obstacle for the ECtHR to expand the Contracting States' responsibility to extraterritorial violations of various existing rights triggered by environmental harm or, if the new protocol were adopted, of the right to a healthy environment, as such.

In this context, the IACtHR and the United Nations Human Rights Committee already rely on a test pursuant to which the exercise of jurisdiction arises where a State of origin has effective control over the activity that caused or that risks causing damage to the environment and the consequent infringement of human rights of persons outside its territory, where the infringement in question was reasonably foreseeable to the State of origin at the time of its acts or omissions.[85] This special, "cause-and-effect jurisdiction" based on the test of effective control over the activity harmful to the environment and to people, is therefore available for the ECtHR to be appropriated.

It is uncertain, however, that the Court is open to have the notion of jurisdiction expanded in this way. Firstly, it has expressly stated, albeit in the context of instantaneous acts, that Article 1 does not admit a cause-and-effect notion of jurisdiction.[86] In more elaborate terms, it has held that the wording of that provision does not accommodate the theory that "anyone adversely affected by an act imputable to a Contracting State, wherever in the world that act may have been committed or its consequences felt, is thereby 'brought within' the 'jurisdiction' of that State for the purpose of Article 1 of the Convention." [87] Secondly, the Court has recently excluded spatial jurisdiction where damage to individuals and properties occurred outside of a State-controlled territory from which destructive action originated.[88] Thirdly, cases concerning climate change and environmental degradation may likely involve

---

[81] *Carter* (n 72), §§ 131–35; and *Rantsev v. Cyprus and Russia*, no. 25965/04, §§ 243–44, 7 January 2010.

[82] *Ukraine and the Netherlands* (n 61), §§ 573–74; and *Güzelyurtlu and Others v. Cyprus and Turkey* [GC], no. 36925/07, § 188, 29 January 2019.

[83] *Ukraine and the Netherlands* (n 61), § 575, with further references; and *Rantsev* (n 81), §§ 243–44.

[84] Keller and Heri (2022), p. 160; Londras and Dzehtsiarou (2018), p. 129; and Raible (2016), p. 161.

[85] Sacchi (n 49), §§ 10.7–10.12; and Advisory Opinion (n 35), paras 101–04.

[86] *Banković* (n 70), § 75; *Medvedyev* (n 67), § 64; and *M.N.* (n 47), § 112.

[87] *Georgia v. Russia (II)* (n 72), § 134.

[88] Firing of weapons, see *Ukraine and the Netherlands* (n 61), § 699.

numerous compounding or cascading events, and potentially large numbers of victims. The situations in question are also largely governed by various sectoral international conventions and agreements (e.g. United Nations Framework Convention on Climate Change and Paris Agreement), even though many of such instruments lack implementation mechanisms. Owing to very similar factors, in a case concerning active hostilities during an international armed conflict, the Court did not consider itself in a position to develop its case-law beyond the present understanding of the notion of "jurisdiction".[89] It further observed that the Contracting Parties would need to provide the necessary legal basis for the change of the current norm.[90] As observed by Lea Raible, the current understanding of extraterritorial jurisdiction by the Strasbourg Court is not likely "to close the accountability gap between major [greenhouse gas] emitters and geographically distant victims".[91] For this author, the protocol on the right to a healthy environment specifically amending Article 1 would be necessary if the aim of the legislator were for the ECtHR to rule on extraterritorial violations of the right to a healthy environment or of collateral rights.[92] Whether this conclusion is true may, in the near future, transpire from the Court's ruling in the case of *Duarte Agostinho and Others v. Portugal and 32 member States,* a Grand Chamber case that raises a specific question of State responsibility for extraterritorial greenhouse gas emissions causing climate change. Alternatively, it may stem, indirectly, from the future judgment on the merits in the case of *Ukraine and the Netherlands v. Russia*, for which the Court reserved the question of whether, in the absence of spatial jurisdiction, there was a State agent authority and control in respect of acts of indiscriminate shelling that had allegedly caused harm to numerous victims outside of the State-controlled territory.[93] This type of extraterritorial jurisdiction based on control over a person does not exactly match the possible complaints about transboundary degradation of the environment or about global climate change. Nevertheless, if the new standard based on the "effective control over the harmful activity" were to be established, it may be pertinent whether it is conceptually associated with the "personal" or "spatial" concept of jurisdiction. In principle, only the former would permit the attribution of extraterritorial responsibility of State Parties to actions or omission impacting victims in the countries outside of the 46 member States of the CoE.

[89] *Georgia v. Russia (II)* (n 72), §§ 126, 132–33, 137–38 and 141; *Ukraine and the Netherlands* (n 61), § 557.

[90] *Georgia v. Russia (II)* (n 72), § 142.

[91] Raible, Extended summary (n 7), pp. 34–35.

[92] Raible, Extended summary (n 7), p. 35.

[93] *Ukraine and the Netherlands* (n 61), § 700.

3.2 The Court's Subject-Matter Jurisdiction

At present, the right to a healthy environment falls outside of the scope of the Court's jurisdiction *ratione materiae*. Environmental protection or conservation can, however, be of an indirect concern to the Strasbourg Court where environmental harm or measures of environment or climate protection cause (or risk causing) negative impacts on the right to life, on the right to respect for private and family life, and for home (comprising life quality, well-being and health), on the right to property, or on participatory (procedural) rights (comprising access to information, participation in a decision-making process and access to a court).[94]

The projected additional protocol to the ECHR would necessarily broaden the Court's subject-matter jurisdiction significantly and in ways than one.

3.2.1 Environment as *Sui Generis* Subject-Matter of Human Rights Protection

The most obvious logical consequence of the projected autonomous right would be the redundancy of looking for a causal connection between an environmental harm complained of and an interference with a human right other than the right to a healthy environment itself.[95] An instance of environment's degradation or deregulation would thus constitute the actual subject-matter of human rights protection and not, as it is now, merely the material event triggering interference with other rights. Put differently, given that the proposed right to a healthy environment is not of an ancillary nature, perceiving it as dependent on whether or not the facts at issue fall "within the ambit" of a conventional first-generation right would be a logical contradiction.[96] Such an approach could also perpetuate the current situation where only the environmental harms that extend as far as to seriously impact people come within the purview of the ECtHR. It is true that the proposed text does not preclude a limiting interpretation pursuant to which only the environmental harm of a certain severity would come under the jurisdiction of the Court.[97] This, however, is clearly not the same as requiring impact on people. Insisting on the latter could render the new right inoperative and undermine the declared intention to have "much more effective", "ambitious" and "enhanced" environmental protection.[98]

---

[94] Winisdoerffer (2003); Kobylarz (2018); and Krenc (2022).

[95] Knox, Extended summary (n 7), p. 27; Lambert, Extended summary (n 7), pp. 43–44; and The Right in Ireland (n 6), p. 16.

[96] Contrast with Article 14 of the ECHR which has a relative autonomy only, see ECtHR's Guide on Article 14 para 5–18. See also, Suprema Corte de Justicia de la Nación [National Supreme Court of Justice, Mexico] no. 649/2019, 11 March 2020, pp. 7 and 8.

[97] Keller, Extended summary (n 7), p. 4.

[98] Recommendation (n 7), §§ 1 and 3.1.

As explained in Sect. 2, the projected right to a healthy environment has a paradigmatic duality to itself. Where the right is viewed in its subjective-anthropocentric dimension, interference may stem from the loss of benefits derived from the environment by humans,[99] possibly also leading to collateral interference with other rights guaranteed by the ECHR and its Protocols. Where the right is viewed in its objective-ecocentric dimension, interference may result from significant harm to elements or functions of ecosystems, even in absence of impact on people.[100] While complaints regarding the former, anthropocentric type of ecological damage may, to some extent, resemble grievances already falling within the scope of indirect rights-based protection, complaints regarding the latter, ecocentric type of ecological damage, would constitute a complete novelty for the Strasbourg system.

3.2.2 Animals as Subject-Matter of Human Rights Protection

The projected additional protocol does not posit that animals or other elements of nature, other than humans, should be considered the holders of the right. Nevertheless, it is undisputed that animals do form part of the environment. Under the anthropocentric approach, they are a critical indicator of the health of the world's biodiversity that benefits humans.[101] Under the ecocentric approach, they also have an intrinsic value. It follows that, even without advancing the rights of nature, the projected right to a healthy environment could theoretically be understood as extending the protection of the right to a healthy environment to issues such as cruel treatment or killing of animals.[102] A clarification may be required from the legislator as to whether the right is intended to protect animals as species only, or also, as singular entities. Theoretically speaking, the former interpretation would have to *prima facie* enable complaints about deficiencies in the protection of wild animals threatened with significant decline or extinction. The latter interpretation could additionally bring into the right's scope *ratione materiae* issues such as the ill-treatment or killing of livestock, or of domestic or captive animals. Having individual animals fall within the scope of the right to a healthy environment would not, however, lead to their normative recognition as "persons" for the purposes of Article 34 of the ECHR (see Sect. 3.3 below). It follows that applications brought on behalf of animals would, in principle, continue to be incompatible *ratione personae* with the Convention. Exactly for that reason, the Court has in the past rejected two

---

[99] Article 1247 of the French Civil Code.

[100] Ibid.

[101] IUCN Red List, https://www.iucnredlist.org/about/background-history and Separate Opinion of Judge Pinto Albuquerque in *Herrmann v. Germany* [GC], no. 9300/07, 26 June 2012.

[102] Compare with *Akkum and Others v. Turkey,* no. 21894/93, §§ 276–77, ECHR 2005-II; and *Chagnon and Fournier v. France*, nos. 44,174/06 et al., 15 July 2010.

applications brought in the name of a captive chimpanzee by the animal's caretaker and by an animal protection activist, who claimed that apes had human rights.[103]

Irrespective of the above considerations, it is of utmost importance to reiterate that the exercise of the right, as advanced, is open to permissible restrictions, for example for the protection of health or of the rights and freedoms of others.[104] Owing to this feature, the right to a healthy environment could not engender any outright prohibition, for example, on eliminating invasive species or on the culling of livestock that might be necessary for containing pandemics or for reducing the carbon footprint of husbandry. Lastly, even pursuant to the principle *in dubio pro natura* or to ecological ethics inherent in the ecocentric dimension of the proposed right, humans are still considered part of nature, whereas animals are not subject to "humanization".

In the light of these considerations, it does not appear conceivable that a normative consequence of the right to a healthy environment would be an unqualified prohibition on raising livestock or hunting game for subsistence; on the management of game or stray animals; on the use of animals for work, medical tests or religious rituals; or on keeping pets. Accepting the use of animals in service of the humans, for example, for food,[105] necessary medical experimentation,[106] with caveats, for religious practice,[107] as well as managing game[108] or animals for human safety,[109] has been implicit in the jurisprudence of the Strasbourg Court and would still be normatively possible. It is true that the right to a healthy environment has in some jurisdictions led to a ban on hunting or fishing for leisure[110] or on subjecting animals to unnecessary cruel treatment.[111] Any such consequence, however, appears to incidentally align with the outcomes already happening under the existing framework insofar as the ECtHR has considered animal protection a public interest,[112] and

[103] *Stibbe v. Austria*, no. 26188/08 and *Balluch v. Austria*, no. 26180/08, both declared inadmissible on 15 January 2010.

[104] Recommendation (n 1), Article 7.

[105] Compare with *Cha'are Shalom Ve Tsedek v. France* [GC], no. 27417/95, § 77, ECHR 2000-VII.

[106] Compare with *Tierbefreier e.V. v. Germany*, 45192/09, 16 January 2014.

[107] *Cha'are Shalom* (n 105), § 77.

[108] *Chassagnou and Others v. France* [GC], nos. 25,088/94 et al., ECHR 1999-III; *Herrmann* (n 101).

[109] *Georgel and Georgeta Stoicescu v. Romania,* no. 9718/03, § 59, 26 July 2011; and *Chagnon* (n 102).

[110] Corte Constitucional [Constitutional Court, Colombia], Sentencia C-045-19, 6 February 2019.

[111] Corte Suprema de Justicia, Sala Constitucional [Constitutional Court, Costa Rica], Resolución 01754—2021, 12/10/2021.

[112] Among many others, *Verein gegen Tierfabriken Schweiz (Vgt) v. Switzerland* (no. 2) [GC], no. 32772/02, § 92, ECHR 2009; *Hashman and Harrup v. the United Kingdom* [GC], no. 25594/94, § 28, ECHR 1999-VIII; *Bladet Tromsø and Stensaas v. Norway* [GC], no. 21980/93, §§ 63 and 73, ECHR 1999-III; and *Barthold v. Germany*, 25 March 1985, § 58, Series A no. 90.

has declined to recognize hunting[113] or unregulated ritual slaughter[114] as a right derived from private life or freedom of religion, respectively. Moreover, conscientious objection to cruelty towards animals or to their killing has, in theory, been integrated by the ECtHR into the right to freedom of thought and conscience protected by Article 9 of the ECHR. To this end, although no examinations on the merits have taken place in this context, the Court has already accepted, for example, that opposition on ethical grounds to hunting constitutes "convictions" that attain a certain level of cogency, cohesion and importance and that are therefore "worthy of respect in a democratic society".[115] Overall, these outcomes appear compatible with the proposed right to a healthy environment. In the near future, the ECtHR may further elaborate on the place of animal welfare in the Convention, when it rules in the cases of *Executief van de Moslims van België and Others v. Belgium* concerning a ban on the ritual slaughter of animals without stunning introduced under the national law for reasons of the protection and welfare of animals.[116] In a philosophical sense, if the ecocentric approach were confirmed, the welfare or the very existence of animals would, conceptually, no longer be viewed as derivative of human ethics and sensitivity,[117] but rather as an end in itself based on the intrinsic value of the environment and its components.[118] Summing up, as observed by Judge Pinto Albuquerque in his separate opinion in the case of *Herrmann v. Germany* and, elsewhere, by Tom Sparks, the Court's evolving case-law shows that a synergistic interaction between the ECHR and the early stages of animal law is normatively and philosophically possible.[119]A consequence of the ecocentric right to a healthy environment would possibly be to ensure consolidation and to give legitimacy to this ongoing, albeit haphazard, evolution of jurisprudence.

3.2.3 Minimum Severity Threshold

While Article 1 of the draft additional protocol to the ECHR does not qualify environmental degradation for the purposes of the proposed right, Article 2 explicitly refers to "irreparable and irreversible damage to life on Earth". Moreover, experts participating in the CoE's preparatory works have agreed that a certain "minimum severity" of environmental harm may legitimately be required for the right to be operational.[120] They advocated, however, that the assessment of such threshold

---

[113] *Friend and Countryside Alliance v. the United Kingdom* (dec.), nos. 16,072/06 et al., §§ 43–44, 24 November 2009; and *Chassagnou* (n 108), § 113.

[114] *Cha'are Shalom* (n 105), § 77.

[115] *Chassagnou* (n 108), § 114.

[116] *Executief van de Moslims van België and Others v. Belgium*, no. 16760/22 and 10 other cases.

[117] Sparks (2018), p. 8.

[118] Corte Constitucional [Constitutional Court, Colombia], Sentencia C-045-19, 6 February 2019.

[119] Separate opinion of Judge Pinto de Albuquerque in the case of *Herrmann* (n 101); and Sparks (2018), p. 13.

[120] Keller, Extended summary (n 7), p. 5; Knox, Extended summary (n 7), p. 31.

should not be hampered by the "beyond reasonable doubt" standard of proof which is unattainable in the context of multifactor or slow onset ecological damage. Instead, causality based on probabilities should be admitted and the burden of proof should be reversed where information about the arguable environmental harm is beyond the applicant's reach.[121] The experts also stressed the importance of basing adjudication on science.[122]

Pursuant to the ECtHR's well-established case-law, the Convention is not engaged every time environmental harm (pollution, noise, disturbance, etc.), or the risk of thereof, occurs.[123] The applicability of various rights offering some level of indirect protection of the environment is conditional on the alleged environmental harm (treated as a material event triggering interference with another right) being "specific", "serious", "severe", "clearly identifiable" or "imminent".[124] Applicants must produce reasonable and convincing evidence of the probability of harm which cannot be "hypothetical", "tenuous" or "too remote".[125] Compatibility *ratione materiae* hinges on all these notions,[126] although, in environment-related cases, the Court often makes this assessment together with the merits of a case.[127]

The Court therefore qualifies environmental harm for the applicability of the relevant provisions offering indirect protection (mainly, Articles 8 and 6). Again, as the ECHR and its Protocols only guarantee rights that are of subjective nature, the qualifying factors pertain to the intensity or duration of the negative impacts on individual applicants. The notions employed have a degree of inherent flexibility and, in practice, they very much depend on the particular circumstances of a given case, taking into consideration the nature of the substantive right at stake, factual complexity, and any evidentiary difficulties involved.

[121] Keller, Extended summary (n 7), p. 6; and Lambert, Extended summary (n 7), p. 44.

[122] Keller, Extended summary (n 7), p. 10; and Lambert, Extended summary (n 7), p. 44.

[123] *Jugheli and Others v. Georgia*, no. 38342/05, § 62, 13 July 2017; and *Çiçek and Others v. Turkey*, (dec.), no. 44837/07, § 22, 4 February 2020.

[124] In respect of harm produced, see for example, *Tauria and 18 others v. France* (dec.), no. 28204/95, 4 December 1995; *Guerra and Others v. Italy*, 19 February 1998, § 57, Reports of Judgments and Decisions 1998-I; *López Ostra v. Spain*, 9 December 1994, § 51, Series A no. 303 C; *Fadeyeva v. Russia*, no. 55723/00, § 68, ECHR 2005-IV; *Băcilă v. Romania*, no. 19234/04, § 64, 30 March 2010. In respect of risk of harm, see for example, *Balmer-Schafroth e.a v. Switzerland* [GC], no. 22110/93, § 40, 26 August 1997; *Asselbourg and 78 others and Greenpeace Association-Luxembourg v. Luxembourg* (dec.), no. 29121/95, ECHR 1999-VI; *Athanassoglou and Others v. Switzerland* [GC], no. 27644/95, § 51, ECHR 2000-IV; *Folkman and Others v. the Czech Republic* (dec.), no. 23673/03, 10 July 2006; *Budayeva and Others v. Russia*, nos. 15339/02 et al., § 137, ECHR 2008 (extracts); *Ivan Atanasov v. Bulgaria*, no. 12853/03, § 66, 2 December 2010; *Vecbaštika and Others v. Latvia* (dec.), no. 52499/11, §§ 69–70, 19 November 2019; and *Thibault v. France* (dec.), nos. 41892/19 et al., §§ 40–48, 14 June 2022.

[125] *Aly Bernard and 47 others and Greenpeace—Luxembourg v. Luxembourg* (dec.), no. 29197/95, 29 June 1999; *Luginbühl v. Switzerland* (dec.), no. 42756/02, 17 January 2006.

[126] *Fadeyeva* (n 124), § 79–88.

[127] *Denisov v. Ukraine* [GC], no. 76639/11, § 92, 25 September 2018; *Kapa and others v. Poland*, nos. 75031/13 et al., § 153, 14 October 2021; and compare with *Thibaut* (n 124), § 48.

For the Court, the severity threshold can be reached even if the pollution or nuisance complained of are only occasional,[128] and even if the applicable national limit values have not been exceeded.[129] On the other hand, non-compliance with national standards for permissible pollution does not automatically lead to a finding that the minimum threshold has been met.[130] Similarly, the fact that an activity harmful to the environment was illegal under the national law, may—but not automatically or in and of itself—lead the Court to conclude that there was an interference.[131] Lastly, any detriment that is "negligible in comparison to the environmental hazards inherent in life in every modern city", in principle, falls below the minimum severity threshold.[132] As a result of the latter, negative environmental impacts occurring in urbanized, industrialized or otherwise degraded areas may currently escape the Court's scrutiny.

The onus is on the applicant to prove that the threshold of severity has been reached.[133] Where there was no evidence about environmental or health hazards in respect of the applicant, the Court has stressed its subsidiary role vis-à-vis the national authorities and courts and the fact that it must be cautious in taking on the role of a tribunal of fact. Where domestic proceedings have taken place, it is, as a rule, not the Court's task to substitute its own assessment of the facts for that of the domestic courts. By contrast, where there have been no national proceedings addressing the key issue in the case, and this was not due to the lack of appropriate remedies in domestic law, the Court held that the lack of a fact-finding exercise at the domestic level and the lack of proof that the activity unduly interfered with the applicant's private life, prevented it from concluding that the nuisance of which the applicant complained reached the minimum level of severity.[134] In assessing evidence, the general principle has been to apply the standard of proof "beyond reasonable doubt". Such proof may follow from the coexistence of sufficiently strong, clear and concordant inferences or of similar unrebutted presumptions of fact.[135] In

[128] *Zammit Maempel v. Malta*, no. 24202/10, § 38, 22 November 2011.

[129] For example, *Frankowski and Others v. Poland* (dec.), no. 25002/09, 20 September 2011; *Calancea and Others v. Moldova* (dec.), no. 23225/05, § 29, 6 February 2018, *Kožul and Others v. Bosnia and Herzegovina*, no. 38695/13, § 35–38, 22 October 2019; *Dmitriyev v. Russia*, no. 17840/06, §§ 33 and 34, 1 December 2020.

[130] *Fägerskiöld v. Sweden* (dec.), no. 37664/04, 26 February 2008; and *Martínez Martínez and Pino Manzano v. Spain*, no. 61654/08, § 46, 3 July 2012.

[131] For example, *Furlepa v. Poland* (dec.), no. 62101/00, 18 March 2008; *Galev and Others v. Bulgaria* (dec.), no. 18324/04, 29 September 2009; *Oluić v. Croatia*, no. 61260/08, §§ 52–62 and 65, 20 May 2010; *Mileva and Others v. Bulgaria*, nos. 43449/02 and 21475/04, § 91, 25 November 2010; *Darkowska and Darkowski* (dec.), no. 31339/04, § 67, 15 November 2011; and *Çiçek* (n 123), § 29. Compare with *López Ostra* (n 124), §§ 16–22; and *Guerra* (n 124), §§ 25–27.

[132] For example, *Fadeyeva* (n 124), § 69; *Dubetska and Others v. Ukraine*, no. 30499/03, § 105, 10 February 2011; *Apanasewicz v. Poland,* no. 6854/07, § 96, 3 May 2011; *Marchiş and Others v. Romania* (dec.), no. 38197/03, 28 June 2011; *Jugheli* (n 123), § 62.

[133] For example, *Atanasov* (n 124), § 75.

[134] *Galev* (n 131).

[135] *Fadeyeva* (n 124), §§ 79 and 88.

the event of scientific uncertainty, the Court has accepted that a sufficiently close link can be established on the basis of a cumulation of factors such as statistics and reports on general causation.[136] The Court has also relied on the presumption that, even assuming that the pollution did not cause any quantifiable harm to the applicants' health, it inevitably made them more vulnerable to various illnesses and adversely affected their quality of life at home.[137] As regards health impairment, the Court has acknowledged that it is hard to distinguish the effect of environmental hazards from the effects of other relevant factors, such as, age, profession or personal lifestyle.[138]

If the projected right to a healthy environment were to operate with a "minimum severity" requirement, depending on the circumstances of the case, such a threshold, in theory, could either be assessed subjectively—similarly to what the Court is doing now—referring to the impacts of the impugned environmental damage on applicants, or objectively, qualifying such damage in terms of ecology only. The latter would be an unfamiliar exercise for the ECtHR, and some of the standards that have so far developed in its environment-related case-law may become obsolete. For example, the right to a healthy environment, similarly to the CoE's Landscape Convention, may likely impose the same level of protection to sites of "outstanding beauty" as to "everyday or degraded landscapes" insofar as the latter can be restored and their ecological quality improved.[139] As a consequence of the right to a healthy environment, the new minimum severity test may have to be based on objective factors, requiring a greater reliance on technical or scientific information.

3.2.4 Relationship Between the Autonomous Right to a Healthy Environment and Other Rights

While the self-standing right to a healthy environment does not suppose, as a necessary condition, an infringement of another fundamental right, collateral human rights violations may nevertheless occur. This phenomenon stems from the inherent dependence of people on the healthy environment and from the indivisibility of human rights.[140] The right to a healthy environment should, by no means, be misconstrued as an extension of the scope of any of the rights that presently offer a level

[136] For example, *Tătar v. Romania*, no. 67021/01, §§ 105–111, 27 January 2009; and *Dzemyuk v. Ukraine*, no. 42488/02, §§ 82 and 83, 4 September 2014.

[137] For example, *Jugheli* (n 123), § 71; *Ledyayeva and Others v. Russia*, nos. 53,157/99 et al., § 90, 26 October 2006; *Wałkuska v. Poland* (dec), no. 6817/04, 29 April 2008. See also, ECHR Guide on Environment, para 79.

[138] For example, *Ledyayeva* (n 137), § 90; and *Dzemyuk* (n 136), § 79.

[139] CoE Landscape Convention of 20 October 2000, as amended by 2016 Protocol, Preamble and Article 2.

[140] Lhaka Honhat (n 35), para 243. See also Daly and May (2019), p. 175.

of indirect protection to the environment.[141] It is a *sui generis* right, with a distinct object of protection—the environment. This does not exclude, however, a certain overlap, similar to what already and frequently occurs, for example, between Articles 2 and 3, Articles 3 and 8, Articles 6 and 8, Articles 8 and 10, or Articles 10 and 11 of the ECHR. With an important caveat that the right to a healthy environment is fully autonomous, the resemblance may also be found with the interplay between Article 14 and any of the rights within whose ambit the facts of a given case have to fall.

Litigating the right to a healthy environment, along individualized grievances about the breach of the right to life, private life, home, health, food or culture, is indeed a dominating trend in the rights-based litigation regarding environmental degradation and climate change.[142] A good example of such indivisibility of rights is the case of *Lhaka Honhat* decided by the IACtHR. In this case, the Argentine authorities failed to recognise land ownership by indigenous communities and tolerated various activities by other residents of the area, such as illegal logging, livestock husbandry and fencing, which had resulted in decreasing forest resources and biodiversity. This, in turn, has interfered with the way the applicant groups traditionally use land, and seek food and water. On the facts of the case, the IACtHR found a violation of the right to a healthy environment, as well as of the rights to communal property, food, water, and to cultural identity.[143] While these specific rights are not guaranteed by the ECHR or its Protocols, vast environment-related jurisprudence developed by the ECtHR, the IACtHR, the African Commission of Human and Peoples' Rights and by the UN Human Rights Committee, supports the conclusion that civil and political rights can be engaged in cases concerning environmental harm.[144] Litigating for the healthy environment as a uniquely ecocentric substantive right, seems to only occur in jurisdictions which either expressly recognise the rights of nature or which allow public interest litigation.[145]

3.2.5 Substantive and Procedural Limbs of the Right to a Healthy Environment

The right to a healthy environment could typically come into play in applications concerning deforestation, urban development, or other forms of significant environmental alterations affecting ecosystems and their living entities. Disturbance or

---

[141] Contrast this with Article 14 of the ECHR and Article 1 of Protocol No. 12 to the ECHR, see ECtHR Guide on Article 14 and on Article 1 of Protocol No. 12 to the Convention, paras 10–19.

[142] Peña Chacón (2018, 2020, 2022).

[143] Lhaka Honhat (n 35), para168 and 289; and Corte Suprema de Justicia, Sala Constitucional [Constitutional Court, Costa Rica] no. 24513—2019, 6/12/2019.

[144] Kobylarz (2022).

[145] Peña Chacón (2022); and Cedars Protected Forest Case, Corte Constitucional del Ecuador [Constitutional Court, Ecuador] no. 1149-19-JP/21, 10 November 2021.

destruction of ecosystems could engage the substantive and/or the procedural aspects of the right.

As regards the substantive aspect, with the projected additional protocol, the Court, in principle, could no longer reject cases like that of *Ogloblina v. Russia,* in which the applicant complained that, by authorizing a housing development in a local forest, the public authorities allowed deforestation compromising environmental safety. Not disposing of the right to a healthy environment, the Court examined the case under the right to respect for private life. But as the applicant could not show any prejudice to her wellbeing or health, the Court declared this complaint inadmissible.[146] Under the projected regime, the logging of an entire forest, unlike falling of several trees only, could objectively seriously alter or destroy an ecosystem, possibly reaching the minimum severity threshold.

Having the right to a healthy environment at its normative arsenal would also, incidentally, allow the ECtHR to release tensions that nowadays accompany the adjudication of claims based on expansive interpretations of Articles 2 or 8 of the ECHR, or of Article 1 of Protocol No. 1 to the ECHR.[147] *Kyrtatos v. Greece*[148] is a landmark example of a complaint that was considered to go beyond the limits of Article 8. This application was brought by individuals living on land adjacent to wetlands which were a natural habitat for various protected species of birds, fish and turtles. The area was damaged as a result of construction of a tourist complex which, despite being declared unlawful by domestic courts, was not demolished. The applicants complained that they were affected by the destruction of the original wilderness area.[149] The majority of judges in this case were of the view that there was no interference with the applicants' private life. They reasoned that the applicants had not shown that the alleged damage to the species living in the swamp was of such a nature as to directly affect their right. They then intuitively added that "[i]t might have been otherwise if, for instance, the environmental deterioration … had consisted in the destruction of a forest area in the vicinity of the applicants' house, a situation which could have affected more directly the applicants' well-being."[150] In his partly-dissenting opinion to this judgment, Judge Zagrebelsky saw "no major difference between the destruction of a forest and the destruction of the extraordinary swampy environment the applicants were able to enjoy near their house." As he wrote, "the importance of the quality of the environment and the growing awareness of that issue … should induce [the Court] to recognise the growing importance of environmental deterioration on people's lives."[151] The approach taken by the majority-judges in this case might have protected Article 8 from deformation, but at

---

[146] *Valentina Viktorovna Ogloblina v. Russia* (dec.), no. 28852/05, § 28, 26 November 2013.

[147] The Right in Ireland (n 6), p. 16.

[148] *Kyrtatos v. Greece*, no. 41666/98, ECHR 2003-VI.

[149] *Kyrtatos v. Greece*, no. 41666/98 (dec.), 11 January 2000 and (dec.), 13 September 2001.

[150] *Kyrtatos* (n 148), §§ 46 and 53.

[151] Ibid, partly dissenting opinion of Judge Zagrebelsky.

the price of what Katalin Sulyok calls "epistemic arbitrariness".[152] As she explains, in the field of environmental litigation, when judges evaluate facts established in the science of ecology with a non-scientific method, namely common sense or everyday logic, what occurs is epistemic trespassing that may lead to a result that lacks epistemic legitimacy.[153] In *Kyrtatos,* the judicial reasoning did not engage with the concept of ecosystem services or ecology to assess the degree to which the applicants' well-being might have directly been affected, and, therefore, whether the case fell within the subject-matter of Article 8. The Court has shown similar reluctance to expand Article 2 of the ECHR to cover the right to "dignified life" in pollution cases,[154] or Article 1 of Protocol No. 1 to the ECHR to protect collective enjoyment of the right to property by members of indigenous communities based on the recognition of their special attachment to and dependency on uncontaminated or unspoiled land.[155] By comparison, examples of such expansive interpretation of these two rights may be found in the jurisprudence of the IACtHR[156] and the works of the UN Human Rights Committee.[157]

A normative consequence of the proposed right to a healthy environment would be that scenarios in which serious objective ecological damage risked occurring or had already occurred, without materializing into damage subjectively affecting humans, would not have to push their way into the anthropocentrically limited scope of the existing first-generation rights. The Court would be called to rule under the *sui generis* right that inherently combines subjective and objective functions of the environment. In the event both types of ecological damage were at stake, the Court would have a choice of ruling, either simultaneously under the right to a healthy environment and Article 8, for example, or considering that there is no separate issue under the latter provision.[158]

---

[152] Sulyok (2021), pp. 40 and 65.

[153] Ibid, pp. 40 and 65.

[154] *Di Sarno and Others v. Italy*, no. 30765/08, §§ 94–96, 10 January 2012; *Hardy and Maile v. the United Kingdom*, no. 31965/07, §§ 183–84, 14 February 2012; and *Cordella and Others v. Italy*, nos. 54414/13 et al., §§ 93–94, 24 January 2019.

[155] *G. and E. v Norway* (dec.), nos. 9278/81 and 9415/81, 3 October 1983.

[156] For example, IACtHR, Mayagna (Sumo) Awas Tingni Community v Nicaragua, Judgment of August 31, 2001; Yakye Axa Indigenous Community v Paraguay, Judgment of June 17, § 162; Sawhoyamaxa Indigenous Community v. Paraguay, Judgment of March 29, 2006, §§ 150–78 and concurring opinion of Judge García Ramírez, §§ 18–23; Saramaka People v Suriname, Judgment of November 28, 2007; Xákmok Kásek Indigenous Community v Paraguay, Judgment of August 24, 2010; Kichwa Indigenous People of Sarayaku v Ecuador, Judgment of June 27, 2012; and Kaliña and Lokono Peoples v Suriname, Judgment of November 25, 2015. For a commentary on the IACtHR's interpretation of the right to dignified life and right to property of indigenous peoples, see Grant (2019), pp. 80–91 and 67–80.

[157] General comment No. 36 (CCPR/C/GC/36), para 3, 3 September 2019; View in Portillo Cáceres et al. v Paraguay (CCPR/C/126/D/2751/2016), paras 7.2–7.5, 20 September 2019; and Oliveira Pereira and Sosa Benega, Indigenous Community of Camp Agua'ẽof the Ava Guaraní People v Paraguay, no. 2552/2015, CCPR/C/ 132/D/2552/2015, paras 8.2–8.4, 14 July 2021.

[158] Compare with, among many others, *Connors v. the United Kingdom*, no. 66746/01, 97 and 103, 27 May 2004.

By the same token, the ECHR provisions that currently enable individuals and/or civil society a degree of participation in environmental matters, would no longer have to be engaged, given that the proposed right to a healthy environment has specific procedural components.[159] Under the current doctrine, while the right to access to information in environmental matters is effectively recognized under Articles 2, 8 and 10 (freedom of expression) of the ECHR, the applicability of these provisions is not automatic or unconditional.[160] The two former provisions require material causality with a given instance of environmental harm personally affecting the applicant. The subject-matter of Article 10, in turn, may, under certain conditions, cover access to information relating to an important subject of general interest of which the public should be informed.[161] The Court has thus allowed concerned individuals and NGOs exercising the role of "public watchdog" to rely on this provision to challenge the refusal or the quality of information concerning projects harmful to the environment.[162] It may be anticipated that the proposed additional protocol would lead to the emergence of a separate, specialized legal regime of environmental participatory rights. The ECtHR's assessment of the exercise of such rights by stakeholders would inevitably gain in coherence, given that the norms would be generated by a unique right to a healthy environment instead of, what is currently the case, by the rules scattered across several rights, with distinct characteristics, none of which is specifically designed for the purposes of environmental governance.

As submitted by Judge Serghides in his separate opinion in the recent case of *Pavlov and Others v. Russia,* the principle of effectiveness that permits the Court's case-law to evolve in the light of present-day conditions, does not go as far as to institutionalize environmental protection under the ECHR. Serghides, and, separately, also John H. Knox, are of the view that the gaps, that have been described above, can therefore only be filled if a right to a healthy environment is included in the Convention by way of a new protocol.[163]

In addition to the above-mentioned normative consequences, the proposed right to a healthy environment would, as a matter of cause, expand the scope of Article 6 § 1 of the ECHR. The procedural guarantees ensured by that provision, such as access to a court,[164] enforcement of final domestic judgments,[165] and fairness of decision-making process, including, but not limited to, equality of arms and neutrality of experts, are crucial for the exercise of the right to a healthy environment.[166]

---

[159] Recommendation (n 7), Article 6.

[160] *Guerra* (n 124), *Magyar Helsinki Bizottság v. Hungary,* no. 18030/11, §§ 149–156, 8 November 2016.

[161] *Cangı v. Turkey*, no. 24973/15, § 34, 29 January 2019; and *Association Burestop 55 and Others v. France*, nos. 56,176/18 et al., §§ 84–90, 1 July 2021.

[162] *Cangı* (n 161), § 35; and *Burestop* (n 161), §§ 78–90.

[163] Separate opinion of Judge Serghides in the case of *Pavlov and Others v. Russia*, §§ 18–22; and Knox, Extended summary (n 7), p. 27.

[164] *Stichting Landgoed Steenbergen and Others v. the Netherlands*, no. 19732/17, 16 February 2021.

[165] *Kyrtatos* (n 148), § 30; and *Taşkın and Others v. Turkey*, no. 46117/99, §§ 135–38, ECHR 2004-X.

[166] Kwiędacz-Palosz (2023).

Under the current doctrine, the applicability of Article 6 to environmental litigation primarily depends on whether the right to a healthy environment or some of its substantive or procedural components are justiciable under the domestic law.[167] Where this is the case, the dispute is, in principle, considered over "a civil right" for the purposes of Article 6 § 1.[168] For example, because the Constitution of Türkiye guarantees justiciable right to live in a healthy and balanced environment, the Court accepted that individual applicants could effectively complain of the non-enforcement of judicial decisions annulling the operating permits of a mine or a thermal power plant,[169] or about the alleged unlawfulness of the environmental impact assessment reports.[170] Similarly, where French environmental associations complained about the shortcomings of proceedings challenging authorizations of the expansion of a nuclear fuel factory or of an underground storage of nuclear waste, the Court accepted that the domestic law recognized an actionable right of the public to information and to participation in the decision-making process in environmental matters.[171] The applicants from these member States could therefore arguably claim to have a right under the respective domestic legal frameworks to substantive or procedural protection against environmental damage caused by industrial activities. The Court thus examined their applications on the merits. Conversely, where the domestic law does not recognize the right to a healthy environment or does not establish it as an individually justiciable right, the applicability of Article 6 is not straightforward and may even be hindered.[172] For example, in Poland, the right to a healthy environment does not operate as a subjective right and is explicitly excluded from the category of personal rights protected by civil law.[173] The practical consequences of such a situation are visible, for example, in the sector of clean air. Leaving aside the doctrine of direct effect of the European Union ("EU") law and the Ambient Air Quality Directive 2008/50/EC,[174] pursuant to the Polish law, neither residents of areas in which authorized limits of air pollutants are exceeded nor environmental NGOs ("eNGOs") are regarded as entitled to challenge air quality programmes before an administrative court. They are considered to lack

---

[167] ECtHR Guide on Environment, paras 39 and 40.

[168] *Taşkın* (n 165), §§ 133; *Okyay and Others v. Turkey*, no. 36220/97, §§ 66–67, ECHR 2005-VII; *Collectif national d'information et d'opposition à l'usine Melox—Collectif Stop Melox and Mox v. France* (dec.), no. 75218/01, 28 March 2006; *Folkman* (n 124); *Sdružení Jihočeské Matky v. Czech Republic* (dec.), no. 19101/03, 10 July 2006; *Atanasov* (n 124), § 91; *Burestop* (n 161), § 57. Contrast *Karin Andersson and Others v. Sweden*, no. 29878/09, § 46, 25 September 2014. See also ECtHR Guide on Environment, para 40.

[169] *Taşkın* (n 165), §§ 130–134; *Okyay* (n 168), §§ 65–68; and *Bursa Barosu Başkanlığı and Others v. Turkey*, no. 25680/05, §§ 126–128, 19 June 2018.

[170] *Çöçelli and Others v. Türkiye*, no. 81415/12, 11 October 2022.

[171] *Melox* (n 168); and *Burestop* (n 161), §§ 56–57.

[172] *Stichting Landgoed Steenbergen and Others v. the Netherlands*, no. 19732/17, § 30, 16 February 2021.

[173] Constitution of Poland, Articles 5, 68 and 74; and Supreme Court of Poland, judgment of 28 May 2021, III CZP 27/20; see also Kwiędacz-Palosz (2023).

[174] Darpö (2021), § 4.3.

the required legal interest to apply for judicial review, given that air quality programmes are only addressed to the organs of public administration and do not set out any obligations for the citizens.[175] Civil law remedies are also unavailable to individual claimants under the Polish law, unless they have suffered a tangible damage to health or to other personal rights (e.g. dignity).[176] Similar doctrines persist in several other member States of the EU or in State Parties to the Aarhus Convention.[177] By comparison, pursuant to the well-established case-law of the Court of Justice of the EU ("CJEU"), natural or legal persons directly concerned with non-compliance with measures or value limits imposed by EU directives, for example, for air or water quality, must have the option of relying upon binding rules within national law in order to exercise their rights and to require the national authorities to take measures to combat pollution.[178] Moreover, although Article 9(3) of the Aarhus Convention (that guarantees members of the public access to judicial review of acts and omissions contravening provisions of national law relating to the environment), does not, as such, have direct effect, with the result that that provision cannot compel the referring court to disapply a national provision which is contrary to it, Article 47 of the Charter of Fundamental Rights of the EU "is sufficient in itself and does not need to be made more specific by provisions of EU or national law" in order to confer on individuals a right to an effective remedy.[179] The Aarhus Convention Compliance Committee, in turn, has observed, in the context of national "impairment of rights" doctrines, that "[w]hile what is considered a public or private interest or an objective or subjective right may vary among Parties and jurisdictions, [pursuant to Article 9 (3) of the Aarhus Convention] access to a review procedure must be provided for all contraventions of national law relating to the environment."[180] Within the Strasbourg system, owing to the existing requirements for the applicability of Article 6 § 1, domestic legal frameworks that do not allow citizens to seek judicial review concerning activities dangerous to the environment unless serious individual impacts on other rights have already been produced, may fall in the ECHR's blind spot and, consequently, escape the Court's scrutiny.

The projected additional protocol would therefore offer an opportunity to create a level playing field among the member States of the CoE insofar as each and every single one of them would have to guarantee the right to a healthy environment in its

[175] *Kuczma v. Poland,* no. 35361/18; see also Kwiędacz-Palosz (2023).

[176] Kwiędacz-Palosz (2023).

[177] EU Commission Notice on access to justice in environmental matters (2017/C 275/01).

[178] Commission of the European Communities v. Federal Republic of Germany, judgment of 30 May 1991, C-361/88 (ECLI:EU:C:1991:224); Commission of the European Communities v Federal Republic of Germany, judgment 17 October 1991, C-58/89 (ECLI:EU:C:1991:391); and Dieter Janecek v. Freistaat Bayern, C-237/07, Preliminary Ruling CJEU, 25 July 2008, § 42 (ECLI:EU:C:2008:447).

[179] Deutsche Umwelthilfe e v Federal Republic of Germany, C-873/19, Preliminary Ruling CJEU Grand Chamber, 8 November 2022, §§ 78 and 79, with further references (ECLI:EU:C:2022:857).

[180] ACCC, Communication ACCC/C/2008/31 (Germany), 20.12.2013.

national law and offer a remedy in the event of the right's violations.[181] Inserting the right to a healthy environment into the ECHR framework would fill the current Article 6 § 1 gap, enabling the Court to set out uniform standards of fairness for proceedings undertaken by citizens in defense of the environment and to ensure compliance with these standards. Whether the Court can achieve this outcome even in absence of the proposed protocol may become known in the near future, given that a series of applications concerning the lack of access to judicial review of air quality programmes have been lodged with the Court.[182]

Another requirement for the applicability of Article 6 § 1 of the ECHR is that the outcome of the proceedings complained of must be "directly decisive" for the applicant's civil right. In this context, the current case-law has produced opposite results, depending on whether procedural or substantive rights were at stake. Where eNGOs complained about the shortcomings of court proceedings related to the exercise of their right to participation in a decision-making process or to access to information about environmentally hazardous projects, Article 6 was found applicable, because the proceedings in question were considered directly decisive for the applicant associations' statutory interests and procedural rights.[183] The Court reached such a conclusion even though the object of the impugned procedures was essentially the defense of the general interest. Conversely, where the applicant was a private individual relying on a substantive right, like in the case of *Ogloblina v. Russia*, discussed above, the Court considered itself estopped from examining the complaint about the shortcomings of the proceedings regarding the project allegedly harmful to the environment,[184] considering that there was no link between the deforestation and the applicant's right to adequate protection of her physical integrity. Given that the applicant could not show that the authorities' actions exposed her personally to danger, the Court found Article 6 § 1 of the ECHR *ratione materiae* inapplicable. The consequence of the right to a healthy environment would possibly be to bring complaints concerning shortcomings of judicial review in environmental matters under the Court's scrutiny regardless of whether the applicant is a natural or legal person and whether the notion of "civil right" for the purpose of the applicability of Article 6 is viewed in substantive or procedural terms.

[181] Recommendation (n 1), Article 6 (d). See also, The Right in Ireland (n 6), pp. 2–4 and 7; and Sébastien Duyck, Extended summary (n 7), p. 16.

[182] *Stowarzyszenie Miasto Jest Nasze and Jędrzejczyk v. Poland*, no. 1138/23; *Marczak and Nowak*, no. 1188/23; *Uliasz v. Poland*, no. 2082/23; *Sitarz-Wójcicka v. Poland*, no. 29053/22; and *Kuczma* (n 175); see also Client Earth, Warszawski smog zaskarżony do Strasburga, 20 December 2022, https://www.clientearth.pl/najnowsze-dzialania/artykuly/warszawski-smog-zaskarzony-do-strasburga/.

[183] *Burestop* (n 161), § 57; and *Melox* (n 168).

[184] *Ogloblina v. Russia* (n 146), §§ 20, 22 and 28.

3.3 The Court's Personal Jurisdiction

The projected right to a healthy environment is guaranteed to "everyone",[185] which is a term typically describing the holders of the rights inscribed in the ECHR and its Protocols. The novelty is the qualification of the right as that "of present and future generations.[186] Moreover, the proposed protocol considers the environment not only as a right, but also as a duty to protect the environment and biodiversity, to prevent irreversible damage to life on Earth, to ensure sustainable use of natural resources, and to remedy environmental damage.[187] Explicitly inherent in the right is therefore the responsibility on the part of the right's holders.[188] Beyond the language described above, the proposed additional protocol does not hint what rules should govern the issues of compatibility *ratione personae.*

Positing the healthy environment as a right[189] and a duty[190] would, in and of itself, require a new approach to the "victim status" and to the legal standing before the ECtHR. The "right aspect" conceptually operates in the sphere of both substantive and procedural limbs. All civil and political rights that are already guaranteed by the ECHR and its Protocols share this feature. The novelty lies in the fact that the material triggering events, such as environmental degradation, pollution or climate change, may cause both singular, as well as wide-ranging and diffuse impacts on the proposed right to a healthy environment and, possibly also, on collateral rights. Moreover, such impacts may be slow-onset and cascading. In turn, the "duty aspect" of the right, as such, effectively concerns participatory rights. The projected responsibilities towards the environment, nature and future generations become not only a valid, but, for the first time, justiciable interest for the members of the society. Article 2 of the proposed draft makes the above-mentioned responsibilities shared and intergenerational, whereas Article 6 reformulates them into individual and specific participatory rights. The proposed participatory rights differ from their current perception by the ECtHR in that the right to access to information, to consultation and, implicitly, to judicial review are not linked with an individual, but with a general interest in protecting the environment and biodiversity. In practice, depending on the circumstances of the case, it may be difficult and, frankly, unnecessary to distinguish conceptual nuances between the right and the duty aspects of the proposed right to a healthy environment. The Aarhus Convention, for example, recognizes the three pillars of participatory rights as essential to "both the achievement of the right to a healthy environment and … the possibility for individuals to fulfill their responsibilities towards others, including future generations."[191] An overriding

[185] Recommendation (n 1), Articles 5 and 6.

[186] Ibid, Article 1.

[187] Ibid, Article 2.

[188] Ibid.

[189] Ibid.

[190] Ibid, Articles 5 and 6.

[191] ECtHR Guide on Environment, p. 30.

conclusion is that the right cannot be fully exercised without the right holders being able to effectively realize their rights-and-duties in national courts and before the ECtHR.[192] Therefore, for the purposes of legal standing tests, a more useful dividing line may be drawn between claims brought by natural persons and those brought by legal entities, with the latter being granted easier access to the ECtHR.

The Court's personal jurisdiction is currently regulated by Article 34 of the ECHR. In the environmental context, a "victim" within the meaning of this provision denotes a "person, non-governmental organization or group of individuals" affected by the alleged violation of their Convention rights.[193] In light of the Court's case-law, to claim to be a victim of a violation, a person must either be affected directly—that is to say, have already suffered the harm,[194] or indirectly (potentially)—that is to say, would likely suffer the harm or would have a valid and personal interest in seeing the potentially harmful situation brought to an end.[195] Moreover, the doctrine of "direct harmful effect" effectively operates with reference to the spatial and temporal proximity of the harm, as well as its severity.

Although the project does not posit that the right to a healthy environment should have unconditional applicability *ratione personae* or offer a remedy to contest, in the abstract, policies, laws or measures regarding industrialization, energy use, the pace of low-carbon transition, or biodiversity conservation, a series of cross-cutting normative consequences for the legal standing of natural persons and civil society are to be anticipated.

3.3.1 The Status of "Future Generations"

The term "future generations" in the projected additional protocol is, first and foremost, a prism helping to define the goal of sustainable environment[196] and the principle of environmental non-discrimination.[197] This concept is already used in international and constitutional law to effectively denote an indefinitely long or open-ended time-dimension of the environment.[198] Additional formulations

---

[192] The duty of environmental protection is recognized by the doctrine, international law and many national constitutions, as integral part of the right to a healthy environment, see Preamble of the 1998 Aarhus Convention; Article 1246 of the French Civil Code; or Article 2 of the French Charte de l'environnement, 01/03/2005. See also, *Chassagnou* (n 58), §§ 113 and 125.

[193] ECHR Guide on Environment, pp. 78–80; *Balmer-Schafroth* (n 124), § 26; *Crash 2000 Ood and Others v. Bulgaria* (dec.), no. 49893/07, § 84, 17 December 2013; *Fadeyeva* (n 124), § 88; and *Cordella* (n 154), § 101.

[194] *Centre for Legal Resources on behalf of Valentin Câmpeanu* [GC], no. 47848/08, § 96, with further references, ECHR 2014; *Cordella* (n 154), § 101; and *Crash* (n 191), § 84.

[195] *Vallianatos and Others v. Greece* [GC], no. 29381/09, § 47, 7 November 2013.

[196] Recommendation (n 1), point 3.1, Preamble.

[197] Ibid, Article 3.

[198] For example, 1972 Stockholm Declaration, Preamble and Principle 1; 1982 World Charter for Nature, Preamble and Article 10; 1992 Rio Declaration, Principle 3; 1992 Convention on Biological Diversity, Preamble and Article 2; Gabčikovo (n 52), § 53; German Basic Law [Constitution],

proposed in the draft, such as “the right of present and future generations”[199] or “duty … to ensure the right of subsequent generations to live in a … [healthy] environment”[200] appear to further construct “future generations” as the holders of the right to a healthy environment. A similar language (“the right of every person of present and future generations”) is employed in the Aarhus Convention, but only to formulate this regulation’s goal[201] and to stress that intergenerational equity is a parameter of the responsibility to protect and improve the environment.[202] The Aarhus Convention, by no means, sets out norms pursuant to which a representative of future generations would have a legal entitlement to have access to information, to public participation in decision-making or to have access to justice in environmental matters. The language of both the proposed additional protocol to the ECHR and the Aarhus Convention confirm that a confusing overlap exists between the concepts of future generations and intergenerational equity.[203] This overlap, as observed by Aoife Daly, is further complicated by the inherently dynamic intersection between future and present generations.[204] The joint term of “present and future generations” is indeed employed throughout the proposed text.[205]

In the current normative framework, in special circumstances, the ECtHR may grant standing to a person who complains on behalf of a direct or indirect victim of an alleged violation. Special considerations may arise in the case of victims of alleged breaches of Articles 2, 3 or 8 of the ECHR at the hands of national authorities, with regard to those victims’ vulnerability on account of their age, sex or disability, when such a factor has rendered them unable to lodge a complaint with the Court.[206] In exceptional circumstances, to prevent respondent State from escaping international accountability for a serious allegation of a violation, an association can act as a representative of a victim, even in the absence of a power of attorney and notwithstanding that the victim may have died before the application was lodged.[207] The ECtHR can also entertain applications lodged by minor children, either directly or through their parents or legal guardians.[208]

By comparison, outside of the Strasbourg system, interests of future generations have been defended by an Ombudsperson for Future Generations entitled, under the

---

Article 20 (a); Constitution of Norway, Article 112; 1995 Constitution of Georgia, Article 37; 1991 Constitution of Poland, Article 74; Constitutional Court of Lithuania, ruling of 14 March 2006, cited in *Kristiana Ltd. v. Lithuania*, no. 36184/13, § 68, 6 February 2018.

[199] Aarhus Convention, Article 1 and Preamble.

[200] Recommendation (n 1), Articles 1 and 2.

[201] Aarhus Convention, Article 1.

[202] UNECE The Aarhus Convention and Implementation Guide (2014), pp. 30 and 42.

[203] Daly (2022), p. 4. For essential reading on intergenerational equity, Brown Weiss (1989, 1990).

[204] Daly (2022), pp. 8, 13, 14, 16–22.

[205] Recommendation (n 1), Preamble and Article 1.

[206] *Câmpeanu* (n 194), § 103.

[207] Ibid, § 112; and *Kondrulin v. Russia*, no. 12987/15, § 31, 20 September 2016.

[208] ECtHR Practical Guide on Admissibility Criteria, para 61; and for example *A. v. the United Kingdom*, no. 25599/94, 23 September 1998.

domestic law, to initiate or take part in judicial proceedings.[209] Intergenerational legal standing was also famously granted by the Supreme Court of the Philippines in the case known as "*Minors Oposa*".[210] That court held that a group of minor children and their parents had active capacity to challenge logging permits leading to massive deforestation of their country, both on behalf of their own generation and generations yet unborn. This was based on the reasoning that each generation was a "trustee and guardian of the environment for succeeding generations."[211]

The Philippine case supports the argument raised by Edith Brown Weiss,[212] Daly,[213] and Knox,[214] that intergenerational equity can materialize through the legal standing of children and the notion of the child's best interests, both of which are already known and operational within the ECHR doctrine.[215] It appears that if the status of future generations remained ambiguous in the projected protocol, the right to a healthy environment would not produce any change for the norms regulating the Court's jurisdiction *ratione personae*. Intergenerational equity, in turn, would be important for the judicial examination of the merits of environment-related cases. Firstly, it would necessarily play a role in defining positive obligations of member States in the context of environmental degradation and conservation. Secondly, it may be a parameter for the assessment of whether fair balance had been struck between conflicting economic and environmental interests at stake. Thirdly, as argued by the applicants in the case of *Duarte Agostinho and Others v. Portugal and 32 other States*, belonging to a young generation must be seen as a ground for discrimination prohibited by Article 14 of the ECHR. The applicants, some of whom are minors, argue that the negative impacts of global warming will be felt by their generation over a longer time. To this end, they submit that there is no objective and reasonable justification for placing the burden of climate change on younger generations through the adoption of belated measures to reduce global warming. In this context, the applicants refer to the UN Convention on the Rights of the Child which requires that any decision concerning them must be based on the paramount consideration of the best interests of the child.[216] The principle of the best interests of the child, together with intergenerational equity and sustainable development are already well-established in intentional law. It follows that, even without the right to a healthy environment being adopted, the above-mentioned principles and concepts

[209] The office of Ombudsman for Future Generations was created by the Hungarian Parliament in 2007. https://www.ajbh.hu/web/ajbh-en/the-role-of-the-ombudsman#:~:text=The%20mandate%20of%20the%20he,the%20interests%20of%20future%20generations.

[210] Minors OPOSA v. Secretary of the Department of Environment and Natural Resources, International Legal Materials, vol. 33 No. 168 (1994).

[211] Ibid.

[212] Brown Weiss (1994), p. 22.

[213] Daly (2022), p. 12.

[214] Knox, Extended summary (n 7), pp. 58 and 59.

[215] For example, *X and Others v. Bulgaria* [GC], no. 22457/16, § 192, 2 February 2021.

[216] *Duarte Agostinho* (n 46), communication report of 13 November 2020, https://hudoc.echr.coe.int/eng?i=001-206535. See also Sandvig et al. (2021).

will guide the evolutive interpretation of the ECHR. The forthcoming Grand Chamber ruling in the *Duarte Agostinho* case, may provide a more definite answer as to whether the right to a healthy environment is a necessary input for the recognition of the intergenerational interest in the healthy environment under the ECHR.

3.3.2 *Actio Popularis*

Conceptually, the right to a healthy environment belongs both to first and second generations of human rights, insofar as it may concern, all at once, individual, diffuse and collective interests.[217] This chameleon-like character of the right allows it to fit (although, sometimes imperfectly) into diverse mechanisms of judicial protection. Those range from pure *actio popularis,*[218] to individual-interest litigation, passing through actions by designated commissions of guardians, ombudspersons or eNGOs. For example, in the African system of human rights protection the right of "all peoples" to a healthy environment is directly inscribed in the African Charter on Human and Peoples' Rights,[219] an instrument that guarantees civil and political, as well as economic, social and cultural rights. The African Commission and the African Court of Human and Peoples' Rights allow applications from individuals and from accredited NGOs.[220] In the landmark case concerning environmental degradation and health problems resulting from the contamination produced by the operations of oil companies, the applicants were two NGOs, whereas the actual direct victims were the Ogoni people on whose territory the contamination occurred.[221] In the inter-American system the right to a healthy environment operates as a right derived from the right to progressive economic, social and cultural development, which is explicitly guaranteed in the American Convention on Human Rights. This system excludes *actio popularis*. The first, and, so far, the only, contentious case engaging the right to a healthy environment was brought by an indigenous community whose members were direct victims of the environmental degradation complained of.[222] In Europe, the Aarhus Convention considers the

---

[217] Knox, Extended summary (n 7), p. 29; and Lambert, Extended summary (n 7), p. 43.

[218] For example, in India access to the Supreme Court is open to persons "genuinely interested in the protection of society on behalf of the community", Public Interest Litigation can also be triggered by the Supreme Court, of its own motion; see Subhash Kumar v. State of Bihar (1991), 1 SCR 5 and M.C. Mehta v. Union of India, SCR 861991 SCC (2) 353 (1991) [Supreme Court of India]. In the Philippines, "writ of Kalikasan" can be pursued on behalf of persons whose constitutional right to a balanced and healthful ecology is violated, or threatened if it involves "environmental damage of such magnitude as to prejudice the life, health or property of inhabitants in two or more cities or provinces"; see May and Daly (2019), p. 38. In Costa Rica environmental amparo can be brought virtually by anyone, including minors; see Peña Chacón (2003).

[219] African Charter of Human and Peoples' Rights, Article 24.

[220] Articles 6 and 34(6) of the Protocol establishing the ACtHPR, with a caveat that the State against which such application is filed has made a special declaration accepting this procedure.

[221] SERAC (n 56).

[222] Lhaka Honhat (n 35).

protection of "the right of every person … to live in an environment adequate to his or her health and well-being" as an objective to be achieved through the three pillars of participatory rights (each pillar, with its own nuanced rules on legal standing).[223] Similarly to what is advanced by the proposed additional protocol to the ECHR, that convention is based on the premise that environmental law protects not only individual interests, but everyone and the environment itself.[224] In the context of access to justice (third pillar), Parties to the Aarhus Convention are required to ensure legal standing for members of the public seeking to enforce environmental law to prevent or redress ecological damage.[225] It is not required, however, that a national law must grant *actio popularis* in its most extensive form, that is to say, standing for anyone to challenge any decision, act or omission relating to the environment. Access to justice can be restricted, as long as the law does not effectively exclude all or almost all members of the public from challenging such acts or omissions that contravene national law.[226] Similar requirements are laid down in EU directives and regulations.[227] As for the CoE's member States, some (e.g. France) take an intermediary approach to public standing, opening it only to registered eNGOs, while others (e.g. Latvia and Portugal) take the most extensive approach, granting pure *actio popularis*.[228]

Article 34 of the ECHR excludes from the Court's jurisdiction any *actio popularis*, that is to say, any public-interest applications that would not have any effect on the applicant's individual rights.[229] Under the current regime, a personal interest is required both for substantive and for procedural grievances concerning any right guaranteed by the ECHR or its Protocols.[230]

The proposed right to a healthy environment would logically have to relax some of the current norms regulating *locus standi* before the ECtHR, especially to enable

---

[223] Aarhus Convention, Articles 1 and 9.

[224] Access to Justice in EU Law. A Legal guide on Access to Justice in environmental matters (Client Earth 2021), p. 29.

[225] Article 9(3) of the Aarhus Convention, see Aarhus Implementation Guide, p. 198. Communication ACCC/C/2006/18 (Denmark), 29.04.2008.

[226] ACCC, Communication ACCC/C/2006/18 (Denmark), § 29, 29.04.2008; Communication ACCC/C/2010/50 (Czech Republic), 29.06.2012; Aarhus Guide (n 225), pp. 197 and 198 and Access to Justice (n 224), p. 38.

[227] For example, Articles 12 and 13, Directive 2004/35/CE on environmental liability with regard to the prevention and remedying of environmental damage, 21/04/2004, and Articles 10 and 11, Regulation (EC) No 1367/2006 on the application of the provisions of the Aarhus Convention (…) to Community institutions and bodies, 6/09/2006, as amended by Regulation (EU) 2021/1767, 6/10/2021; see also Access to Justice (n 224), p. 38 et al.

[228] Mikosa (2020), pp. 79–80 and 86–102; and Darpö (2013).

[229] *Câmpeanu* (n 194), § 101; *Roman Zakharov v. Russia* [GC], no. 47143/06, § 164, ECHR 2015; *Bursa* (n 169); *Crash* (n 191), § 84; *Di Sarno* (n 154), § 80; and *Cordella* (n 154), § 100.

[230] X. *v. Federal Republic of Germany* (dec.), no. 7407/76, 13 May 1976; *Ünver* (n 182); *Ogloblina* (n 146), §§ 20–22; *Kyrtatos* (n 148), §§ 46, 52 and 53; *Hamer* (n 36), § 79; *Turgut* (n 39), § 90; and *Dubetska* (n 132), § 105.

the realization of the general duty to protect the environment and biodiversity[231] through participatory rights.[232] As for complaints concerning the substantive components of the right, however, the proposed text does not posit that standing should be based on an unlimited public interest. It must be stressed that the ongoing political process of anchoring the right to a healthy environment within the CoE also postulates the adoption of an additional protocol to the European Social Charter which is already equipped with the mechanism (albeit limited in range) of a collective complaints procedure.[233] In sum, even though the projected right to a healthy environment would bring the complaints "aiming at the defense of the environment in general"[234] within the Court's subject-matter jurisdiction, there is no contradiction to require that litigating this right before the ECtHR be reserved, as a matter of personal jurisdiction, only to those with the "victim status" regulated by Article 34 of the ECHR.[235] Not to defeat the right's purpose, however, the provision in question would have to be interpreted in the light of new criteria that would adequately accommodate the premise that having the right to a healthy environment respected, protected and fulfilled is a diffuse, preventive-and-precautionary, and long-term legal interest.[236]

3.3.3 Legal Standing of Natural Persons

Complaints regarding the subjective aspect of the right to a healthy environment may, at the first glance, resemble those concerning toxic pollution or noise that already fall within the system of indirect rights-based protection. This is a misconception, because, as explained in Sect. 2, the right to a healthy environment has its own object of protection. A violation of this *sui generis* right may, but does not have to, lead to an additional violation of the right to life or the right for respect of private life that protects a person's well-being or health. The objective aspect of the right to a healthy environment distinguishes the project under discussion even more from the current doctrine. No other right that is currently guaranteed by the ECHR or its Protocols has a similar objective feature.

Pursuant to the well-established jurisprudence of the ECtHR, legal standing for cases concerning subjectively-felt environmental damage or nuisance is, in principle, only conferred by the risk or the actual deprivation of the most obvious (tangible), direct and individualized environmental benefits in relation to people's wealth,

---

231 Recommendation (n 1), Article 2.

232 Ibid, Article 6; Duyck, Extended summary (n 7), p. 13.

233 Additional Protocol to the European Social Charter Providing for a System of Collective Complaints, 9 November 1995.

234 *Kyrtatos* (n 148), §§ 46 and 53.

235 Knox, Extended summary (n 7), p. 31; and The Right in Ireland (n 6), pp. 16, 18 and 31.

236 The Right in Ireland (n 6), pp. 16–18.

life, health or well-being.[237] As illustrated by the case of *Kyrtatos v. Greece,* in the context of Article 8, impacts on the applicant's well-being or health have so far never been argued or assessed in scientific terms, with reference to "ecosystem services".[238] If the right to a healthy environment were recognized, ecosystem services could become a criterion for establishing legal standing of an individual applicant acting in defense of the environment. This useful and practicable concept could incidentally also allow the Court to determine whether there was an interference with the subjective aspect of the right to a healthy environment. For the right to be fully operational, however, ecosystem services would necessarily have to be grasped comprehensively including indirect, wider-range, and longer-term benefits.[239] As Knox submits, the point of the right to a healthy environment is to enable complaints about a broader range of environmental harms.[240] A broadly interpreted notion of ecosystem services is a basis of the legal test for standing in the Mexican *amparo* procedure for the protection of the right to a healthy environment. In that system, active legal capacity to claim protection of diffuse interests in a healthy environment and to realize the constitutional duty to protect it, is linked with the special personal situation that the claimant has with specific ecosystem benefits in his or her "adjacent surroundings".[241] Such a test would have likely rendered the case of *Kyrtatos v. Greece* compatible *ratione personae*.

An alternative and simpler test for legal standing for complaints about the right to a healthy environment, may be a spatial nexus. In Chile, where action of *amparo* is constitutionally open to anyone on behalf of everyone, even in absence of personal injury,[242] courts require, for the active capacity to bring action, that the claimant resides in the place where environmental harm has or risks to occur.[243] Pursuant to such a test of a geographical relationship with the radius of action of the emissions affecting the environment, the applicant in the case of *Ogloblina v. Russia,*

[237] *X.* (n 228); *Ogloblina* (n 146), §§ 20–22; *Kyrtatos* (n 148), §§ 46, 52 and 53; *Hamer* (n 36), § 79; *Turgut* (n 39), § 90; and *Dubetska* (n 132), § 105.

[238] Ecosystem services fall within four categories: (1) provisioning services, which include food, water, medicine, timber, and fiber; (2) regulating services, which are concerned with the regulation of climate, floods, disease, waste and air and water quality; (3) cultural services, which include recreation, aesthetic enjoyment, and spiritual fulfillment; and (4) supporting services, which are responsible for soil formation, photosynthesis, and nutrient cycling. See, for example, IUCN Global Ecosystem Typology 2.0: descriptive profiles for biomes and ecosystem functional groups (March 2021); and IPBES (2019): Global assessment report on biodiversity and ecosystem services of the Intergovernmental Science-Policy Platform on Biodiversity and Ecosystem Services.

[239] Lambert (2020), p. 4.

[240] Knox, Extended summary (n 7), p. 30.

[241] Liliana Cristina Cruz Piña y otra c. actos del presidente municipal de Tampico, estado de Tamaulipas, y otras autoridades, Suprema Corte de Justicia de la Nación [Supreme Court, Mexico] no. 307/2016, 14 November 2018, paras 147–173.

[242] Constitution of Chile, Article 20 in conjunction with Article 19 (8).

[243] Corte Suprema [Supreme Court, Chile], Sentencia Puchuncaví-Quintero, Rol Núm. 5888-2019, 28/05/2019.

discussed above, could have possibly been granted legal standing to effectively bring her complaints under the Court's scrutiny.

Irrespective of these two possible standards, in some situations, legal standing for complaints regarding the right to a healthy environment could also be borrowed from the current doctrine under Article 6 of the ECHR. In a recent case of *Çöçelli and Others v. Türkiye*, the Court rejected the objection that the complaint about the national authorizations of cement production in the applicants' village was of *actio popularis* nature.[244] The applicants' procedural complaint in respect of the compatibility of the environmental impact assessment reports on the factories with the legal requirements, was thus accepted, based on the criterion that the applicants were party to the proceedings and had their case examined on the merits by the domestic court. In this sense, the applicants were directly affected by the alleged shortcomings in the impugned proceedings, and they could claim to be victims in respect of their Article 6 § 1 complaint.[245] While such a test is more straightforward than the one based on ecosystem services in adjacent surroundings, it could only operate as a default test for cases where national jurisdictions had indeed allowed the applicant's application for a judicial review. Making it a general test, would create an unwanted double standard and a free pass for States that do not allow such access as a matter of their national law.[246]

Summing up, if the proposed additional protocol were to remain silent about the specific criteria for legal standing for applications under the right to a healthy environment, it would be up to the ECtHR to make the necessary adjustments to the current Article 34 norms. As already explained, legal standing for complaints about violations of the right to a healthy environment can, conceptually and normatively, depend on factors limiting the circle of those with a "victim status". Considering, however, that the aims of the proposed additional protocol are a better, more preventive and precautionary environmental protection, it could be concluded that the parameters currently regulating the compatibility *ratione personae* of the ECHR's provisions, such as Article 8, could not be simply transposed unchanged into the framework of the new right. The proposed additional protocol does not posit that the norms regulating legal standing for the existing rights should change, given that it creates a separate legal framework especially designed for environment-related cases. Put differently, while an interference with the right to a healthy environment may be found to exist, other rights, such as the right to private life would be collaterally engaged only where environmental harm produced, as required today, severe effects on the applicant's well-being or health. As shown above, various options exist for formulating a new special test for individual legal standing that would be both pragmatic and reflective of the premise that the proposed right to a healthy

[244] *Çöçelli* (n 170), § 37.

[245] Ibid, § 40; see also *Bursa* (n 169), §§ 106–112 and 117.

[246] In a similar vein on standing of NGOs, Keller, Extended summary (n 7), p. 8.

environment necessarily entails the ability of the members of "human community" to demand effective protection of the environment.[247]

3.3.4 Legal Standing of NGOs

The projected additional protocol to the ECHR does not specify the role of legal entities in the enforcement of the right to a healthy environment. In the works leading up to the proposed draft, however, Elisabeth Lambert submitted that, owing to the diffuse or collective interests in the realization of the right and to its objective-ecocentric dimension, the project should postulate a wider and easier access of eNGOs to the Strasbourg Court.[248]

Pursuant to the current doctrine of "personal interest", the ECtHR does not entertain applications in which a legal entity exclusively relies on a Convention right, such as the right to life or the right to respect for home and private life (whether in terms of nuisance or impacts on health). Such grievances are considered to be inherently attributable to natural persons only.[249] The sole fact that an NGO considers itself a guardian of the collective interests of its members does not suffice to make it a victim within the meaning of Article 34 of the ECHR, owing to the prohibition of *actio popularis*.[250] On this basis, the Court has recently rejected the case of *Yusufeli İlçesini Güzelleştirme Yaşatma Kültür Varlıklarını Koruma Derneği v. Turkey*, brought, under Article 8 of the ECHR, by a local NGO campaigning against a planned hydroelectric power plant and dam whose construction would entail the submersion of a nearby town and villages.[251]

It is only where an association can demonstrate that the impugned measure—apart from infringing the individual substantive rights of its members—has had a significant impact on itself, the Court may recognize it as a victim. One example is the case of an association that was allowed to rely on the right to freedom of assembly (Article 11 of the ECHR) to complain that the disruption of a public demonstration by violence inflicted on the association's members breached its own corporate

---

[247] Suprema Corte de Justicia de la Nación [Supreme Court of Mexico] no. 649/2019 (11 March 2020), p. 20.

[248] Lambert, Extended summary (n 7), pp. 43, 44 and 50; see also Keller, Extended summary (n 7), p. 6.

[249] ECtHR Guide on Environment, paras 192 and 193; *Federation of Heathrow Anti-noise Group v. the United Kingdom* (dec.), no. 9310/81, 15 March 1984; *Association des Résidents du Quartier Pont Royal, la commune de Lambersart and Others v. France* (dec.), no. 18523/91, 8 December 1992; *Asselbourg* (n 124); *Aly Bernard* (n 125); *L'Association des Amis de Saint-Raphaël et de Frejus and Others v. France*, no. 45053/98, 29 February 2000; *Maatschap Smits and Others v. Netherlands* (dec.), 2001; *Greenpeace e. V. and Others v. Germany* (dec.), no. 18215/06, 12 May 2009; and *Yusufeli İlçesini Güzelleştirme Yaşatma Kültür Varlıklarını Koruma Derneği v. Turkey* (dec.) no. 37857/14, §§ 38, 41 and 43, 7 December 2021.

[250] *Kalifagiannis and Prospert v. Greece* (dec.), § 50, no. 74435/14, 9 June 2020.

[251] *Yusufeli* (n 249), §§ 38–44.

interest in disseminating messages.[252] As already mentioned, eNGOs can, under certain conditions, rely on the right to freedom of expression to challenge the refusal to provide them with adequate information concerning projects harmful to the environment of which the public should be informed.[253] As regards Article 6 of the ECHR, the Court has granted victim status to eNGOs in respect of proceedings where the substance of the association's claim before the domestic courts concerned its members' interests in respect of their private lives, families and homes and their right to participate in the decision-making process.[254] To distinguish such applications from *actio popularis*, the Court required that the applicant NGO was set up with the specific purpose of defending its members' interests before the courts, that its members were directly affected by the impugned measures and that they had been granted legal standing in the domestic proceedings. In other cases, the Court has granted legal standing to eNGOs on the grounds that the impugned proceedings had a sufficient link with the organizations' own right to enable the public to be informed and to participate in the decision-making process in environmental matters.[255]

More generally, the Court has recognized that, in modern-day societies, when citizens are confronted with particularly complex administrative decisions, recourse to collective bodies such as associations is one of the accessible means—sometimes the only means—available to them whereby they can defend their particular interests effectively. The Court has also attached importance to the fact that the standing of associations to bring legal proceedings in defense of their members' interests is recognized by the legislation of most European countries.[256]

Overall, while the Court shows some understanding about the modern-day civil society and is sensitive to contemporary strategic litigation, it still considers that, where substantive provisions are at stake, an association only plays an intermediary role in the pursuit of its members' rights and lacks legal standing before the Court. By comparison, the Aarhus Convention Compliance Committee requires that where national legal frameworks make access to court dependent on "impairment of a subjective right", eNGOs must be considered affected in their subjective rights by acts or omissions of public authorities or private persons which contravene national laws relating to the environment, as they have the right to demand compliance with

---

[252] *Identoba and Others v. Georgia*, no. 73235/12, §§ 47–48, 12 May 2015.

[253] *Burestop* (n 161), §§ 84–90.

[254] For example, *Gorraiz Lizarraga and Others*, v. Spain, no. 62543/00, §§ 9–10, 32 and 36–39, 27 April 2004; *L'Erablière A.S.B.L. v. Belgium,* no. 49230/07, §§ 28–29, ECHR 2009 (extracts); *Bursa* (n 169), §§ 114–116. ECHR Guide on Environment, para 194.

[255] *Erablière* (n 254), § 26; and *Melox* (n 168). Compare and contrast with *Lesoochranarske zoskupenie Vlk v. Slovakia* (dec.), no. 53246/08, §§ 77, 78, and 88, 2 October 2012, and *Burestop* (n 161), §§ 57–61.

[256] *Gorraiz* (n 254), §§ 38 and 39; and *Beizaras and Levickas v. Lithuania*, no. 41288/15, § 81, 14 January 2020.

such laws.[257] In the context of the EU law (prior to the amendments to the Aarhus Regulation),[258] the Compliance Committee considered that requiring NGOs to demonstrate a "direct and individual concern" to have standing to challenge decisions of EU institutions before the EU courts, was in breach of Article 9 of the Aarhus Convention.[259] The CJEU, itself, has interpreted Article 9(3) of the Aarhus Convention, read in conjunction with Article 47 of the Charter of Fundamental Rights, as precluding a situation where an environmental association, authorized to bring legal proceedings in accordance with national law, cannot challenge before a national court an administrative decision granting or amending EC type-approval which may be contrary to the EU law that has direct effect.[260] It has also stressed that such an association is part of the "public concerned", within the meaning of Article 2(5) of the Aarhus Convention, and as such is deemed as having an interest in environmental decision-making.[261] As for national jurisdictions, standing of legal persons, as plaintiffs or interveners in climate litigation, was accepted within respective legal frameworks by the Dutch Supreme Court in the case of *Urgenda,*[262] by the Brussels First Instance Court in the case of *Klimaatzaak,*[263] as well as in the French cases: *Affaire du Siècle*, decided by the Paris Administrative Court and *Grande-Synthe*, decided by the Council of State.[264]

Summing up, the proposed right to a healthy environment may be necessary to enable a broad interpretation of the notion of "subjective right" in the context of impacts of climate change, environmental degradation or pollution, allowing a legal entity, on the basis of its protective purpose, to make a third-party interest its own concern for the purpose of legal standing before the ECtHR.[265] Legal standing of applicants like those in the case of *Yusufeli,* mentioned above, would thus be based

[257] ACCC, Communication ACCC/C/2008/31 (Germany), 20.12.2013; see also on 5 September 2013 the Federal Administrative Court (BVerwG 7 C 21.12).

[258] Regulation (EC) No 1367/2006 on the application of the provisions of the Aarhus Convention (...) to Community institutions and bodies, 6/09/2006, as amended by Regulation (EU) 2021/1767, 6/10/2021.

[259] ACCC/C/2008/32 (European Union), 14/04/2011, ECE/MP.PP/C.1/2011/4/Add.1, para. 94; and ACCC/C/2008/32 (European Union),17.03.2017, ECE/MP.PP/C.1/2017/7, para. 121.

[260] Deutsche Umwelthilfe (n 179), §§ 46–81.

[261] Ibid. § 62.

[262] Urgenda Foundation v. the Netherlands, Supreme Court of the Netherlands, 20/12/2019, §§ 5.9.2–3; see also Vereniging Milieudefensie et al. v. Royal Dutch Shell PLC, Rechtbank Den Haag, C/09/571932/HA ZA 19-379, 26/05/2021, § 4.2.5. with a caveat that the NGOs' statutory and actual activity had to be within the specific subject and geographical area. As for the doctrine that individual claimants do not have to be uniquely affected, see Urgenda § 5.3.1; Asbl Klimaatzaak v. Belgium, Tribunal de 1ere instance de Bruxelles, 17/06/2021, p. 51, and Neubauer (n 80), §§ 110 and 131.

[263] Klimaatzaak (n 262), pp. 51–55.

[264] Oxfam France, Association Notre Affaire à Tous et al., Tribunal Administratif de Paris, nos. 1,904,967 et al., 3/02/2021, §§ 10–15 and 40–45; and Conseil d'État no. 427301, 19/11/20, Article 3 and 1/07/2021, § 6.

[265] Federal Administrative Court (BVerwG 7 C 21.12), § 45, 5 September 2013.

on a new logic, namely that an eNGO has an interest in preventing or seeking remediation of objective type of environmental harm. The right could also make emerge a new doctrine, pursuant to which eNGOs with a recognized status under domestic law, or as proposed by Keller, NGOs accredited with the CoE,[266] may automatically be deemed as having a valid and justiciable interest in environmental decision-making procedures. Such a norm is already present, for example, in the EU Council Directive 85/337/EEC on the assessment of the effects of certain public and private projects on the environment,[267] and, for example, in the French environmental law.[268]

3.3.5 Status of Potential Victim

The proposed additional protocol on the right to a healthy environment recognizes the prevalent multi-factor, wide-spread and often, slow onset character of environmental harm and climate change. It institutionalizes preventive and precautionary judicial intervention by expressly imposing obligations to prevent environmental damage and to take precautionary action where "threats of severe damage to the environment or to human, animal or plant health, lack scientific certainty".[269] To achieve these projected goals, the determination of whether the applicant's complaint is *ratione personae* compatible and, linked to this, whether there has been an interference with the right to a healthy environment, should not be as restrictive as it is today, under the norms that regulate the applicability of the provisions offering a level of indirect protection of the environment through human rights. It has thus been postulated that the existing norm of a "direct and immediate link" between the triggering material event and the applicant should be relaxed.[270] Keller argues that not only prevention, but also the precautionary principle should play a much more prominent role in the assessment of environmental risks.[271]

The risk of future harm has, to a certain extent, been incorporated by the ECtHR in the concept of a "potential victim" within the meaning of Article 34 of the ECHR. To claim such status, applicants must present plausible and cogent evidence of the probable materialization of a violation which will personally affect them.[272] The accompanying requirement of an "immediate link"[273] or of "imminent danger"[274]

[266] Keller, Extended summary (n 7), p. 7.

[267] Article 1 (d) and (e).

[268] French Environment Code, Article L142-1.

[269] Recommendation (n 1), Article 4; and Lambert, Extended summary (n 7), p. 44.

[270] Knox, Extended summary (n 7), p. 30; and Kobylarz (2018), pp. 112–113.

[271] Keller, Extended summary (n 7), p. 5.

[272] ECHR Practical Guide on Admissibility Criteria, paras 30–31; ECtHR Guide on Environment, paras 197 et al. *Asselbourg* (n 124); see also *Aly Bernard* (n 125); and *Vecbaštika* (n 124), §§ 79–84.

[273] *Atanasov* (n 124), § 66 in fine; *Balmer-Schafroth* (n 124), § 40; *Athanassoglou* (n 124), § 51; and *Folkman* (n 124).

[274] *Balmer-Schafroth* (n 124), § 40; *Tauria* (n 124); *Asselbourg* (n 124); *Athanassoglou* (n 124), § 51; and *Budayeva* (n 124), § 137.

is, however, stringent, especially if taken against the Court's own observation that the exercise of the right of individual petition cannot have the aim of preventing a violation of the Convention.[275] This doctrine engendered scholarly disapproval of international human rights litigation in the field of environmental protection, as being deprived of the essential preventive and, even less so, precautionary character.[276] The test of "serious, specific and imminent danger" for the applicability of Article 6 was even criticized as unattainable by some of the Court's own judges.[277] The Court has indeed examined the merits of several cases in which applicants managed to assert, arguably and in a detailed manner, that for lack of adequate precautions taken by the authorities their Convention rights were at, not too remote, risk of being harmed.[278] This approach, however, is considered by the Court exceptional.[279]

The terms such as "imminent" or "immediate" are factual rather than legal. As such, they are inherently susceptible to a case-by-case interpretation informed by scientific information about a given environmental-human rights risk. Recent rulings regarding climate change cases issued by several national courts[280] and the Child Rights Committee[281] show that an international scientific consensus or, in the event of scientific uncertainty, the precautionary principle, can be relied on to justify a conclusion that the risk, for example, of climate change impacts, is "reasonably foreseeable" or sufficiently close to grant the claimants the status of a potential victim of the alleged future violation. Knox proposes an alternative definition linked to science: environmental threat is imminent when it reaches its point of no return, that is to say, in the last moment when preventive action is still possible.[282] Another possible approach is detached from science. It is to view the risks of future impairments of fundamental rights, not in terms of timelined material events, but as something "built into" the current legislation, for example, regarding the trajectory of mitigation of climate change.[283] From this formalist perspective, actionable harm "[is] potentially set into irreversible motion today", even though the actual impairment would only arise as a result of future law.[284] This approach was taken by the German Constitutional Court in its landmark case of *Neubauer*.

---

[275] *Tauria* (n 124); and *Aly Bernard* (n 125).

[276] Boyle (2012), pp. 613–642.

[277] Dissenting opinion of Judge Petiti and six other judges in *Balmer-Schafroth* (n 124); and dissenting opinion of Judge Costa and four other judges in *Athanassoglou* (n 124).

[278] *Asselbourg* (n 124); *Taşkın* (n 165), § 113; *Öçkan and Others v. Turkey*, no. 46771/99, 28 March 2006; Lemke v. Turkey, no. 46771/99, 5 June 2007; *Hardy* (n 154); and *Tătar* (n 136), §§ 93–97.

[279] *Tauria* (n 124); *Asselbourg* (n 124); *Câmpeanu* (n 194), § 101; *Aly Bernard* (n 125); *Luginbühl* (n 125); and *Folkman* (n 124).

[280] Urgenda (n 262), §§ 4.1.1–7 and 5.6.1–8; and Klimaatzaak (n 262), p. 50.

[281] Sacchi (n 49), § 9.4.

[282] Knox, Extended summary (n 7), p. 59.

[283] Neubauer v. Germany, Bundesverfassungsgericht 24/03/2021—1 BvR 2656/18 et al. §§ 130–133.

[284] Ibid.

In the light of the above examples, it may be concluded that the insertion of the right to a healthy environment does not appear indispensable for the Court to be able to give, on a case-by-case basis, a contextually adequate definition to the notions of "immediate link" and "imminent danger". The fundamental principles of prevention and precaution that are well-established in international environmental law, already at present constitute a framework of interpretation for the ECHR.[285] The proposed right would, however, institutionalize such a normative transformation and back it up with political legitimation.

3.3.6 Loss of Victim Status

To reiterate, the proposed right to a healthy environment carries with it a duty to take, first and foremost, preventive action for the protection of the environment and biodiversity. As described in previous sections, this duty would have to become actionable pursuant to a set of changed norms concerning the subject-matter and personal jurisdiction of the ECtHR. The result of this may likely be a higher ratio of complaints seeking to avert the risk of future violations of the right to a healthy environment than those seeking to redress the harm that has already occurred. Given that the right to a healthy environment is, moreover, concerned with diffuse and general impacts and interests, the same environmental threat may become the subject of several sets of proceedings in a member State.[286] It may therefore happen that the situation that has given rise to an application lodged with the ECtHR in a given moment by a particular applicant, may, by the time the Court examines it, be resolved by the domestic authorities. Lastly, as the right to a healthy environment combines substantive and procedural elements, it may also happen that while the impugned decision-making process might have allegedly violated the procedural rights of an applicant, the main substantive issue will be resolved to a general satisfaction.[287]

These hypothetical situations may raise a question of whether the applicants have lost their victim status for the purposes of the admissibility of their applications. This issue has so far not been addressed by the project of the right to a healthy environment.

Pursuant to the Court's well-established case-law concerning individual rights protected by the ECHR and its Protocols, applicants may be considered to have lost their victim status where domestic authorities have found a violation in respect of the applicant and given a decision which constitutes adequate and sufficient redress for the violation in question.[288] In the context of environmental harm that has already

[285] *Demir and Baykara v. Turkey* [GC], no. 34503/97, §§ 76–84, ECHR 2008; and *Tătar* (n 136); Kobylarz (2018), p. 107.

[286] E.g. Oxfam France (n 264) and Conseil d'État (n 264).

[287] *Carême* (n 46).

[288] ECtHR Practical Guide on Admissibility Criteria, paras 36–39.

occurred, the Court has found that the mere fact of putting an end to an instance of pollution or an environmental disturbance, that had for a certain time affected the applicants, is insufficient to deprive them of their victim status.[289] Similarly, the fact that local residents had moved houses to escape the pollution or environmental disturbance to which they had previously been exposed did not deprive them of their victim status for the purposes of the proceedings before the Strasbourg Court.[290]

The loss of a victim status in the context of complaints about threats of future environmental harm may for the first time be addressed by the Court in *Duarte Agostinho* insofar as the complaint about insufficient climate action concerns, among others, several States in which the domestic courts have in the meantime ordered public authorities to improve their mitigation ambition within the meaning of Paris Agreement, and compliance with the domestic emission reduction targets. In this connection, it is also worth taking notice of the recent application, *Engels and Others v. Germany*, which claims that the improved objectives of the German Climate Protection Act, as amended following the 2021 judgment of the German Constitutional Court, are still insufficient to meet the Paris Agreement temperature goals.[291] As to the question of whether a complaint about an alleged violation of participatory rights is actionable even though the proceedings on the merits have ultimately led to the result that was originally sought by the applicant, this may, in the near future, be examined in the case of *Carême v. France.* This case concerns proceedings for declaration that the State did not undertake sufficient climate action to meet its statutory targets regarding greenhouse gas emissions. The applicant—a natural person, was denied legal standing before a national court, but the legal action was pursued to the merits stage by the applicant's fellow claimants—NGOs. The ultimate court ruling was in line with what the applicant and the other claimants intended to achieve.[292]

4 Conclusions

The overall conclusion of the present study is that recognizing the right to a healthy environment in an additional protocol to the ECHR would have significant normative consequences for admissibility requirements. A separate regime specifically designed for the right to a healthy environment would thus be institutionalized, leaving unchanged the general norms insofar as they regulate the existing provisions operating in environment-related cases. Adjudicating under the new specific right

---

[289] *López Ostra* (n 124), § 42; Martínez and Pino Manzano v. Spain, 2012, § 28, Guide § 198.

[290] *López Ostra* (n 124), § 42; *Dmitriyev* (n 129), § 37.

[291] ECtHR Press Release of 9.02.2023, https://hudoc.echr.coe.int/eng-press?i=003-7566368-10398533.

[292] Affaire Grande Synthe, Conseil d'État no. 427301, 19/11/20 and 1/07/2021.

would thus release tensions currently generated by environmental claims based on the expansive interpretations of these provisions.

The most direct and obvious normative consequence of the right would be that an instance of environmental degradation or deregulation would constitute a *sui generis* subject-matter of the Court's protection. Viewed from the anthropocentric perspective, interference could stem from the loss of benefits derived from the environment by humans, possibly also leading to collateral interference with other rights guaranteed by the ECHR and its Protocols. Viewed from the ecocentric perspective, interference could result from significant harm to elements or functions of ecosystems, even in absence of impact on people. Insofar as the right to a healthy environment would be qualified by a minimum severity requirement, such a threshold could consequently be assessed either subjectively or objectively, in both cases necessarily informed by relevant science. Animals that inherently comprise the environment, would become one of the objects (but not the subject) of the protection. In the context of participatory rights, the right to a healthy environment would be the missing tool to enable the Court to set out uniform standards of fairness for proceedings undertaken by citizens in defense of the environment, irrespective of whether a national legal framework recognizes the healthy environment or its components as a justiciable right.

The right to a healthy environment would further require a new approach to the "victim status" and to the legal standing before the ECtHR, especially to enable the realization of the inherent general duty to protect the environment and biodiversity through participatory rights. As a consequence, a regime of default legal standing for certain categories of eNGOs may emerge. It would not be required, however, that standing for complaints concerning the substantive components of the right should be based on a completely unlimited public interest. The question of the victim status could still be regulated by Article 34 of the ECHR, interpreted in the light of new criteria that would have to adequately accommodate the premise that having the right to a healthy environment respected, protected and fulfilled is a diffuse, preventive/precautionary, and long-term legal duty. Various options have been shown for formulating a new special test for individual legal standing that would be both pragmatic and reflective of the right's special character. Owing to the textual and conceptual ambiguity regarding the status of future generations, it appears that the right would not engender any automatic change for the Court's jurisdiction *ratione personae*.

Lastly, while the insertion of the right to a healthy environment may be necessary for the emergence of some new specialized norms, theoretically, it does not appear indispensable for the Court to be able to develop a contextually adequate definition of the notion of "potential victim" of environmental risks. The latter is also true regarding the emergence of a new test regulating extraterritorial effects of environmentally harmful activities of States. As these and other relevant norms are subject to ongoing judicial deliberation, it will soon become clear whether the current doctrine of "green human rights" has reached its interpretative limits and, consequently, that any further expansion could only be triggered through the political process underway.

Closing with the remarks of Robert Spano, a former judge and President of the ECtHR, an additional protocol to the ECHR on the right to a healthy environment would "[give judges] a tool to deal with a different dynamic, where the anthropocentric nature of the right is potentially diminished by opening up the possibility of humans being more involved in taking care ... of their environment."[293] The ensuing normative changes, no matter how innovative and intricate, would be manageable for the ECtHR.[294]

References

Araszkiewicz M, Płeszka K (2015) The concept of normative consequence and legislative discourse. In: Araszkiewicz P (ed) Logic in the theory and practice of lawmaking, Legisprudence library 2. Springer, Cham

Baumann P (2018) Le droit à un environnement sain et la Convention européenne des droits de l'Homme. LDGJ

Boyd D (2012) The environmental rights revolution: a global study of constitutions, human rights, and the environment. University Columbia Press

Boyd D (2018) Catalyst for change. In: Knox J, Pejan R (eds) Human right to a healthy environment. Cambridge University Press

Boyd D (2019) Report of the Special Rapporteur on Human Rights and the Environment, Human rights obligations relating to the enjoyment of a safe, clean, healthy and sustainable environment, A/74/161

Boyd D (2020) Report of the Special Rapporteur on the issue of human rights obligations relating to the enjoyment of a safe, clean, healthy and sustainable environment, Human rights depend on a healthy biosphere, A/75/161

Boyle A (2012) Human rights and the environment: where next? Eur J Int Law 23(3):613

Brown Weiss E (1989) In fairness to future generations. Transnational Publishers Inc.

Brown Weiss E (1990) Our rights and obligations to future generations for the environment. Am J Int Law 84:198–207

Brown Weiss E (1994) In fairness to our children: international and intergenerational equity. Childhood 2:22

Daly A (2022) Intergenerational rights are childrens' rights: Upholding the right to a healthy environment through the UN Convention on the Rights of the Child

Daly E, May JR (2019) Indivisibility of human and environmental rights. In: Daly E, May JR (eds) Human rights and the environment, legality, indivisibility, dignity and geography. Edward Elgar

Darpö J (2013), Effective Justice? Synthesis report of the study on the Implementation of Articles 9.3 and 9.4 of the Aarhus Convention in the Member States of the European Union, 2013-10-11/Final

Darpö J (2021) Can nature get it right? A study on rights of nature in the European context. European Union

de Londras F, Dzehtsiarou K (2018) Great debates on the European Convention on Human Rights. Palgrave

[293] Interview between Philippe Sands and Robert Spano, Journal of Human Rights and the Environment, Vol. 13 Special Issue Human Rights and the Planet, p. 9.

[294] Ibid.

Grant E (2019) American convention on human rights and environmental rights standards. In: Turner SJ et al (eds) Environmental rights: the development of standards. Cambridge University Press

Husain T, Bakr Amin M (1995) Kuwaiti oil fires. Regional Environmental Perspectives. https://doi.org/10.1016/B978-0-08-042418-7.X5000-2

Keller H, Heri C (2022) The future is now: climate cases before the ECtHR. Nord J Hum Rights 40(1):153–174. https://doi.org/10.1080/18918131.2022.2064074

Knox J H (2018) Report of the Special Rapporteur on Human Rights and the Environment, Human rights obligations relating to the enjoyment of a safe, clean, healthy and sustainable environment, A/73/188

Kobylarz N (2018) The ECtHR, an underrated forum for environmental litigation. In: Anker T, Olsen E (eds) Sustainable management of natural resources, legal instruments and approaches. Intersentia

Kobylarz N (2022) Balancing its way out of strong anthropocentrism: integration of "ecological minimum standards" in the ECtHR "fair balance" review. J Hum Rights Environ 13(Special Issue):16–85

Krenc F (2022) La Cour européenne de droits de l'Homme: prétoire pour les contentieux environnementaux et climatique? Liber Amicorum Robert Spano, Anthemis

Ksentini FZ (1994) Final report Commission on Human Rights, Sub-Commission on Prevention of Discrimination and Protection of Minorities, E/CN.4/Sub.2/1994/9

Kwiędacz-Palosz M (2023) Nowe prawo człowieka. Online Think Tank

Lambert E (2020) The environment and human rights introductory report to the high-level conference environmental protection and human rights

May JR, Daly E (2019) Global judicial handbook on environmental constitutionalism, 3rd edn. United Nations Environment Programme

Mikosa Ž (2020) Evolution of procedural Rights and Legal Standing in Environmental Matters in Latvia. In: Baltic Yearbook of International Law, vol 19

Peña Chacón M (2003) La legitimación procesal en el derecho ambiental, el caso de Costa Rica. Revista de Direito Ambiental, año 8, 2003, número 29

Peña Chacón M (2018) Revolución de los derechos humanos ambientales y derechos de la naturaleza. Revista del Programa en Derecho de la Universidad de Costa Rica

Peña Chacón M (2020) Enverdecimiento de las cortes latinoamericanas, últimos avances jurisprudenciales, vol XII. Revista Monfragüe Desarrollo Resiliente

Peña Chacón M (2022) Derechos humanos y derechos de la Naturaleza: Diálogo y fertilización cruzada entre ambas visiones

Raible L (2016) The extraterritoriality of the ECHR: why Jaloud and Pisari should be read as game changers. European Human Rights Law Review 2:161–168

Sandvig J, Dawson P, Tjelmeland M (2021) Can the ECHR encompass the transnational and intertemporal dimensions of climate harm? In: EJIL:Talk! vol 23

Sparks T (2018) Protection of animals through human rights. The case-law of the European Court of Human Rights. MPIL Research Paper Series

Sulyok K (2021) Science and judicial reasoning: the legitimacy of international environmental adjudication. Cambridge University Press

Winisdoerffer Y (2003) La jurisprudence de la CrEDH et l'environnement. Revue juridique de l'Environnement

Young A L (2009) Tactical use of herbicides during the Vietnam War. https://doi.org/10.1007/978-0-387-87486-9_3

Zheng X (2021) Does Fukushima wastewater decision violate our environmental rights? EJIL

Natalia Kobylarz works as a Senior Lawyer at the Registry of the ECtHR in Strasbourg. She also teaches the courses of International Human Rights Law and Human Rights and the Environment at the University of Grenoble Alpes and at the Strasbourg University. She holds an LL.M. degree in

International and Comparative Law from the Southern Methodist University in Dallas, United States; a Master 2 degree in Environmental Law from Strasbourg University, France; and a Master of Law degree from the University of Maria Curie-Sklodowska in Lublin, Poland. In 2016 she was seconded to the Inter-American Court of Human Rights in San José, Costa Rica.

Environmental Law in the Courts of Europe: A Rough Sketch

Luc Lavrysen

Abstract This chapter examens how domestic judiciaries in Europe deal with environmental law. This produces a very varied picture. The way in which domestic courts in Europe engage with environmental law is influenced by various factors: different environmental conditions, pressures and awareness in the various subregions and countries, different legal systems, various degrees of development of environmental law, being a party or not to the Aarhus Convention, membership of the Council of Europe, the European Union (EU) or the European Economic Area (EEA). The intensity with which the concerned judiciaries dialogue with the Court of Justice of the EU (CJEU) or follow the case law of the European Court of Human Rights (ECtHR), as well as the organisation of the judiciary, the investment in it and the degree of specialisation are further factors of influence. It is argued that the legal anchoring of specialized environmental departments or judges in the relevant courts, as applied in some European countries, is a must.

1 Introduction

Several factors influence the way in which domestic courts in Europe engage with environmental law. Apart from the different environmental conditions, pressures and awareness in the various subregions and countries, there are the different legal systems, the degree of development of environmental law, membership and the various degrees of implementation of the Aarhus Convention, membership of the Council of Europe, the European Union (EU) or the European Economic Area (EEA) and, the intensity with which the concerned judiciaries dialogue with the Court of Justice of the EU (CJEU) or follow the case law of the European Court of Human Rights (ECtHR). Further factors of influence are the organisation of the judiciary, the investment in it and the degree of specialisation. In this contribution

L. Lavrysen (✉)
Constitutional Court of Belgium, Ghent, Belgium
e-mail: luc.lavrysen@ugent.be

G. Antonelli et al. (eds.), *Environmental Law Before the Courts*,
https://doi.org/10.1007/978-3-031-41527-2_8

we try to sketch an overall picture of environmental law in the domestic courts in Europe. The picture is far from being homogeneous.

2 Different Environmental Conditions, Pressures and Awareness

With nearly 750 million people, the average population density in Europe is 34 persons/km^2, but with great disparities between the countries, ranging from 508 in The Netherlands, 383 in Belgium, over 281 in the UK, 240 in Germany, 206 in Italy, 119 in France, 94 in Spain, to 25 in Sweden, 18 in Finland and 3 in Iceland.[1]

The economic development shows also great differences. GDP per capita in 2020 ranged from 110,261 USD in Luxembourg to 10,776 in Kosovo, with an EU average of 41,504 USD, compared with 60,236 USD in the US.[2] The *Human Development Report 2021–22*[3] ranks eight European countries amongst the first ten, with a first place for Switzerland. 28 European countries can be found amongst the first 40, with the lowest rank for Moldova (80th). The *2022 Environmental Performance Index*[4]ranks 16 European countries in the highest ranks, with a first place for Denmark, 19 amongst the first 20, another 19 amongst the first 40. Ukraine ranks 52th. The lowest ranks are for Bosnia and Herzegovina (102th), Russia (112th) and Turkey (172th). The *Sustainable Development Report 2022*[5] ranks 18 European countries in the highest ranks, with a first place for Finland, 19 of the 20 first places are occupied by European countries, while 32 European countries can be found among the first 40. The lowest rank here is for Montenegro, with an 86th place.

The European Environmental Agency (EEA),[6] that covers not only the current 27 Member States of the EU[7] and the UK, as a former Member State, but also the other EEA countries,[8] as well as Switzerland and Turkey, and is cooperating with six more countries,[9] 38 countries in total,[10] has determined 11 biogeographical regions

[1] https://www.worldometers.info/world-population/population-by-country/.

[2] https://tradingeconomics.com/country-list/gdp.

[3] United Nations Development Programme (2022).

[4] Wolf et al. (2022), p. 206.

[5] Sachs et al. (2022), p. 508.

[6] https://www.eea.europa.eu/.

[7] Austria, Belgium, Bulgaria, Croatia, Republic of Cyprus, Czech Republic, Denmark, Estonia, Finland, France, Germany, Greece, Hungary, Ireland, Italy, Latvia, Lithuania, Luxembourg, Malta, Netherlands, Poland, Portugal, Romania, Slovakia, Slovenia, Spain and Sweden.

[8] Norway, Iceland, Liechtenstein.

[9] Albania, Bosnia and Herzegovina, Kosovo, Montenegro, North Macedonia, Serbia.

[10] The agency does not cover the following 4 Eastern European countries: Belarus, Moldova, Russia, Ukraine. Formally 4 small countries in Western and Southern Europe are not covered either: Andorra, Monaco, San Marino and Vatican City.

in Europe.[11] Environmental problems cause different impacts in the different subregions of Europe as can be seen by the EEA environmental indicators.

Greenhouse gas emissions in the EU decreased by 31% between 1990 and 2020, exceeding the EU's 2020 target by 11 percentage points. EU greenhouse gas emissions are expected to further decline until 2030.[12] Due to climate change, the Arctic sea ice is declining rapidly. This declining trend is projected to continue.[13] In the period 2000–2019, the proportion of a country's territory suffering from *drought impacts* was highest in Malta. However, the average annual drought impact intensity on vegetation was highest in Czechia, followed by Austria and Sweden, and to a lesser extent Portugal, Spain, Latvia, Slovenia, Germany and Finland. Drought impact on cropland productivity was worst in Portugal, followed by Austria and Finland.[14] Climate change has increased forest fire risk across Europe. In recent years, forest fires coinciding with record droughts and heatwaves have affected regions in central and northern Europe not typically prone to fires.[15]

At the EU level, only 15% of *habitat assessments* have a good conservation status, with 81% having a poor or bad conservation status. At the EU Member State level, the majority of assessments indicate a low number of habitats with a good conservation status. Intensive agriculture, urban sprawl and pollution are the top reported pressures to habitats.[16] At EU level, only 27% of *species assessments* indicate a good conservation status, with 63% having a poor or bad conservation status. Only 6% of all species have improving trends. At Member State level, a large proportion of assessments show few species with a good conservation status. Agriculture, urban sprawl, forestry and pollution are the pressures on species most reported.[17]

More than 70% of EU citizens live in urban areas, where high population densities and economic activities cause *high levels of air pollution*. Between 2005 and 2019, emissions of four pollutants declined considerably in the 27 EU Member States. Decreases in emissions from the energy, industry and transport sectors are largely responsible for these declines. NH_3 emissions also declined, but by only 8% overall, with emissions even increasing slightly between 2013 and 2017. This reflects a lack of progress in the agriculture sector, which is responsible for more than 90% of NH_3 emissions.[18] Overall, progress has been made, with the proportion of the EU urban population exposed to PM, NO_2 and SO_2 above 2005 WHO guideline values decreasing since 2000. In 2019, 36% of the EU urban population was

[11] https://www.eea.europa.eu/data-and-maps/figures/biogeographical-regions-in-europe-2/map_2-1_biogeographical-regions/image_large.

[12] https://www.eea.europa.eu/ims/total-greenhouse-gas-emission-trends.

[13] https://www.eea.europa.eu/ims/arctic-and-baltic-sea-ice.

[14] https://www.eea.europa.eu/ims/drought-impact-on-ecosystems-in-europe.

[15] https://www.eea.europa.eu/ims/forest-fires-in-europe.

[16] https://www.eea.europa.eu/ims/conservation-status-of-habitats-under.

[17] https://www.eea.europa.eu/ims/conservation-status-of-species-under.

[18] https://www.eea.europa.eu/ims/emissions-of-the-main-air.

estimated to be exposed to PM_{10} above the WHO guideline value, 61% to $PM_{2.5}$, 3% to NO_2 and 17% to SO_2.[19]

Environmental noise remains a major problem in Europe, with at least 20% of the EU population living in areas where noise levels are considered harmful to health. Most of the people affected live in urban areas. Road traffic is by far the most dominant source of environmental noise. The number of people exposed to high levels of environmental noise has broadly remained stable since 2012.[20]

In the period up to 2015, good or better (high) ecological status has been achieved for only around 40% of *surface waters* (rivers, lakes, and transitional and coastal waters). The percentage of water bodies with less than good (i.e. moderate, poor or bad) ecological status varies between river basin districts across Europe. Surface water bodies in north-western Europe have the lowest status. In Belgium (Flanders), northern Germany and the Netherlands, the ecological status of more than 90% of surface waters is reported to be less than good. Other problem areas include Czechia, southern England (United Kingdom), northern France, southern Germany, Hungary and Poland, as well as several individual river basin districts in some countries, where 70–90% of surface water bodies are reported to have less than good status. Northern countries, particularly the northern Scandinavian region, Iceland, Ireland and Scotland (United Kingdom), along with Estonia, Romania, Slovakia and several river basins in the Mediterranean region, have a high proportion of water bodies with good or better (high) ecological status or potential.[21]

According to the *Special Eurobarometer Report 2019*,[22] more than nine in ten respondents answered that protecting the environment is important for them and for just over half it is very important, but there are great differences between the member states, ranging from a "very important"—score of 81% in Sweden, to 36% in Estonia and Latvia. A large majority believed that the problems are best handled by the EU together with the Member States. The environment and climate change were viewed in the *2022 Eurobarometer Survey*[23] as the most important issues at EU level by 26% of respondents in the EU as a whole. It was the most mentioned issue in nine countries, with the highest levels seen in Denmark (53%), Sweden (52%) and the Netherlands (50%). It ranked second in Ireland (29%) and Portugal (25%), and joint second with immigration in Luxembourg (28% each). The environment and climate change ranked third in Hungary and Slovenia (both 22%) and joint third with immigration in Czechia (25% each issue) and Estonia (26% each issue). At national level the environment and climate change had risen from the fifth to the fourth position after rising prices/inflation/cost of living, health and the economic situation. The environment and climate change ranked fourth in the EU overall

[19] https://www.eea.europa.eu/ims/exceedance-of-air-quality-standards.

[20] https://www.eea.europa.eu/ims/exposure-of-europe2019s-population-to.

[21] https://www.eea.europa.eu/ims/ecological-status-of-surface-waters.

[22] European Commission (2019) *Special Eurobarometer 51, Attitudes of European citizens towards the environment*, Brussels, p. 144.

[23] European Commission (2022) *Standard Eurobarometer 96, Winter 2021–2022, Public opinion in the European Union First results Fieldwork: January–February 2022*, Brussels, p. 24.

(16%), but was the most mentioned item in Denmark (44%) and Sweden (34%). It was the second most mentioned item in the Netherlands (40%), Malta (30%) and Germany (29%) and the third most mentioned item in France and Latvia (both 17%). However, in the majority of member states it counted for less than 10%. The invasion of Russia in Ukraine in early 2022, the continuing war and its consequences on energy supply and the economic situation have influenced public opinion and the place environmental issues are taking amongst the priorities. In the Autumn of 2022 rising prices, inflation, and cost of living was the main concern of Europeans at the EU level, followed by energy supply, the international situation and the economic situation. Environmental and climate concerns fell down to the fifth position.[24]

3 Variety in Legal and Judicial Systems

Within Europe various legal systems exists. In the first place there are the civil law systems with their variations on the continent. The French civil law system is applied in France and has strongly influenced the legal systems of the Benelux countries, Albania, Italy, Lithuania, Portugal, Romania and Spain. The German civil law system is applied in Germany, and has strongly influenced the legal systems of Austria, Belarus, Bulgaria, Bosnia and Herzegovina, Croatia, the Czech Republic, Estonia, Greece, Kosovo, Latvia, Montenegro, North Macedonia, Russia, Slovakia, Switzerland, Slovenia, Serbia and Turkey. Scandinavian civil law is applied in Denmark, Norway, Sweden, Finland and Iceland. In England and Wales and Ireland common law is applied. Common law also influenced the mixed legal systems of Cyprus, Malta and Scotland. Huge diversity is also observed in terms of budgets (nominal and as percentage of GDP) for the court systems, number of judges and staff per capita, expenditure for prosecution and legal aid, court fees, disposition time, clearance rate, etc. [25]

In most of the national constitutions—but not all—there is one or another reference to environmental protection.[26] There are federal and unitary states. Federal states usually have a specialised constitutional court.[27] That is also the case in most countries that have known a period of authoritarianism.[28] In some countries

---

[24] European Commission (2022) *Special Eurobarometer 531, Autumn 2022, Key Challenges of our Times – Autumn 2022*, Brussels, p. 44.

[25] Council of Europe (2020), p. 138.

[26] Lavrysen (2006a), pp. 82–83.

[27] Austria, Belgium, Germany, Italy and Spain. There are also Constitutional Courts in the unitary states Andorra, France, Luxembourg, and Malta.

[28] Albania, Bosnia and Herzegovina, Bulgaria, Croatia, Czech Republic, Hungary, Latvia, Lithuania, North Macedonia, Moldova, Montenegro, Poland, Portugal, Romania, Serbia, Slovakia, Slovenia, Turkey, Ukraine. There is also a Constitutional Court in Belarus and the Russian Federation.

constitutional review is performed by the whole judiciary and/or the supreme court.[29] In the countries that have no constitutional court, a distinction can be made between countries where the ordinary courts and tribunals have power of constitutionality review. Denmark, Finland, Greece and the United Kingdom fall into this category. The Finnish Constitution, for instance, stipulates that when the application of a law could manifestly come into conflict with the Constitution, and the statute was not adopted in the manner provided for constitutional amendments, the court must give priority to the provision of the Constitution. Moreover, if a provision of a decree or any other legislative rule ranking lower than a statute comes into conflict with the Constitution or another law, it shall not be applied by a court or any other authority. On the other hand, there are also countries where this power is not given to the ordinary courts and tribunals. In the Netherlands, for instance, it is accepted that constitutionality review is prohibited by the Constitution. According to the Constitution, the Dutch Parliament is responsible for the conformity of legislation with the Constitution.

In federal states one can have state and federal courts alongside, as is the case in Germany, while in other states, e.g. Belgium, the judiciary is exclusively or mainly a federal competence. In most continental legal systems, a dual structure has been put in place, with on the one hand ordinary courts and tribunals, that have jurisdiction in civil and criminal cases, and on the other hand administrative courts and tribunals. In Cyprus an Administrative Court has been created, with a possibility to appeal to the Supreme Court.[30] In Estonia there are two administrative courts of first instance. Appeals are handled by administrative law chambers of the courts on appeal and further on by the Administrative Law Chamber of the Supreme Court. Most of the time there is a Supreme Court as apex court of the ordinary judiciary and a Supreme Administrative Court as apex court of the administrative judiciary. In some cases there are administrative law divisions in the ordinary judiciary. That is the case in Romania, Russia, Slovenia and Spain. In The Netherlands, the administrative law sections of the district courts are competent to hear disputes in all areas of administrative law. However, there are exceptions to this. The Administrative Jurisdiction Division of the Council of State administers e.g. justice in first and sole instance in cases concerning spatial planning law and some environmental law cases. It also hears some appeals against decisions of administrative law sections of the district courts.[31] In the UK, planning courts and administrative courts are part of the civil court system. They also oversee the work of the tribunals, some of them dealing with environmental matters (the General Regulatory Chamber of the First Tier Tribunal and the Administrative Appeals Chamber of the Upper Tribunal).[32] In Ireland, judicial review is a competence of the High Court, the Court of Appeal, and

---

[29] Cyprus, Denmark, Estonia, Ireland, Liechtenstein, Monaco, Norway, Switzerland.

[30] Nicolatos et al. (2018), pp. 7–8.

[31] https://www.aca-europe.eu/index.php/en/tour-d-europe-en.

[32] https://www.judiciary.uk/about-the-judiciary/the-justice-system/court-structure/.

the Supreme Court.[33] There can also be some types of specialised courts (military courts, finance courts, labour courts, social courts…), that are supervised by the Supreme Court or by Specialised Supreme Courts (as is the case in Germany). It should be pointed out, however, that the powers of the administrative courts may differ. While in most European countries those courts only have the power to suspend and/or annul administrative decisions, in certain countries such as France, Germany and Spain they have more extensive powers. The administrative courts in Spain, for instance, may impose provisional protective and compulsory measures. However, they cannot take the place of the administrative authorities. In Germany, they are empowered to annul administrative decisions as well as to oblige the authorities to take a decision, and they can also substitute their decision for that of the authorities. In France, administrative courts have distinct powers in cases of extreme urgency and in case of judgment on the merits. In cases of extreme urgency, they can suspend the decision, take any measure necessary to protect the fundamental rights of the petitioners and impose orders, where appropriate with periodic penalty payment. In cases where a judgment is given on the merits, they can annul the challenged decision and order the authorities to pay compensation. In special cases they can substitute their decision for the challenged decision, and they can impose fines and damages. In Portugal, they are empowered to annul administrative acts and to decide on claims for compensation against the authorities.[34] In Denmark all sorts of cases are settled by one court system. In practice, a number of administrative tribunals have been set up for certain specific matters which are empowered to take a final decision in disputes between the State and private individuals. After that, the dispute can be brought before the ordinary courts. Norway does not have specific administrative courts and tribunals either. In Estonia, there are only separate administrative and judicial courts in the first instance, with no separate administrative court at Circuit or Supreme Court level. The Circuit Courts are comprised of civil, criminal and administrative chambers, as is the case with the Supreme Court of Estonia.[35]

Whether this means the same courts and tribunals have jurisdiction in both criminal and civil cases is not a foregone conclusion. In the United Kingdom, Germany, Spain, Hungary and Poland, for instance, there is a distinction between civil courts and criminal courts. In most countries, civil and criminal cases are tried by different divisions or chambers of the ordinary courts, and this in first instance as well as on appeal or on cassation. In some countries, such as Portugal, this specialization has only been implemented at the appeal level. Finally, there are the countries where this distinction is less strictly emphasized, as is the case in Denmark, Ireland and Finland. Norwegian courts have general competence, meaning that they can rule in civil, criminal, administrative and constitutional matters.[36]

---

[33] https://www.courts.ie/judicial-review#:~:text=Judicial%20Review%20is%20a%20mechanism,Rules%20of%20the%20Superior%20Courts.

[34] Lavrysen (2006a), pp. 84–85; EUFJE and Milieu Consulting (2019), pp. 34–35.

[35] EUFJE and Milieu Consulting (2019).

[36] Lavrysen (2006a), p. 85; EUFJE and Milieu Consulting (2019), p.35

4 Environmental Law and the Various Types of Courts

Al those different courts with their different competencies and roles can be confronted with environmental cases to some extent.

Constitutional courts are specialised courts that provide centralised constitutional review, mainly based on the model developed by Hans Kelsen in Austria,[37] as opposed to the US model of decentralised constitutional review exercised by the whole judiciary.[38] Depending on national constitutions and their organic laws, they review in the first place the constitutionality of acts of parliaments, but their competence may include the review of the constitutionality of acts of central, regional or local governments, the constitutionality of judgments (the so-called constitutional complaint procedure), the review of electoral processes and other constitutional issues. Constitutional courts are dealing with environmental issues in the first place from an institutional point of view. They have to judge conflicts of competences between different levels of government (federal, state or regional)[39] and between the legislature and the executive,[40] also when environmental issues are at stake. As substantive law is concerned, most of the time they review environmental related issues from two distinct perspectives.[41] In countries in which there is a constitutional provision on the protection of the environment or the climate, it may happen that legislation is challenged because it provides in view of the complainers insufficient protection.[42] In this respect, a remarkable case is the climate case of the Federal Constitutional Court of Germany. On March 24, 2021 the Federal Constitutional Court, in a case that became known as the *Neubauer et al.* case,[43] struck down parts of the Federal Climate Change Act as incompatible with fundamental rights for failing to set sufficient provisions for emission cuts beyond 2030. The Court found that Article 20a of the Federal Constitution obliges the legislature to protect the climate and aim at achieving climate neutrality. Further, the Court stated that Article 20a "is a justiciable legal norm that is intended to bind the political process in favour of ecological concerns, also with a view to the future generations that are particularly affected". Accepting arguments that the legislature must follow a carbon budget approach to limit warming to well below 2 °C and, if possible, to 1.5 °C, the Court found that the legislature had not proportionally distributed the budget between current and future generations, writing '*one generation must not be allowed to consume large parts of the CO_2 budget under a comparatively mild reduction burden if this would at the same time leave future generations with a radical reduction*

[37] Bezemek (2012), pp. 115–128.

[38] Kelsen (1942), pp. 181–200.

[39] Winter (2012), pp. 59–61; Lavrysen (2009a), pp. 96–97 and 99–100.

[40] E.g. the Conseil Constitutionnel in France; https://www.conseil-constitutionnel.fr/.

[41] Lavrysen (2022), pp. 2–3.

[42] Lavrysen (2008b), pp. 2; Lavrysen (2009a), pp. 103–105.

[43] BVerfG, Order of the First Senate of 24 March 2021, 1 BvR 2656/18, paras. 1–270; https://www.bundesverfassungsgericht.de/e/rs20210324_1bvr265618en.html.

burden... and expose their lives to serious losses of freedom'. The Court ordered the legislature to set clear provisions for reduction targets from 2031 onward by the end of 2022.[44] Constitutional Courts are on the other hand checking the compatibility of environmental legislation with property rights and the freedom of trade and industry in cases introduced by industrial federations, businesses, and landowners.[45]

In contrast with the global trend,[46] there are not many independent specialised environmental courts in Europe. Environmental cases are generally assigned to judges of the ordinary courts and administrative courts. Sometimes, specialisation occurs in practice, because the environmental cases are systematically referred to the same court division and the judges concerned train themselves in environmental law on a voluntary basis.

Criminal courts can in general pass sentences for environmental crimes ranging from fines to imprisonment, where appropriate concomitant with compensation and/or safety measures and remediation measures.[47] There is a lack of reliable, accurate and complete statistical data on environmental crime proceedings. It has been noted that even within the EU, there is no level playing field regarding the enforcement of environmental law. The Directive 2008/99/EC on the protection of the environment through criminal law did not have much effect on the ground: over the past 10 years the number of environmental cases successfully investigated and sentenced remained very low.[48] That absence of enforcement exists with regard to sanctions used by countries, but it is equally absent at the level of the criminal track as a whole, even when the situation seems to improve slowly. A lack of training and specialisation of prosecutors and judges in most countries has been highlighted.[49] In practice, certain larger courts, in Belgium e.g., have developed a limited form of specialization as a result of environmental cases being consistently referred to the same division on the basis of an internal division of tasks, or, more recently on the basis of a specific court regulation. But those divisions are usually also entrusted with other forms of crime, so that this specialization does not go to such an extent

---

[44] Kotzé (2021), pp. 1423–1444; Winter (2022), pp. 209–221.

[45] Aragão (2016), pp. 282–290; Bándi (2016), pp. 188–189; García-Ureta (2016), pp. 294–304; Iwanska and Baran (2016), pp. 248–260; Jancarova et al. (2016), pp. 117–126; Knez (2016), pp. 282–290; Mikosa (2016), pp. 212–217; Montini (2016), pp. 206–207; Lavrysen (2016), pp. 99–107; Ofak (2016), pp. 130–138; Ofak (2021), pp. 85–98; Winter (2016), pp. 4 and 174–184.

[46] According to the UNEP's initial study on Environmental Courts & Tribunals (Pring and Pring (2016), p. 120), the number of Environmental Courts has been exploding since 2000. In 2016, there were over 1200 Environmental Courts and Tribunals in 44 countries at the national or state/provincial level, with some 20 additional countries discussing or planning Environmental Courts & Tribunals. In the Updated Guide 2021 one has observed a steady growth since 2016 with the number of operational ECTs standing at 2116 in 67 countries (United Nations Environment Programme (2022), pp. 64–66).

[47] Lavrysen (2006a), p. 86; Corn and Perilongo (2015), pp. 3–5.

[48] European Commission, Proposal for a Directive of the European Parliament and of the Council on the protection of the environment through criminal law and replacing Directive 2008/99/EC, Brussels, 15.12.2021, COM (2021) 861 final, p. 1.

[49] Billiet (2016), p. 44; Billiet (2020) pp. 25–26.

that the criminal judges can concentrate exclusively on environmental criminal law.[50] In France, six specialised coastal courts exist for marine pollution. Recently one has created 37 specialised environmental courts within the ordinary courts. In the jurisdiction of each Court of Appeal, one court has been designated and is territorially competent for investigation, prosecution, and adjudication of offences of the Environment Code, the Forest Code and some other offences in the Mining Code, the Rural Code and Marine Fishing or illegal trade of wood or wood products, in cases which are or appear to be complex in view of their technicity, the importance of the damage or their geographical scale.[51]

In most countries, the *civil courts* are empowered to award damages, either in kind or by equivalent. Consequently, the civil courts are usually only confronted with environmental cases where damages are claimed. But they can also be competent to impose other measures in case the liability of the state, another authority or a private natural or legal person is at stake, e.g. by giving an order to reduce or prevent the damage. The famous *Urgenda* case in the Netherlands is an example of such a civil court case. The role of civil courts can vary slightly in different countries as can the volume of environmental cases referred to civil courts. While in France e.g. the civil courts have many environmental cases, in Sweden the civil courts are generally not involved in environmental cases. In Denmark, the Netherlands, Ireland and the United Kingdom, the civil courts are charged with two types of environmental cases: in private law (e.g. nuisance) and in public law (e.g. assessment of unlawful acts, omissions and decisions). While the civil courts in most countries are primarily empowered to award damages, in Ireland and the United Kingdom, a judge who is confronted with a particular dispute may also make other orders such as an injunction or declaration on a point of law, a quashing order, mandatory order or prohibition order. In Lithuania, the civil court can also make a range of orders, including granting a particular right, restoration of the relationship, prohibition to perform certain acts.

In Italy and Poland, the civil courts have jurisdiction in relation to issues relating to environmental taxes and challenges to the imposition of administrative sanctions.[52]

In various countries *administrative courts* are the courts that are most confronted with environmental cases as they are courts specializing in administrative law, particularly disputes concerning the exercise of public power. Their role is to ascertain that administrative acts are consistent with the law. As environmental policy is conducted to an important extent through multiple specific administrative acts (environmental licences, EIA, SEA, etc..), judicial review of these acts is a competence of those courts. The French climate cases *Grande-Synthe* and *Notre Affaire à Tous* are example of cases decided by the administrative judiciary. The intensity of the review performed is varying under influence of the legal system in which they operate and according to traditions. In reformatory systems the court decides the case on its

---

[50] Lavrysen (2006a), p. 86.

[51] Lavrysen and Bouquelle (2021), pp. 6–7.

[52] EUFJE and Milieu Consulting (2019), p. 40.

merits and the courts can substitute the administrative decision of its own, as is the case in Sweden and Finland. On the other hand, there are systems where legality control of the environmental decision is very formal, allowing the administration almost full discretion on the substance of law and where the court mostly focuses on procedural aspects, as is e.g. the case in Portugal and Spain.[53] In between are systems in which courts review both the substantive and procedural law aspects of the challenged decision, while respecting the discretionary powers that have been attributed to the administrative authorities.[54] Administrative courts are developing a certain degree of specialization in environmental law since the settlement of virtually all disputes between citizens and public authorities in environmental matters fall within their remit. Where environmental disputes account for a substantial portion of the administrative disputes this leads to a certain kind of specialization as those cases are consistently referred, whether or not on the basis of a legal rule, to the same court chamber or divisions.[55] This type of specialization in courts appears e.g. in Belgium, Bulgaria, Finland, Greece, Germany, Italy and Spain. Often, the appointment of specialized chambers or divisions is based on a court's regulation or a president's decision. Those chambers or divisions are thus not structurally guaranteed but have remained stable for many years. Often, they also handle non-environmental cases, and judges can easily be moved to other chambers. In this regard, the Netherlands is an exception, as, since 2020, the new Environmental Chamber of the Council of State only handles environment and planning cases.[56]

5 Specialized Environment Courts and Tribunals

The second model of specialization in Europe are administrative courts or appeal bodies, whose competences are restricted to appeals against decisions, fines, or permits falling under specifically listed environmental legislation. That is e.g. the case in Belgium, Denmark, Finland, Iceland, Ireland, Malta and the UK.[57] The Flemish Region of Belgium has two specialized environmental administrative courts: the Council for Permit Disputes and the Enforcement College. The Council for Permit Disputes has competence for the annulment and suspension of integrated environmental permits and similar administrative acts. Administrative fines in the field of environmental and planning law can be challenged before the Enforcement College.[58] Denmark has a long tradition of quasi-judicial administrative appeal bodies that have traditionally provided an alternative option for an independent review

[53] Darpö (2019), pp. 96–97; Sulyok et al. (2019), pp. 19–21.

[54] Darpö (2019), pp. 97–98.

[55] Lavrysen (2006a), p. 86.

[56] Lavrysen and Bouquelle (2021), p. 1.

[57] Lavrysen and Bouquelle (2021), p. 2; Pring and Pring (2016), pp. 81–85.

[58] Lavrysen and Bouquelle (2021), pp. 4–6.

of administrative decisions. In environmental matters, the Nature and Environmental Appeals Board dealt with administrative appeals until 2016. Two appeal Boards replaced it in 2016: the Environment and Food Appeals Board and the Planning Appeals Board, covering plans and decisions under the Planning Act. A decision by the Appeals Boards can be appealed to the general courts, generally to the district courts within 6 months, and with the option for a second appeal to the high courts.[59] In Finland, the Vaasa Administrative Court has exclusive jurisdiction to hear appeals under the Environmental Protection Act and the Water Act. As a result, environmental cases represent a significant share of its overall caseload and two divisions of that court deal almost exclusively with environmental cases. The administrative appeal is a reformatory remedy, which means the court can amend the challenged decision. In environmental matters, revision of disputed permit conditions comprises a typical use of this reformatory power. When considering administrative appeals, the administrative court's responsibilities are quite broad, and the review is not restricted explicitly to what has been alleged in the appeal. When an administrative appeal has been lodged, both the procedural and substantial legality of the challenged decision are liable to review, as well as the underlying material and technical findings. The Vaasa Administrative Court has judges with technical and scientific training[60] and legally qualified judges to provide the Court with sufficient expertise for such consideration. Parties may appeal against the decision of the Vaasa Administrative Court to the Supreme Administrative Court (SAC), provided the SAC admits the appeal. When the SAC hears a case, the Court may review the decision of the Vaasa Administrative Court as such or change provisions of the decision. In the Supreme Administrative Court of Finland, all environmental cases are consistently referred to the same court chamber, the First Chamber. The First Chamber of the Supreme Administrative Court also handles cases regarding land use, planning, building, asylum and international protection, data protection, and access to public documents. When recruiting and assigning judges and judge assistants to the chamber, special knowledge on environmental law plays a significant role. When the Supreme Administrative Court hears an appeal, two expert counsellors for the environment (who are qualified engineers or natural scientists) sit on the bench side by side with the lawyer judges.[61]

The Environmental and Natural Resources Board of Appeal of Iceland is an independent body, able to develop its own rules and procedures and provides the possibility to appeal against administrative decisions in matters concerning the environment and natural resources. The Board is comprised of nine members with different expertise, four of these are legal experts, and five are experts in other fields. An appeal can be brought to the Board against the decisions of the National Planning Authority based on Act no. 106/2000 on Environmental Impact Assessment.[62]

---

[59] Lavrysen and Bouquelle (2021), pp. 8–10.

[60] Eklund (2018), pp. 3–4.

[61] Lavrysen and Bouquelle (2021), pp. 10–12.

[62] Lavrysen and Bouquelle (2021), pp. 15–17.

In Ireland, the *An Bord Pleanála* (ABP) is an administrative appellate body empowered to make binding decisions in planning (i.e. land use) disputes. The Board's primary function is to determine appeals from decisions on applications for development consent made by local planning authorities. In the specific case of applications for development consent for strategic infrastructure development (e.g., motorways, airports, hospitals) and strategic housing development, however, the Board is the decision-maker at first instance. The ABP uses scientific-technical experts as decision-makers and can hire experts to provide advice and to assist in the evaluation of evidence presented by the parties. A decision of the ABP may only be challenged by way of judicial review proceedings in the High Court. The Aquaculture Licenses Appeals Board determines appeals against decisions of the Minister for Agriculture, Food, and the Marine on aquaculture license applications. The Forestry Appeals Committee determines appeals from license decisions made by the Minister for Agriculture, Food, and the Marine.[63]

In Malta administrative review of decisions from the Environment and Planning Authority (EPRT) is the competence of the Environment and Planning Review Tribunal. The role of the EPRT is to review planning, and environmental decisions in terms of law and fact within the parameters set out in the EPRT Act. Besides accepting and/or dismissing appeals, the EPRT has the power to order a change of plans and/or variations to planning and environmental decisions or order the Planning Authority to reconsider and decide the case again. The decisions of the EPRT are subject to appeal in front of the Court of Appeal.[64]

In the UK the General Regulatory Chamber (GRC) of the First-Tier Tribunal is part of the unified tribunal structure and is entirely independent of the executive. The GRC is the primary jurisdiction for determining appeals against decisions taken by a wide range of environmental regulators. These include appeals against a fine or notice for an environmental offence, for example, pollution or dumping toxic waste or against some other decisions. The Planning Courts are part of the Administrative Court, a specialist court within the Queen's Bench Division of the High Court of Justice. The Planning Courts are based at the Royal Courts of Justice in London and district registries across England and Wales. Cases at these courts are heard by the planning liaison judge or a high court judge, while a district judge hears cases at district registries. The Planning Courts hear claims for judicial review and statutory challenges regarding decisions made by planning authorities and other public bodies. Its broad jurisdiction encompasses granted planning permission, development consent, compulsory purchase orders, highways and other rights of way, decisions under EU environmental legislation. The "Verderers' Courts" of New Forest/Court of Swainmote, Forest of Dean/Court of Speech House, Epping Forest/Courts of Waltham Forest date back to medieval times. They have certain powers related to

63 Lavrysen and Bouquelle (2021), pp. 17–18.

64 Lavrysen and Bouquelle (2021), pp. 19–20.

the preservation and management of the said forests. The Court of Swainmote has the same status as a Magistrates' Court.[65]

More far-reaching specialization can be found in Sweden. The Environmental Code, which became effective in 1999, established a system of environmental courts in Sweden. The Land and Environment Courts (LEC) and the Land and Environment Court of Appeal, are part of the general court system. There are five regional LEC at the district court level and one LEC of Appeal. The regional environmental courts function both as (1) trial courts (first-instance) on permits for hazardous activities, water developments and environmental damage claims made by individuals, groups, NGOs and government and as (2) appellate courts (second-instance) for appeals of decisions by local and regional bodies on environmental permits, disposal of waste and clean-up orders. The one Environmental Court of Appeal hears appeals of cases from the regional Land and Environment Courts. Its decisions in the category (1) cases can be appealed to the Supreme Court, and its decisions in category (2) are in most instances final.[66] The LEC are also competent for planning and building, and real estate matters. Installations and activities involving a substantial environmental impact must obtain a permit of the LEC, as must most kinds of water operations. They try civil cases related to the environment, decisions appealed from administrative authorities, and handle cases on imposition of conditional fines after application from the supervisory authorities.

They have the power to impose orders, injunctions, prohibitions, and withdrawals of permits, in combination with administrative fines, decide on compensation and order the payment of damages. It should be noted that most decisions from local authorities are first challenged through administrative appeal to the County Administrative Boards and then to the LEC. A LEC consists of one judge trained in law, one environmental "technical expert" (with a science or technical education) and two "lay expert" members. The regional judge and technical expert are fulltime members of the court, and the two lay experts are selected depending on the expertise required in a given case. All four members of the panel are equals in the decision-making process. This multidisciplinary judicial approach acknowledges that environmental adjudication is increasingly based on highly complex scientific and technical projections of uncertain future impacts on intricate social, economic and environmental factors and that law-trained judges do not generally have the scientific-technical training to analyse expert testimony on these issues.[67]

The technical judges must have a long experience and good scientific and technical education. They must also be familiar with judicial interpretation and the procedural code. The advantages of technical judges are:

- Their ability to understand and evaluate what experts/specialists in different fields of science produce as material in the cases;
- Their ability to ask "the right questions" in main hearings;

[65] Lavrysen and Bouquelle (2021), pp. 24–27.

[66] Pring and Pring (2016), p. 27.

[67] Lavrysen and Bouquelle (2021), pp. 22–24.

- Their ability to make an assessment of the environmental effects in the cases;
- Their ability to clarify the scientific reasons for the conditions in the permit for an environmentally hazardous activity;
- Their ability to formulate the conditions of the permit so that the handling of the issue at hand, is correct according to scientific knowledge.

They know the standards for noise, water and air pollution measurements, the most recent guidelines and methods in science to keep up with scientific news because processes and measurements evolve constantly. Technical judges have contributed to a better understanding of EIAs, more transparency at the hearings, and to equalizing the parties' expert reports.[68]

6 The Aarhus Convention

The UNECE Convention on Access to Information, Public Participation in Decision-making and Access to Justice in Environmental Matters was adopted on 25th June 1998 in the Danish city of Aarhus at the Fourth Ministerial Conference in the 'Environment for Europe' process, in the framework of the United Nations Economic Commission for Europe.[69] The Convention, which is in force since 30 October 2001, has been ratified by 47 Parties, including the EU.[70] The GMO-Amendment to the Convention, has been ratified by 32 Parties, including the EU. The PRTR-Protocol has been ratified by 38 Parties, including the European Union and entered into force on 8 October 2009. The Russian Federation, a member state of UNECE from the beginning, is not a party to the Convention, while Ukraine is party to the Convention and the PRTR Protocol. Belarus, has following the invasion of the Russian Federation in Ukraine notified its withdrawal of the Convention on 26 July 2022, with effect on 24 October 2022.[71]

The Aarhus Convention links environmental rights and human rights. It acknowledges that we owe an obligation to future generations. It establishes that sustainable development can be achieved only through the involvement of all stakeholders. It focuses on interactions between the public and public authorities in a democratic context and it is forging a new process for public participation in the negotiation and implementation of international agreements. The Convention is therefore not only an environmental agreement, it is also a convention about government

[68] Schultz (2018), p. 5; Schultz (2019), pp. 118–135.

[69] The UNECE region covers more than 47 million square kilometres. Its member States include the countries of Europe, but also countries in North America (Canada and United States), Central Asia (Kazakhstan, Kyrgyzstan, Tajikistan, Turkmenistan and Uzbekistan) and Western Asia (Israel) (https://unece.org/map-region). UNECE has 56 member States (https://unece.org/member-states-and-member-states-representatives).

[70] The following countries which are not or only partially situated in Europe are a party to the Convention: Armenia, Azerbaijan, Georgia, Kazakhstan, Kyrgyzstan, Tajikistan, and Turkmenistan.

[71] https://treaties.un.org/doc/Publication/CN/2022/CN.259.2022-Eng.pdf.

accountability, transparency and responsiveness. The Aarhus Convention grants the public rights and imposes on parties and public authorities obligations regarding access to information, public participation and access to justice.[72] As its title suggests, the Convention contains three broad themes or 'pillars': access to information, public participation and access to justice.

The third pillar of the Convention (Article 9) aims to provide access to justice in environmental matters in three different contexts: (a) review procedures with respect to information requests, (b) review procedures with respect to specific (project-type) decisions which are subject to public participation requirements, and (c) challenges to breaches of environmental law in general.

Article 9.1 of the Aarhus Convention deals with access to justice concerning *information* appeals. A person whose request for information has not been dealt with to his satisfaction must be provided with access to a review procedure before a court of law or another independent and impartial body established by law. Standing must, under this provision, be granted to "any person who considers that his or her request for information under Article 4 [of the Convention] has been ignored, wrongfully refused, whether in part or in full, inadequately answered, or otherwise not dealt with under the provisions of that article". No additional standing requirements may be imposed.[73]

Article 9.2 of the Aarhus Convention deals with access to justice concerning *environmental decision-making with regard to activities that may have a significant effect on the environment*. The Convention provides for a right to seek review in connection with decision-making on projects or activities covered by Article 6 of the Convention. The review procedure should be organized before a court of law and/or another independent and impartial body established by law and make it possible "to challenge the *substantive* and *procedural* legality of any *decision*, *act* or *omission* subject to the provisions of Article 6". Parties may apply the review procedure to other provisions of the Convention by providing for review in those cases. Those may include decisions covered by Article 7 (plans, programmes and policies relating to the environment) or Article 8 (executive regulations and generally applicable legally binding normative instruments). The review procedure should be open to "members of the public", that is to say "the public affected or likely to be affected, or having an interest in the environmental decision making", including environmental NGOs "meeting any requirements under national law" (Art. 2.5) in so far as they have "a sufficient interest" (notion often used in the legal systems inspired by those of France) or "maintain impairment of a right, where administrative procedural law of a Party requires this as a precondition" (concept used in the legal systems inspired by German law). So, State Parties may impose certain standing requirements for members of the public and environmental NGOs, but their room for manoeuvre in this respect is not unlimited. Article 9.2, subparagraph 2, states: "[w]hat constitutes

---

[72] UNECE (2000), p. 12; UNECE (2004), p. 15; Redgwell (2007), pp. 156–157; Lavrysen (2010), p. 651.

[73] UNECE (2000), p. 126; UNECE (2004), pp. 190–193.

a sufficient interest and impairment of a right shall be determined in accordance with the requirements of national law and *consistent with the objective of giving the public concerned wide access to justice within the scope of this Convention.* To this end *the interest of any non-governmental organization* meeting the requirements referred to in Article 2, paragraph 5, *shall be deemed sufficient* for the purpose of subparagraph (*a*) above. Such organizations shall also be deemed to have rights capable of being impaired for the purpose of subparagraph (*b*) above". According to Article 9.2, third subparagraph, this provision on access to justice shall not exclude the possibility of a preliminary review procedure before an administrative authority and shall not affect the requirement of exhaustion of administrative review procedures prior to recourse to judicial review procedures, where such a requirement exists under national law. The administrative appeal system is not intended to replace the opportunity of appeal to the courts, but it may in many cases resolve the matter expeditiously and avoid the need to go to court.[74]

Article 9.3 concerns violations of environmental law in general. The Convention requires Parties to provide access to administrative or judicial procedures to challenge *acts and omissions* by *private persons* and *public authorities* which *breach laws relating to the environment.* The Convention introduces in so doing a form of direct citizen enforcement which can be used not only against administrative acts, but also against material acts and omissions. Omissions include the failure to implement or enforce environmental law with respect to other public authorities or private entities.[75] The Convention uses the terms "which contravene provisions of its national law relating to the environment." In the absence of a specific definition of "national law", it includes not only domestic law (both federal and regional), but also European and international law that is binding on the Member States, in particular those provisions of international or European law that have direct effect. Such access is to be provided to members of the public 'where they meet the criteria, if any, laid down in national law'. In other words, the issue of standing is primarily to be determined at the national level, as is the question of whether the procedures are judicial or administrative. Members of the public include natural or legal persons, and, in accordance with national legislation or practice, their associations, organizations or groups (Art. 2.4).

Art. 9.4 and 9.5 are setting minimum requirements concerning access to justice which should be provided for under Art. 9.1, 9.2 and 9.3 of the Aarhus Convention. Article 9.4 stipulates that these procedures should provide *adequate and effective remedies, including injunctive relief as appropriate*, and be fair, equitable, timely and not prohibitively expensive. Decisions under this article shall be given or recorded in writing. Decisions of courts, and whenever possible of other bodies, shall be publicly accessible. Injunctive relief is a remedy to prevent or remedy injury. The Convention requires injunctive relief and other remedies to be "adequate and effective". Adequacy requires the relief to fully compensate past damage,

[74] UNECE (2004), pp. 193–196.

[75] UNECE (2004), p. 197.

prevent future damage, and may require it to provide restoration. The requirement that the remedies should be effective means that they should be capable of efficient enforcement.[76] Article 9.5 prescribes that in order to further the effectiveness of the provisions of Article 9, each Party shall ensure that information is provided to the public on access to administrative and judicial review procedures and shall consider the establishment of appropriate assistance mechanisms to remove or reduce financial and other barriers to access to justice.[77]

Art. 9 of the Aarhus Convention is of particular relevance for the national judiciaries.[78] In most of the countries the legislation uses a rather vague formula in describing the conditions to have standing. As has been indicated the EU is a party to the Aarhus Convention. The EU has also adopted some implementing legislation of the Aarhus Convention. As EU Member States are concerned, the Court of Justice of the European Union (CJEU) plays a crucial role in the interpretation and implementation of EU law, including international agreements to which the EU is a party, as is the case with the Aarhus Convention (see below paragraph 8). An intensive dialogue between national judges and the CJEU has been established in relation to access to justice in environmental matters, while the European Commission is also bringing some infringement cases to the CJEU. That gave the CJEU the possibility to interpret different aspects of Art. 9 of the Aarhus Convention or of its EU implementing legislation.[79] Through more than two dozen judgments, the Court gave a broad interpretation of Art. 9 in relation to decisions, acts and omissions by public authorities of the Member States. The CJEU has held that it is for the national courts, in order to ensure effective judicial protection in the fields covered by EU environmental law, to interpret its national law in a way which, to the fullest extent possible, is consistent with objectives laid down in the Aarhus Convention. There is clear evidence that under the influence of the Aarhus Convention, the Findings and Recommendations of its Compliance Committee[80] and the case law of the CJEU on the Aarhus Conventions and its implementing EU provisions, judges in various member states are reinterpreting their national provisions on access to justice, when the legislator is not taking the requested action, to bring them in line with the Aarhus Convention.[81]

But in general, implementation of the access to justice provisions of the Convention remains the most difficult pillar for the Parties to the Convention. Two of the main issues mostly reported are: (a) the regulation of the rights of environmental NGOs to seek judicial or administrative remedies in environmental cases (standing); and (b) financial barriers. Parties to the Convention are according to the

---

[76] UNECE (2004), p. 200.

[77] UNECE (2004), pp. 199–205; Lavrysen (2010), pp. 661–668.

[78] L. Lavrysen (2008b), pp. 7–10.

[79] Lavrysen (2014), pp. 7–11; European Commission (2017) Commission Notice on access to justice in environmental matters (2017/C 275/01), *OJ* C 275, 18 August 2017.

[80] https://unece.org/environment-policy/public-participation/aarhus-convention/compliance-committee; Andrusevych and Kern (2016), p. 231.

[81] Lavrysen (2014), pp. 12–13.

latest Implementation Report aware of these difficulties, and the efforts reported demonstrate how keen Parties are to promote implementation of this Convention pillar. Some Parties amended their legislative provisions as a result of developments in the case law or on the basis of recommendations by the Aarhus Convention Compliance Committee. During the most recent reporting cycle (2017–2020) under the Convention several positive trends have been identified, namely: (1) increasing admissibility of public interest litigation in environmental matters; (2) increasing review by courts and other review bodies of the substantive legality of challenged decisions, acts and omissions; (3) measures introduced to remove or reduce financial barriers; and (4) promotion of awareness-raising and specialization of the judiciary and other legal professionals in environmental matters. The Implementation Report is recommending for those Parties that did not do so, to strive for full implementation of the Convention's access to justice provisions, in particular, by: (1) ensuring the clarity of legislation on access to justice and the compliance of practice with such legislation with the requirements of the Aarhus Convention; (2) speeding up the process of adoption of relevant amendments to national legislation with regard to standing, scope of review, burden of proof, timeliness in review procedures, especially with regard to information cases, remedies, including injunctive relief; (3) continuing to develop and promote electronic information tools and e-justice initiatives; (4) reducing or eliminating financial and other related barriers to review procedures; (5) ensuring that necessary and sufficient legal framework and assistance mechanisms are provided and are available in practice for members of the public and NGOs wishing to exercise their rights under the Convention, in particular the right of access to justice; (6) continuing to raise awareness of the public, judiciary, judicial institutions, prosecutors, judicial training institutions, other review bodies and legal professionals, about the third pillar of the Convention.[82]

7 The European Convention on Human Rights and the European Court of Human Rights

The Convention for the Protection of Human Rights and Fundamental Freedoms, better known as the European Convention on Human Rights (ECHR),[83] was opened for signature in Rome on 4 November 1950 and came into force on 3 September 1953. Since its adoption in 1950 the Convention has been amended a number of times and supplemented with many rights in addition to those set forth in the

[82] UNECE (2021) Meeting of the Parties to the Convention on Access to Information, Public Participation in Decision-making and Access to Justice in Environmental Matters, Seventh session, Geneva, 18–20 October 2021, Item 7 (a) of the provisional agenda, Procedures and mechanisms facilitating the implementation of the Convention: reporting mechanism Synthesis report on the status of implementation of the Convention, Report by the secretariat, ECE/MP.PP/2021/6.

[83] https://www.echr.coe.int/.

original text.[84] It had been ratified by the 47 Member States of the Council of Europe. Following the invasion of Ukraine by the Russian Federation the Committee of Ministers decided on 16 March 2022 that the Russian Federation is no longer a member of the Council of Europe,[85] while Belarus has never been a member. As a consequence the Russian Federation ceases to be a Party to the Convention on 16 September 2022.[86]

Even though the ECHR does not enshrine any right to a healthy environment as such, the European Court of Human Rights (ECtHR) has been called upon to develop a case-law in environmental matters on account of the fact that the exercise of certain Convention rights may be undermined by the existence of harm to the environment and exposure to environmental risks. The environmental jurisprudence of the ECtHR is based on the Right to life (Article 2 ECHR), the Prohibition of inhuman or degrading treatment (Article 3 ECHR), the Right to a fair trial (Article 6 ECHR), the Right to respect for private and family life and home (Article 8 ECHR), the Freedom of expression/Freedom to receive and impart information (Article 10 ECHR), the Freedom of assembly and association (Article 11 ECHR), the Right to an effective remedy (Article 13 ECHR) and the Protection of property (Article 1 of Protocol No. 1 ECHR).[87] Cases can only be brought to the ECtHR after domestic remedies have been exhausted. In other words, individuals complaining of violations of their rights must first have taken their case through the courts of the country concerned, up to the highest possible level of jurisdiction. The applicant must be, personally and directly, a victim of a violation of the Convention, and must have suffered a significant disadvantage. Protocol No. 16 to the ECHR allows the highest courts and tribunals of a State Party to request the Court to give advisory opinions on questions of principle relating to the interpretation or application of the rights and freedoms defined in the Convention or the protocols thereto. Protocol No. 16 came into force on 1 August 2018 in respect of the States which have signed and ratified it.[88]

The ECHR and the caselaw of the ECtHR serve in most of the countries concerned as an interpretative aid for national judges as part of a set of fundamental principles used by courts in environmental case-law. In some countries the case law has a direct (decisive) impact. In e.g. Finland, Hungary and Belgium the right

[84] https://www.echr.coe.int/Documents/Convention_ENG.pdf.

[85] https://search.coe.int/cm/Pages/result_details.aspx?ObjectID=0900001680a5da51.

[86] Resolution of the European Court of Human Rights on the consequences of the cessation of membership of the Russian Federation to the Council of Europe in light of Article 58 of the European Convention on Human Rights, 22 March 2022, https://echr.coe.int/Documents/Resolution_ECHR_cessation_membership_Russia_CoE_ENG.pdf

The Court remains competent to deal with applications directed against the Russian Federation in relation to acts or omissions capable of constituting a violation of the Convention provided that they occurred until 16 September 2022.

[87] ECtHR, Press Unit, *Environment and the European Convention on Human Rights*, Factsheet – Environment and the ECHR, https://www.echr.coe.int/documents/fs_environment_eng.pdf.

[88] https://www.coe.int/en/web/conventions/full-list?module=signatures-by-treaty&treatynum=214.

appears to be used directly as a legal reference by judges. In some countries, references to this right can be made, when no national act appears to provide the corresponding type or level of protection (e.g. France), or alternatively through a direct reference to the ECtHR case-law within constitutional adjudication or criminal case-law (e.g. Belgium).[89]

8 The EU and the Court of Justice of the EU

The European Union counts 27 member states, since the UK has left the EU on 31 January 2020. EU law follows a dual integration approach. In the first approach, the so-called negative integration, the Treaty on the Functioning of the European Union (TFEU) imposes a series of restrictions on the policies of the Member States, with a view to the institution and proper functioning of the internal market. In turn, there are exceptions to those restrictions. The second approach is the so-called positive integration. By enacting common rules in the form of regulations, directives and decisions in certain areas, the laws of the Member States are approximated and a common policy takes shape. The EU has developed since the start of its environmental policy program in 1973—at that time it was called the European Economic Community (EEC) with 9 Member States—and after different changes in its founding treaties, that has strengthened gradually its environmental competences and expanded its membership, an impressive corpus of legal texts (regulations, directives, decisions, international agreements) covering the whole area of environmental and climate policy.[90] The Directory of Legal Acts[91] contains more than 750 consolidated texts under the theme "environment", distributed as follows: general provisions and programmes: 11; pollution and nuisances: 21; nuclear safety and radioactive waste: 31; water protection and management: 38; monitoring of atmospheric pollution: 50; prevention of noise pollution: 64; chemicals, industrial risk and biotechnology: 70; space, environment and natural resources: 86; management and efficient use of space, the environment and natural resources: 88; conservation of wild fauna and flora: 90; waste management and clean technology: 101; international cooperation: 116. Of course not all these acts or of equal importance as both basic legislation and detailed executive measures are counted. Those legal texts are not only binding for the EU Member States, they keep in a transitionally period their relevance for the UK[92] and most of them are through the Treaty on the European Economic Area[93] also binding for Norway, Iceland and Liechtenstein. Not surprisingly, EU environmental law increasingly directs national environmental

[89] Caliceti et al. (2016), pp. 19–21.

[90] Jans and Vedder (2012), p. 570; Kingston et al. (2017) p. 558; Van Calster and Reins (2017), p. 384; Thieffry (2021), p. 520.

[91] https://eur-lex.europa.eu/homepage.html?locale=en.

[92] (Macrory 2019)

[93] https://www.efta.int/eea.

and climate legislation in the member states. We can say that 80–100% of national environmental legislation, depending on the Member State, is directly influenced by or the result of implementation of European environmental law.

The CJEU[94] has the task of ensuring respect for the law by explaining and applying the treaties and secondary law. Two of the competencies of the Court are of great importance for environmental law: (1) the Court delivers judgements in disputes between the European Commission and the Member States or between Member States themselves on the failure of a member state to fulfil its European obligations. That is the case among other things for violations of directly operating provisions of the treaties (e.g. with regard to free movement of goods, with regard to the advisory obligation for radio-active discharge) or on failure to comply with directives or to implement them. If the CJEU finds that a Member State has failed to fulfil an obligation under the Treaty, the State shall be required to take the necessary measures to comply with the judgement and under some conditions sanctions can be imposed. The Court can also impose interim measures in this context. (2) The Court is also competent to judge by means of preliminary rulings on questions referred by national judges on the interpretation of the treaties and secondary European law and on the validity of secondary law. In the final instance the national judges are obliged in certain cases to refer such a question if this arises in a pending case. Preliminary rulings represent in recent years more than two third of the judgments handed down by the Court.[95]

It may be safely concluded that the CJEU plays a crucial role in giving practical effect to EU environmental law. Since the judgement of 7 February 1985 in the case of *L'Association de défense des brûleurs d'huiles usagés,*[96] the Court acknowledges that protection of the environment is a 'mandatory requirement' which, in the absence of exhaustive harmonisation of legislation, may justify trade-restricting measures of Member States. Over the years, the Court has adopted an increasingly benevolent attitude on this point, so that the Member States, in the absence of fully harmonised common environmental rules, retain sufficient discretion to conduct an ambitious domestic environmental policy. The importance of environmental protection is thus soundly balanced against the economic interests that sometimes conflict with it[97] The general doctrines developed by the Court concerning the enforcement of regulations[98] the transposition and practical application of directives, the direct effect of directives,[99] consistent interpretation,[100] and Member State liability for

[94] https://curia.europa.eu/jcms/jcms/j_6/en/.

[95] Court of Justice on the European Union (2022).

[96] ECJ, 7 February 1985, Procureur de la République v Association de défense des brûleurs d'huiles usagées (ADBHU), Case 240/83, *European Court Reports*, 1985, p. 531.

[97] Lavrysen (2006b), p. 447; Lavrysen (2008a), pp. 78–79.

[98] Billiet and Meeus (2009), pp. 278–306

[99] Krämer (2013), pp. 53–71.

[100] Macrory et al. (2013), p. 37–50.

violations of EU law[101] also contribute to the effective application of European environmental law.[102] What is also noteworthy is that the authority that the Court has received to impose in certain circumstances a penalty (lump sum and daily fines) in non-compliance cases was applied for the first time in cases involving the failure to apply an environmental directive in practice[103] and that the environment is still the area in which the Court has used this competence in the most intensive way.[104] In recent times the possibility to impose interim measures has been applied more frequently in environmental cases, in particular in biodiversity cases.[105] In one of those cases, the concerned member state has been ordered to pay to the European Commission a penalty payment of EUR 500000 per day, until that Member State complies with the interim order.[106] As regards the interpretation of environmental regulations and directives, we can be brief. Provisions that protect the environment are usually interpreted extensively, while provisions that constitute an exception to those rules are usually given a restrictive interpretation. We can only agree with Ludwig Krämer where he wrote: '*The Court of Justice record in environmental matters over the last twenty-five years is almost entirely positive*'[107] and '*The Courts have almost always tried to interpret existing legislation in a way which is favourable to the environment and to formalise the concept of environmental law which has often been rather general and vague*'.[108] Twenty years later, those observations are more than ever valid.

Application of EU Environmental Law is however in the first place a task of national judges. On this point, we may endorse the words of Gil Carlos Rodríguez Iglesias, former President of the European Court of Justice, '*All national judges – tens of thousands of them – are competent to apply E[U] law on an everyday basis. They apply it directly; they interpret their national laws in conformity with it, if at all possible; if not, they must leave aside national laws that are contrary to E[U] law, because it is the duty of national judges to guarantee the rights provided for in the treaty and in E[U] legislation. In other words, individuals may rely upon provisions of [Union law before national courts without any implementing element of*

---

101 Moreno Molina (2013), pp. 99–103.

102 Lavrysen (2006b), P. 447; Lavrysen (2008a), pp. 78–79.

103 ECJ, Case C-387/97, *Commission v. Greece;* Case C-278/01, *Commission v Spain.*

104 Hedemann-Robinson (2017), p. 100.

105 Case C-76/08 R, Order of the President of the Court of 24 April 2008, *Commission v Malta* and Case C-573/08 R, Order of the President of the Court of 10 December 2009, *Commission v Italy* (order to refrain from applying hunting rules in derogation of the Birds Directive); Case C-441/17 R, Order of the Court (Grand Chamber) of 20 November 201, *Commission v Poland* (order to cease immediately active forest management operations in Białowiesa Forest) and Case C-121/21R, Order of the Vice-President of the Court of 21 May 2021, *Czech Republic v Poland (Mine de Turów)* (order to cease immediately lignite mining activities at the Turów mine).

106 Case C-121/21R, Order of the Vice-President of the Court of 20 September 2021, *Czech Republic v Poland (Mine de Turów)*.

107 Krämer (2003a), p. 86.

108 Krämer (2003b), p. 45.

domestic law, the only requirement being that the provisions relied upon should be sufficiently clear and unconditional to create such rights. The co-operation between the Court of Justice and the national courts through the preliminary reference procedure has been decisive to ensure the proper application of [Union] law and the protection of individual rights created by the [EU] legal order. The Court's jurisprudence in the area of environmental protection shows particularly well the important role that national judges play in the implementation and enforcement of obligations created by [EU directives].'[109]

The role of the national judge in the application of EU Environmental law is not more complicated than in the application of domestic law in the ideal situation that Directives are transposed timely and correctly in domestic law by legally binding rules in a way that is consistent with both provisions of primary European law, taking into account the latest case law of the CJEU, and with other pieces of domestic law and that the member states have taken on time the necessary complementary provisions to Regulations. However, such an ideal situation seems not to be real. It is sufficient to look at statistics of the European Commission to understand that the transposition of European environmental law in domestic law is not a success story. DG Environment still has the highest number of open infringement cases.[110] Also the high number of condemnations of Member States by the ECJ for bad application of European Environmental law show that the situation is far from ideal.

So the reality seems to be that in various member states one is confronted with relatively bad or delayed transposition of some Environmental Directives and bad application of certain Environmental Regulations. In such circumstances the role of the national judge in upholding European Environmental Law is crucial, but at the same time complicated. He has to look in a critical way to his domestic law. He has to make an in depth analysis of European Environmental Law, taking into account the ever growing case law of the CJEU. Has the rule of European Environmental Law which is thought to be violated, direct effect or not? Is the party who is arguing that the provisions of a directive are violated, entitled to raise this argument, taking into account that Directives do not produce horizontal or third-party effect, but can produce on the other hand horizontal side-effects of vertical direct effect. If there are some differences between domestic and European Environmental Law he has to ask himself if such differences are allowed by European law. He must look at the nature of the Directive. Is the Directive providing for minimum or uniform harmonization, or for a mixed or another form of harmonisation? In case the Directive provides for minimum harmonisation, is the domestic law in concordance with the minimum requirements of the Directive? If domestic law is going further than such a Directive, are these further going requirements compatible with primary European law? When the Directive is providing for uniform harmonization, is there nevertheless room for further going requirements, on the basis of secondary or primary European law? If

[109] Rodríguez Iglesias and Riechenberg (2002), p. 31

[110] European Commission (2021), *Monitoring the Application of European Union Law. 2021 Annual Report. General Overview*, Brussels, https://ec.europa.eu/info/sites/default/files/general_overview_en.pdf.

the differences between domestic and European law are not allowed by European law, can these differences be faded away by interpreting domestic law in a way that domestic law is becoming consistent with European law? Can this be done within the boundaries set by the CJEU in its case-law? If consistent interpretation is not possible, to which extent domestic law must be set aside, or, if that is allowed by national law, annulled? In case of annulment, what should be the scope of it and should this operate with full retro-active effect or not? Is there still room for balancing interest and to what extent? If there is question of bad application of European Environmental law, can the *Francovich liability* be applied, and how should it be applied within the domestic procedural framework?.[111] Different types of questions on the validity and the interpretation of European Environmental Law may thus arise before national judges. Judges may feel it in such a situation necessary to refer such cases for preliminary ruling to the CJEU, and, if such a question arises for the highest national judges they may be obliged to refer the question. It seems however that the willingness for raising such questions varies considerably from one member state to another. The same holds true for the further follow-up.[112] In the field of the environment a quick search in the database of the CJEU[113] resulted in 371 preliminary rulings classified under the term "environment". These cases have been referred from the following member states: Italy: 66; Germany: 62; Belgium: 38; Netherlands: 34; France: 33; UK: 26; Austria: 22; Ireland and Sweden: 10; Finland and Spain: 9; Greece, Hungary and Slovakia: 6; Czech Republic: 5; Bulgaria, Denmark and Poland: 4; Croatia and Luxembourg: 3; Estonia, Latvia, Lithuania, Portugal and Romania: 2; Slovenia: 1. There were no such references from Cyprus and Malta. Maybe these figures can, with some caution, be interpreted as an indicator of the willingness of national judges to give precedence to European environmental law over domestic law. Some caution is indeed needed. The better the implementation situation is, the less the need to consult the CJEU.[114] There is only a partial match with the number of decided infringement cases. In the same period 790 environmental infringement cases have been decided, concerning the following member states: Italy: 110; Greece 74; Spain: 64; Belgium: 63; France: 60; Germany: 51; UK: 46; Portugal; 45; Ireland: 44; Austria: 40; Luxembourg and Poland: 35; Netherlands: 21; Finland:19; Malta: 13; Sweden: 10; Estonia: 9; Romania: 8; Czech Republic, Slovakia and Slovenia: 7; Denmark and Cyprus: 5; Bulgaria: 4; Croatia and Hungary 2; Latvia and Lithuania: 0. Of course other factors or influencing this result. Recent member states will have witnessed less cases that older ones, but also the availability of Commission staff familiar with the environmental law of the concerned member states is influencing the number of started infringement cases.

It may also be assumed that many cases are settled by national courts without it being necessary to refer questions to the CJEU for a preliminary ruling, either

[111] Lavrysen (2009b), pp. 137–138.

[112] Squintani and Annink (2018), pp. 147–170; Squintani and Kalisvaart (2020), pp. 931–961.

[113] Situation on August 18th 2022.

[114] Lavrysen (2009b), pp. 138–139.

because the court is of the opinion that there is no reasonable doubt about the validity or interpretation of the provisions of European law relied upon (the so-called "*acte claire*"), or because those questions could be solved on the basis of the existing case-law of the CJEU (so-called "*acte éclairé*").[115] As regards the existing national judicial culture in the various Member States, it seems that although the level of knowledge of EU law in general and the preliminary ruling procedure in particular is considered high among the highest courts of the member states, still, there seems to be room for improvement at the level of lower courts.[116]

9 Conclusion

Under the influence of an ever growing corpus of environmental law, the rising awareness of environmental problems, the growing biodiversity and climate crises and the relaxation of standing requirements, the judiciaries of Europe, in their different forms and to a variable extent, are more and more dealing with environmental issues. Over the past half century, environmental law has gradually developed into a very extensive, complex, rapidly evolving, sometimes very technical-scientific and multi-layered area of law, in which national law is strongly influenced and guided by international and, in the greater part of Europe, EU environmental law. It may be expected that this evolution has not yet come to a halt, certainly not when adjacent (spatial planning, water law) and partly overlapping (climate law) areas of law are taken into account. Environmental law is characterized by a combination of laws and regulations. It consists of different branches of law, environmental policy principles, policies and soft law and a particularly dynamic European case law. Generalist judges who are confronted with an environmental case from time to time, still the reality in the bigger part of Europe, are in an uncomfortable position. This applies all the more if they have not received thorough and up-to-date training in environmental law and if the parties do have specialized lawyers at their disposal. The legal anchoring of specialized environmental departments or judges in the relevant courts, as applied in some countries, is a must. Specialization within the existing courts can only be guaranteed structurally and in the long term if it is enshrined in law and is mandatory. A legal status for environmental judges would make it possible to attract specialized and motivated environmental judges who have the prospect of being able to hold the position for a sufficiently long period of time and also to pursue a career in it. Environmental judges should receive basic training in environmental law (international, European and national) and basic training in environmental sciences that enable them to understand the scientific and technical data involved in environmental court cases, with the help of subject matter experts if necessary. In addition to initial training, continuous training is also a must, given the

[115] Lavrysen (2009b), pp. 139–140.

[116] Squintani and Kalisvaart (2021), pp. 1 and 29.

continuous evolution of environmental problems, science, technology, policy, legislation, jurisprudence and legal doctrine.[117]

References

Andrusevych A, Kern S (2016) Case law of the Aarhus Convention Compliance Committee (2004–2014), 3rd edn. RACSE, Lviv

Aragão A (2016) Property and environmental protection in Poland. In: Winter G (ed) Property and environmental protection in Europe. Europa Law Publishing, Groningen, pp 282–290

Bándi G (2016) Property and environmental protection in Hungary. In: Winter G (ed) Property and environmental protection in Europe. Europa Law Publishing, Groningen, pp 188–189

Bezemek C (2012) A Kelsenian model of constitutional adjudication. The Austrian Constitutional Court. Zeitschrift für öffentliches Recht 67:115–128. https://doi.org/10.1007/s00708-012-0127-5

Billiet CM (2016) Sanctioning environmental crime (WG4). Prosecution and judicial practices. Interim report. ENPE, Brussels

Billiet CM (2020) Sanctioning environmental crime (WG4), Final report. Key observations and recommendations, 2016–2020. ENPE, Brussels

Billiet CM, Meeus R (2009) Europese verordeningen als wetgevingsvraagstuk: het voorbeeld van de milieuverordeningen. TVW 2009:278–306

Caliceti E, Corn E, Penasa S, Perilongo GF, Pulice E (2016) Human Rights and the enforcement of environmental law. Summary report: analysis of the questionnaires, 2016 EUFJE conference, Bucharest. https://www.eufje.org/images/docConf/buc2016/sumbuc2016.pdf

Corn E, Perilongo GF (2015) The protection of the environment through criminal law. Summary report. Analyses of the questionnaires, Bolzano, 2015 EUFJE conference. https://www.eufje.org/images/docConf/bol2015/sum%20bol2015.pdf

Council of Europe (2020) European judicial systems. CEPEJ Evaluation report 2020 evaluation cycle (2018 data), Strasbourg. https://rm.coe.int/evaluation-report-part-1-english/16809fc058

Court of Justice on the European Union (2022) Annual report 2021 judicial activity, Luxembourg

Darpö J (2019) Understanding the nuts and bolts: scientific and technical knowledge in environmental litigation – national solutions, EU requirements and current challenges. In: Squintani L, Darpö J, Lavrysen L, Stoll PT (eds) Managing facts and feelings in environmental governance. Edward Elgar Publishing, Cheltenham, pp 82–102

Eklund J (2018) The role of technical judges in the Finnish administrative courts. Presentation at the occasion of the 2018 EUFJE annual conference. https://eufje.org/images/docConf/so2018/Eklund_Sofia.pdf

EUFJE and Milieu Consulting (2019) Summary report. Questionnaire relating to the training and specialisation of magistrates in environmental law, p 58. https://www.eufje.org/images/docConf/so2018/EUFJE_Report_on_Training_and_Specialisation_in_Environmental_Law_2019.pdf

García-Ureta A (2016) Property and environmental protection in Spain. In: Winter G (ed) Property and environmental protection in Europe. Europa Law Publishing, Groningen, pp 294–304

Hedemann-Robinson M (2017) Enforcement of European Union environmental law. Legal issues and challenges, Chapter 4 A. Routledge. https://static.routledge.com/9780415659598/Chapter_4A.pdf

[117] Lavrysen and Bouquelle (2022), pp. 556–565.

Iwanska B, Baran M (2016) Property and environmental protection in Poland. In: Winter G (ed) Property and environmental protection in Europe. Europa Law Publishing, Groningen, pp 248–260

Jancarova I, Hanak J, Vomavka V (2016) Property and environmental protection in Czechia. In: Winter G (ed) Property and environmental protection in Europe. Europa Law Publishing, Groningen, pp 117–126

Jans JH, Vedder HHB (2012) European environmental law: after Lisbon, 4th edn. Europa Law Publishing, Groningen

Kelsen H (1942) Judicial review of legislation: a comparative study of the Austrian and the American Constitution. J Polit 4(2):182–200. https://doi.org/10.2307/2125770

Kingston S, Heyvaert V, Čavoški A (2017) European environmental law. Cambridge University Press

Knez R (2016) Property and environmental protection in Slovenia. In: Winter G (ed) Property and environmental protection in Europe. Europa Law Publishing, Groningen, pp 282–290

Kotzé LJ (2021) Neubauer et al. versus Germany: planetary climate litigation for the Anthropocene? Ger Law J 22:1423–1444. https://doi.org/10.1017/glj.2021.8

Krämer L (2003a) The future role of the ECJ in the development of European environmental law. In: Jans JH (ed) The European convention and the future of European environmental law. The Avosetta series I. Europa Law Publishing, Groningen

Krämer L (2003b) EC environmental law, 5th edn. Sweet & Maxwell, London

Krämer L (2013) Direct effect and consistent interpretation: strengths and weaknesses of the concepts. In: Jans JH, Macrory R, Moreno Molina AM (eds) National courts and EU environmental law. Europa Law Publishing, Groningen, pp 53–73

Lavrysen L (2006a) The role of national judges in environmental law. In: Ormond T, Führ M, Barth R (eds) Environmental law and policy at the turn to the 21st century – Umweltrecht und -politik an der Wende zum 21. Jahrhundert. Gedenkschrift/Liber Amicorum Betty Gebers, Lexxion, Berlin, pp 81–96

Lavrysen L (2006b) The European Court of Justice and the implementation of environmental law. In: Macrory R (ed) Reflections on 30 years of EU environmental law. A high level of protection. The Avosetta series, nr. 7. Europa Law Publishing, Groningen, pp 417–448

Lavrysen L (2008a) The European Court of Justice and the implementation of environmental law. In: Postiglione A (ed) The role of the judiciary in the implementation and enforcement of environmental law. Bruylant, Brussels, pp 25–80

Lavrysen L (2008b) National judges and the convention – how the judiciary can further the implementation of the third pillar. The Aarhus Convention: how are its access to justice provisions being implemented? Conference Paper, Brussels

Lavrysen L (2009a) Chapter 2: Belgium. In: Kotzé LJ, Paterson AR (eds) The role of the judiciary in environmental governance: comparative perspectives. Kluwer Law International, Alphen aan den Rijn, pp 85–122

Lavrysen L (2009b) Application of European environmental law by national courts. In: Bándi G (ed) The impact of ECJ jurisprudence on environmental law. Szent István Társalut, Budapest, pp 135–141

Lavrysen L (2010) The Aarhus Convention: between environmental protection and human rights. In: Liège, Strasbourg, Bruxelles: parcours des droits de l'homme, Liber Amicorum Michel Melchior. Anthemis, Limal, pp 647–672

Lavrysen L (2014) Access to justice in environmental matters: perspective from the European Union forum of judges for the environment. Global symposium on the environmental rule of law, proceedings, https://biblio.ugent.be/publication/4426506/file/4426507.pdf

Lavrysen L (2016) Property and environmental protection in Belgium. In: Winter G (ed) Property and environmental protection in Europe. Europa Law Publishing, Groningen, pp 99–107

Lavrysen L (2022) Environmental cases before the Belgian Constitutional Court, Workshop organised by the Constitutional Council, Courts faced with new public health, technological and environmental challenges. Conference of the heads of the Supreme Courts of the European Union Member States, Paris. https://www.const-court.be/public/stet/n/stet-2022-001n.pdf;

https://www.conseil-constitutionnel.fr/actualites/conference-des-chefs-des-cours-supremes-des-etats-membres-de-l-union-europeenne-le-lundi-21-fevrier

Lavrysen L, Bouquelle F (2021) ECT regional report Europe, 2021. Ghent University

Lavrysen L, Bouquelle F (2022) Waarom zijn milieurechters noodzakelijk? Ars Aequi 2022:556–565

Macrory R (2019) Environmental law in the United Kingdom post Brexit. ERA Forum 2019(19):643. https://doi.org/10.1007/s12027-018-0531-6

Macrory R, Madner V, Mayr S (2013) Consistent interpretation of EU environmental law. In: Jans JH, Macrory R, Moreno Molina AM (eds) National courts and EU environmental law. Europa Law Publishing, Groningen, pp 37–52

Mikosa Z (2016) Property and environmental protection in Latvia. In: Winter G (ed) Property and environmental protection in Europe. Europa Law Publishing, Groningen, pp 212–217

Montini M (2016) Property and environmental protection in Italy. In: Winter G (ed) Property and environmental protection in Europe. Europa Law Publishing, Groningen, pp 206–207

Moreno Molina AM (2013) Direct effect and state liability. In: Jans JH, Macrory R, Moreno Molina AM (eds) National courts and EU environmental law. Europa Law Publishing, Groningen, pp 53–72

Nicolatos MM, Parparinos L, Hadjiprodromou M (2018) Administrative justice in Europe. The Supreme Court of Cyprus. https://www.aca-europe.eu/en/eurtour/i/countries/cyprus/cyprus_en.pdf

Ofak L (2016) Property and environmental protection in Croatia. In: Winter G (ed) Property and environmental protection in Europe. Europa Law Publishing, Groningen, pp 130–138

Ofak L (2021) The approach of the Constitutional Court of the Republic of Croatia towards the protection of the right to a healthy environment. J Agric Environ Law 16(31):85–98. https://doi.org/10.21029/JAEL.2021.31.85

Pring G, Pring C (2016) Environmental courts & tribunals. A guide for policy makers. UNEP, Nairobi

Redgwell C (2007) Access to environmental justice. In: Francioni F (ed) Access to justice as a human right. Oxford University Press, pp 153–175

Rodríguez Iglesias GC, Riechenberg K (2002) Sustainable development in the European Union – environmental law before the European Court of Justice, Contribution to the global judges symposium. UNEP, Johannesburg

Sachs JD, Lafortune G, Kroll Ch, Fuller G, Woelm F (2022) From crisis to sustainable development: the SDGs as roadmap to 2030 and beyond. Sustainable development report 2022. Cambridge University Press. https://s3.amazonaws.com/sustainabledevelopment.report/2022/2022-sustainable-development-report.pdf

Schultz M (2018) The role of the technical judge in Sweden. EUFJE annual conference 2018, presentation, https://eufje.org/images/docConf/so2018/so2018_presMS.pdf

Schultz M (2019) Scientific evidence in Swedish courts: the use of technical judges for better integration of scientific data in environmental decision-making. In: Squintani L, Darpö J, Lavrysen L, Stoll PT (eds) Managing facts and feelings in environmental governance. Edward Elgar Publishing, Cheltenham, pp 118–135

Squintani L, Annink D (2018) Judicial cooperation in environmental matters: mapping national courts' behaviour in follow-up cases. J Eur Environ Plan Law 15(2):147–170

Squintani L, Kalisvaart S (2020) Environmental democracy and judicial cooperation in environmental matters: mapping national courts behaviour in follow-up cases. Eur Pap 5(2):931. https://doi.org/10.15166/2499-8249/415

Squintani L, Kalisvaart S (2021) Report on judicial cooperation in follow up judgments in environmental matters. 2021 EUFJE annual conference, cooperation between CJEU and national judges in environmental cases, pp 1 and 29. https://www.eufje.org/images/docConf/visio2021/Report_based_on_EUFJE_questionnaires_on_follow-up_judgments_-_Definitive_06-11-2021.pdf

Sulyok K, Bögös F, Paloniitty TM, Eliantonio M (2019) The role of science in environmental adjudication summary report: analysis of the questionnaire, 2019 EUFJE conference. https://www.eufje.org/images/docConf/hel2019/Summary_report_Questionnaire_EUFJE2019.pdf
Thieffry P (2021) Handbook of European environmental law, 2nd edn. Larcier, Brussels
UNECE (2000) The Aarhus Convention: an implementation guide. United Nations, New York and Geneva
UNECE (2004) The Aarhus Convention: an implementation guide. 2nd ed, United Nations, Geneva
United Nations Development Programme (2022) Human development report 2021–22. Uncertain Times. Unsettled lives: shaping our future in a transforming world the next frontier. Human Development and the Anthropocene, New York. https://hdr.undp.org/system/files/documents/global-report-document/hdr2021-22pdf_1.pdf
United Nations Environment Programme (2022) Environmental courts and tribunals-2021: a guide for policymakers. UNEP, Nairobi. https://wedocs.unep.org/handle/20.500.11822/40309
Van Calster G, Reins L (2017) EU environmental law. Edward Elgar Publishing, Cheltenham-Northampton
Winter G (2012) Environmental governance in Germany. In: Alberton M, Palermo F (eds) Environmental protection in multi-layered systems: comparative lessons from the water sector. Martinus Nijhoff Publishers, Leiden
Winter G (2016) Property and environmental protection in Germany. In: Winter G (ed) Property and environmental protection in Europe. Europa Law Publishing, Groningen, pp 174–184
Winter G (2022) The intergenerational effect of fundamental rights: a contribution of the German federal constitutional court to climate protection. J Environ Law 34(1):209–211. https://doi.org/10.1093/jel/eqab035
Wolf MJ, Emerson JW, Esty DC, de Sherbinin A, Wendling ZA et al (2022) 2022 environmental performance index. Yale Center for Environmental Law & Policy, New Haven. https://epi.yale.edu/

Luc Lavrysen is Emeritus Professor at the Centre for Environmental and Energy Law, Ghent University, Belgium.

How German Judges Decide Environmental Cases Under Public Law

Matthias Keller

Abstract Environmental cases address problems known across the globe: climate change, air pollution, impact assessment and access to justice. Therefore, the German way to decide them under public law may be of interest for foreign lawyers. A proper understanding necessitates a look at the background. This includes German history, the impact of the Basic Law on Administrative law, Europeanisation by EU Directives and Internationalisation by the Aarhus Convention. Other important features are standard rules on interpretation, the court system, the style of legal reasoning and the active role of the German judge in administrative court procedures. A landmark case is the Climate Decision of the German Constitutional Court that establishes a new fundamental right for young people. EU Directives regulate how environmental impact assessments are to be carried out. Court cases in this context raise questions of access to justice. The Trianel judgment of the CJEU is fundamental for German NGO-standing. In the field of air quality it is the Janecek judgment of the CJEU that made air quality plans judicially enforceable. A recent judgment by the CJEU in a not yet finished Dieselgate case affirms the right of an environmental NGO to challenge product-approvals. Against this backdrop the author suggests to reconsider the German nineteenth century legal postulates on the concept of individual rights and to include NGO-rights.

1 Introduction

The goal of this contribution is to provide an insight on how German judges decide environmental cases under public law. English-reading legal practitioners and academics from abroad are addressed. The author is a judge with a German legal background and has been engaged in judicial training on environmental law for judges and prosecutors from all EU Member States. This experience may explain the

M. Keller (✉)
Administrative Court, Verwaltungsgericht Aachen, Aachen, Germany
e-mail: matthias.keller@vg-aachen.nrw.de

G. Antonelli et al. (eds.), *Environmental Law Before the Courts*,
https://doi.org/10.1007/978-3-031-41527-2_9

chosen style and focus. Legal contexts are explained and—like in a guide for foreign jurists—supplemented by information on German history, legal culture and judicial practices. Before this background, cases, controversies and concepts are discussed.

To draw is to omit, as the painter Max Liebermann once said. A special focus is put on the—already famous—Climate Decision of the German Federal Constitutional Court from March 2021. Furthermore, it is shown how the present-day compound of international and supranational law, namely the Aarhus Convention and EU law Directives, play a crucial role in deciding German environmental cases under public law. The environmental impact assessment of projects (EIA), the enforcement of EU air quality standards and access to justice of an NGO challenging a product approval ('Dieselgate') are presented and explained.

Last but not least, the author wants to provoke discussions on the doctrine of NGO standing and argues for an adjustment of the traditional German concept of subjective-public right by including NGO rights.

2 Background

2.1 German History

Do nations with a bad history have a good public law? Obviously, the question is far too general. An application for taking evidence on such a question would be dismissed by any court. However, when it comes to German public law and its culture, one is confronted with the sharp contrast between the past and the present-day legal order. Being aware of this phenomenon is a good door-opener to understand German public law and its underlying narratives.[1] The 'good public law' is shaped by the German constitution from 1949, known as Basic Law (*Grundgesetz—GG*). Its Article 1 provides:

> (1) Human dignity shall be inviolable. To respect and protect it shall be the duty of all state authority.
> (2) The German people therefore acknowledge inviolable and inalienable human rights as the basis of every community, of peace and of justice in the world.
> (3) The following basic rights shall bind the legislature, the executive and the judiciary as directly applicable law.

The powerful provisions of Article 1 of the Basic Law put the individual person and his or her inalienable human rights in the centre of the German legal order. It is a reaction to 'the evil', namely the terrible aberrations of state power that happened from 1933 to 1945 under the German Nazi-regime. Holocaust and World War II lead to the biggest catastrophe in German history and legal culture.

---

[1] Cover (1983) 'Once understood in the context of the narratives that give it meaning, law becomes not merely a system of rules to be observed, but a world in which we live.'

Sadly enough, in one part of post-war Germany totalitarian state power continued to exist. By establishing a Soviet-style communist government and the will to overcome the evils of capitalism, militarism and fascism, the German Democratic Republic—GDR—(*Deutsche Demokratische Republik—DDR*) abandoned the Western ideas of individual freedom, liberty and property and severely oppressed all persons who did not think along the line of the communist party which required strict obedience enforced by the Ministry for State Security (*Staatssicherheit—Stasi*).

After the fall of the Berlin Wall on 9 November 1989 and the following reunification (*Wiedervereinigung*) of Germany, the Basic Law became the Constitution of present-day Germany.

2.2 Constitutionalisation

Under the constitutional provisions of the Basic law from 1949 a new era started for the method of applying German public law: The legal method to construe specific provisions of Administrative law (on public order, social benefits, assemblies, construction permits, urban planning etc.) 'in the light' of fundamental rights. Fritz Werner, President of the Federal Administrative Court from 1958 to 1969 and Law Professor, framed the concept and saw Administrative law as 'concretized constitutional law' (*konkretisiertes Verfassungsrecht*).[2] This highly respected judge and esteemed university lecturer, who did a lot to rebuilt a German Administrative judiciary with independent judges and a mission to protect fundamental rights in all fields of Administrative law, had not always been a decent man. Even before 1933 he joined the Nazi-movement not only as a member of the Nazi-party (NSDAP) but also as member of its paramilitary wing (SA).[3]

Examples of the process of Constitutionalisation[4] can be found in many fields of Administrative law. The most prominent feature is the incorporation of the principle of proportionality as a limit for the exercise of discretion by public administration. Another example is the process of widening judicial review referring to the guarantee of Article 19 (4) of the Basic law and recognizing more and more 'subjective-public rights' (*subjektiv-öffentliche Rechte*)[5] which are enforceable individual rights vis-à-vis the state.

---

[2] Werner (1959).

[3] Bunke (1997), p. 93.

[4] Wahl (2003), p. 411.

[5] Entitlement to social welfare: BVerwG 1, 159.

2.3 Europeanisation

The Europeanisation[6] of German public law has been a result of an increasing European integration. Its main steps are marked by the Treaty of Rome of 1957, the Single European Act of 1986, the Treaty of Maastricht of 1992 and the Treaty of Lisbon of 2007. In the field of environmental law many Directives had to be transposed into national law. Just to name a few: the Birds Directive,[7] the Habitats Directive,[8] the Water Framework Directive,[9] the Industrial Emissions Directive,[10] the Directive on Environmental Impact Assessment.[11]

The EU has a limited number of administrative bodies. Examples are the EU Commission and the EU Agencies. In general, it is the task of the Member States and their administrative bodies to enforce EU law. This reflects the principle of 'indirect administration'[12] within the EU. The Member States are free to establish their own administrative institutions and regulate administrative procedures. They enjoy so-called 'procedural autonomy'. However, when applying EU law the procedural autonomy is limited.

2.4 Internationalisation

The Aarhus Convention and its implementation are an example of the ongoing process of internationalisation[13] with an impact on deciding environmental cases by German courts. The Convention was signed on 25 June 1998 under the aegis of the United Nations Economic Commission for Europe (UNECE). Its full name is Convention on Access to Information, Public Participation in Decision-Making and Access to Justice in Environmental Matters.[14]

---

[6] Wahl (2003), p. 411.

[7] Directive 2009/147/EC of the European Parliament and of the Council of 30 November 2009 on the conservation of wild birds.

[8] Council Directive 92/43/EEC of 21 May 1992 on the conservation of natural habitats and of wild fauna and flora.

[9] Directive 2000/60/EC of the European Parliament and of the Council of 23 October 2000 establishing a framework for Community action in the field of water policy.

[10] Directive 2010/75/EU of the European Parliament and of the Council of 24 November 2010 on industrial emissions (integrated pollution prevention and control).

[11] Directive 2014/52/EU of the European Parliament and of the Council of 16 April 2014 amending Directive 2011/92/EU on the assessment of the effects of certain public and private projects on the environment.

[12] Dubey (2003).

[13] Wahl (2003), p. 411.

[14] The text of the Convention is available at: https://www.unece.org/fileadmin/DAM/env/pp/documents/cep43e.pdf. The Aarhus Implementation Guide is a useful tool for a proper understanding

The Convention legally binds the EU and its Member States. For Member States the legal obligations are twofold. First, they exist under international law being a (direct) Party of the Convention. In Germany, after being submitted to the legislator and ratified, the Convention has the legal status of an ordinary statute, cf. Article 59 (2) of the Basic Law and is to be applied as such. Secondly, they exist under the supranational law of the EU. The European Union has adopted Aarhus rules which then have become an integral part of EU law. This 'Aarhus-influenced' EU law is mostly to be applied by the Member States that—on their turn—must guarantee the EU standards (principles of equivalence and effectiveness[15]) and the compliance with the case law established by the judgments of the CJEU.

The Aarhus Convention and its three pillars (information, participation and access to justice) are a new approach of international law to address environmental matters. This approach can be traced back to Principle 10 of the Rio Declaration[16] that was adopted already in 1992 at the United Nations Conference on Environment and Development. It states:

> Environmental issues are best handled with the participation of all citizens concerned, at the relevant level. (…)
>
> Effective access to judicial and administrative proceedings, including redress and remedy, shall be provided.

It is worthwhile to mention that the Aarhus Convention implemented a so-called Compliance Committee.[17] This quasi judicial body considers complaints ('communications') from the public relating to the compliance of the Parties to the Convention. The Committee has been quite successful to remind the Parties of the Convention (among them the EU[18]) of their obligations under international law. In the course of time, traditional restrictions to access to justice in environmental matters were set aside. Germany, to give an example, has established access to justice for NGOs in environmental matters that require public participation and/or an environmental impact assessment.[19]

For many Member States and the EU itself the approach of the Aarhus Convention marked nothing less than a 'legal turn' as to their concept of rights to information, participation and legal review in environmental matters. The Aarhus Convention tackles deficiencies in the enforcement of existing environmental law by making the civil society, namely affected persons and NGOs (*the public concerned*),[20] a

---

and available at: http://www.unece.org/fileadmin/DAM/env/pp/Publications/Aarhus_Implementation_Guide_interactive_eng.pdf.

[15] See Case 33/76 Rewe v Landwirtschaftskammer für das Saarland [1976] ECR 1989, para. 5.

[16] Available at: https://un.org/en/development/desa/population/migration/generalassembly/docs/globalcompact/A_CONF.151_26_Vol.I_Declaration.pdf; accessed 2 February 2023.

[17] More information available at: https://unece.org/environment-policy/public-participation/aarhus-convention/compliance-committee; accessed 2 February 2023.

[18] See Regulation (EU) 2021/1767—adopted by the European Union on 6 October 2021—introducing amendments to the Aarhus Regulation (EC) No 1367/2006.

[19] Under the German Environmental Appeals Act (Umwelt-Rechtsbehelfsgesetz—UmwRG).

[20] Article 2 (5) Aarhus Convention.

watchdog for the proper implementation of environmental law. Rights (*to information and participation*)[21] and remedies (*access to justice*)[22] are guaranteed in order to make the watchdog 'bark and bite' in the field of environmental law.

2.5 Interpretation

German judges conceive it as a self-evident part of their legal professionalism to decide cases and controversies by exploring the 'true meaning' of the relevant legal provisions, even if they appear in quite specific and technical contexts. Deference[23] to the fact-finding and/or the opinion of the deciding administrative authority is not totally unknown, but considered as a seldom exception which has to be thoroughly justified being an interference to the constitutional guarantee of effective legal protection.[24]

Judges may go beyond the literal meaning of a single word, a specific sentence or paragraph in a legal provision. Under Article 20 (3) of the Basic law, judges shall be bound 'by law <u>and</u> justice' [emphasis by the author]. Methods of interpretation are meant to avoid personal preferences and arbitrariness.

The standard rules of interpretation are a legacy of the nineteenth century. An important century in which Germany failed politically (1848), but gained a reputation of excellency in many fields of science and culture. Friedrich Carl von Savigny[25] (1779–1861), an influential German jurist and scholar of Roman law, established the historical school of jurisprudence and is considered to be the founder of 'legal science' (*Rechtswissenschaft*) in Germany. His brilliant work, especially his 'Treatise on Possession' (*Das Recht des Besitzes*) combined historical and systematic as well as doctrinal and practical legal thinking in a unique way. For him, systematic and historic legal thinking and the 'spirit of the people' (*Volksgeist*) were essential elements of law. Jurists with a deep and solid academic education (also known as doctors and professors of law), but not politicians (let alone pastors, priests or philosophers) are in the centre of the legal system and have to unfold the 'sense' behind the legal texts (*Rechtsgedanke*). If the author is not mistaken, neither Savigny's 'spirit of the people' (*Volksgeist)* nor Hegel's 'world spirit' (*Weltgeist*) have made it into colloquial English. However, the word *zeitgeist* is understood and used from time to time. The *zeitgeist* during the nineteenth century in Germany embraced the rule of law, but not the democratic State,[26] a political ideal of the West

[21] Articles 4–8 Aarhus Convention.

[22] Article 9 Aarhus Convention.

[23] US example: US Chevron U.S.A., Inc. v. Natural Resources Defense Council, 467 U.S. 837 (1984).

[24] Article 19 (4) of the Basic Law.

[25] Savigny is presented by J. Rückert (2001), p. 555; from a French perspective: Champeil-Desplats (2014), p. 82.

[26] The failure of the 1848 revolution has been interpreted by some historians as a crossroads of history that directed Germany on a 'special path' (Sonderweg), paved with scientific, technical and

that was so brilliantly expressed by the words of Abraham Lincoln in his Gettysburg Address[27] (1863): 'Government of the people, by the people, for the people'.

The four rules on interpretation, established by Savigny, are still a widely accepted method to explore the meaning of legal texts: A grammatical (literal), systematic (context-related), historical and teleological method of interpretation. The four rules have become canons of legal interpretation and have stood the test of time. Today, they belong to the basics in legal methodology that every examined practitioner of law is supposed to know.

Rudolf von Jhering[28] (1818–1892), another German genius in legal scholarship, criticized a pure legal approach to the understanding of law as too 'formalistic'. He fought for a new method that takes the conflicts of human societies and their solutions into account. The sociological essence of his jurisprudence is well reflected in the title of his best known book 'The Struggle for Law'[29] (*Der Kampf ums Recht*). His work can be seen as a step towards present-day methods of interpretation.

Today, a sound legal interpretation gives Savigny's canons a new focus. Jurists have to explore the 'underlying values of the law' and prioritise them in the same way as the legal system does. The concept is called 'value-oriented jurisprudence' (*Wertungsjurisprudenz*).[30] Therefore, under the case law of the Federal Constitutional Court[31] the Basic law and its fundamental rights and freedoms represent not only individual rights, but also a comprehensive 'system of values'[32] with human dignity on top. Under this constitutional doctrine the German legal order in its entirety has to be interpreted 'in the light of the Basic law'. A transformation that has not been praised in every field of law and caused, especially in contract law, some uncertainties.[33]

However, all in all, it can be seen as a success for the rule of law in Germany. Judges of all jurisdictions and all ranks have gained an awareness of the constitutional values at stake when they interpret specific legal provisions in one way or the other.

---

economic excellence and political backwardness, see Chapoutot (2014), p. 39.

[27] 'Nicolay Copy' of the Gettysburg Address, 1863, available at: www.loc.gov/exhibits/gettysburg-address/exhibitionitems/Assets/Nicolay2_enlarge.jpg; accessed 2 February 2023.

[28] Jhering presented by Falk (2001a), p. 334.

[29] Available in English at: https://archive.org/details/struggleforlaw00jher/page/8/mode/2up; accessed 2 February 2023.

[30] On the German narrative of an evolution from 'Begriffs-, Interessen-' to 'Wertungsjurisprudenz': Meder (2021), p. 449.

[31] All cases are cited in the original German way 'BVerfGE'. This acronym literally means 'Federal Constitutional Court Decisions' followed by volume and page number where they were published in the official collection of cases.

[32] BVerfGE 7, 198 (Lüth) on the 'objective function' of fundamental rights embodying an 'objective value system' that applies to all areas of law.

[33] For an in-depth analysis: Ruffert (2012).

2.6 Court System

Apart from the Federal Constitutional Court (*Bundesverfassungsgericht*) and State Constitutional Courts, the court system in Germany is quite different from that of other federations, such as the United States of America. Generally, Federal Courts (*Bundesgerichte*) are—as Courts of Cassation—at the top of the court hierarchy and assure the uniform application of federal law by the lower courts. Lower courts are State Courts of first or second instance. The German judiciary has five branches: the ordinary courts having jurisdiction in civil and criminal matters and four court systems with specialized jurisdiction—in labor, social security, tax and administrative matters. As a general rule, environmental cases under public law are administrative matters and therefore decided on the state level by Administrative Courts (*Verwaltungsgerichte*) at first instance, by Higher Administrative Courts (*Oberverwaltungsgerichte/Verwaltungsgerichtshöfe*) at second instance and last but not least on the federal level by the Federal Administrative Court (*Bundesverwaltungsgericht*) at third instance. It should be mentioned that the German legislator does not hesitate to change this well proven court structure when attempting to 'accelerate' legal proceedings concerning important infrastructural projects. Consequently, cases concerning the development consent on important infrastructure projects like motorways are to be decided by the Federal Administrative Court as court of first and last instance.[34] Controversies against bigger installations like windfarms are to be decided by Higher Administrative Courts as courts of first instance.[35] The complexity of the allocation of cases to different courts does not harm an effective access to justice. Bringing an action 'to the wrong court' is without major consequences under the German law of court procedure. It is up to the respective judge to determine the competent court and take a decision of referral (*Verweisungsbeschluss*)[36] which is, as a rule, binding for the court that receives it. Decisions of referral are possible between all German courts.

2.7 Style

Decisions of German courts are written in a particular style. Like in France, a 'judicial institution' speaks to the parties. Therefore, the style is impartial and objective. The German judge never speaks in the first person. Not even the judge-rapporteur, who prepares the case by drafting an opinion, is allowed to do so. Concurring or dissenting opinions are only permitted at the Federal Constitutional Court. The courts have to provide reasons for their decisions. German judgments begin with a

[34] Cf. Section 50 Code of Administrative Court Procedure (VwGO).

[35] Concerning windfarms cf. Section 48 (1) No 3a Code of Administrative Court Procedure (VwGO).

[36] Section 83 Satz 1 Code of Administrative Court Procedure (VwGO) in conjunction with Section 17a (2) Courts Constitution Act (GVG).

preamble that shows the parties, the object matter and the deciding judges (*Rubrum*), followed by the operative part (*Tenor*), the facts of the case (*Tatbestand*), the grounds of the decision (*Entscheidungsgründe*) and, last but not least, the instructions on the rights to appeal (*Rechtsmittelbelehrung*).

The grounds of the decision (Entscheidungsgründe) are written in a standardized style which is quite difficult to master properly. The basic idea of a judgment is easy to understand: the decision puts or—more technically—subsumes a concrete case under a general rule and draws a conclusion.

The specific German style of writing judgments (*Urteilsstil*) always starts with the case related conclusion (e.g.: 'The appeal is inadmissible'). Then the legal reasoning introduces the general rule applied (e.g.: 'Appeals have to be brought within 2 months.'). It continues with a cascade of legal requirements to present a complete picture of the logic structure of the judgment which is meant to be a logic 'deduction' from pertinent law. (e.g.: 'The time limit of 2 months begins with the notification of the appealed decision.') In this way legal rules are concretized and approximated to the facts of the case. (e.g.: 'The time limit of 2 months ends on the expiry of the day of the last month which, in its designation or its number, corresponds to the day on which the decision was notified'). The categorical gap between facts ('is') and law ('ought') can be bridged. (e.g.: 'Here, the decision was notified on 1 April 2022, the time limit for bringing an appeal ended 2 month later on 1 June 2022. The claimant's appeal that reached the court on 1 July 2022 was too late and inadmissible'). Deviations from this neutral and, admittedly, a bit dry style are frowned upon. If pertinent provisions are not referred to, the logic 'deduction' will be seen as incomplete and a judgment in an important matter will be criticised by the legal community in forums, law reviews etc., even though none of the parties pleaded the provision that was omitted by the judge.

2.8 Role of the Judge

The role of the German judge in court procedure can be described as 'active'. Deciding (environmental) cases German judges have to take into account all (!) legal requirements, irrespective of whether they are a part of national, international or EU law. As a rule, German courts apply legislation whether or not it was pleaded by one of the parties. The court has to consider the pertinent law and its principles ex officio, it follows the maxim of 'the court knows the law' (iura novit curia).

For a lawyer with a common law background it may be a surprise that the German Administrative Courts (not Civil Courts) have the power to gather all relevant facts on their own motion. The so called inquisitory principle[37] (*Inquisitionsmaxime* or *Amtsermittlungsgrundsatz*) might appear quite unattractive or even unlawful to lawyers who see the adversarial principle as a safeguard of due process and the rule of

[37] Section 86 (1) Code of Administrative Court Procedure (VwGO).

law. However, in deciding environmental cases, the inquisitory principle, applied to the gathering of the relevant facts, has important benefits.

First of all, it does not prevent the parties from putting forward facts and filing an application to the court to take evidence just like in a civil law procedure. Secondly, it is a crucial tool to protect public interests like air quality, water quality, protection of flora and fauna, noise control etc. Defendants, mostly the public agency that issued the challenged decision, and their legal representatives want to 'win the case' and avoid to pay the costs of the adversarial attorney. Therefore, the agency may have no interest in giving notice that relevant facts were not explored. The claimant, in construction law typically a person with private (property-) interests who wants to realize a project as quickly as possible, has no interest in raising unnoticed environmental aspects that could make the realisation of the project even more complicated. If neighbours are involved, they may not have a sufficient expertise to detect the facts missing, even if they are represented by an attorney. Thirdly, in daily practice, the inquisitory principle results in a quite trivial, though important court order for the functioning of the German administrative justice system: At an early stage of the court procedure the court requests the public agency to produce all the relevant administrative files.[38] From then on the files of the preceding administrative procedure are part of the files of the court case. Consequently, they can be inspected by the attorney of the claimant. The judge will prepare the case by reading every detail of the file concerning the preceding administrative procedure which gives him or her a solid factual basis for conducting the audience and writing the judgment.

3 Cases, Controversies and Concepts

3.1 Climate Decision

3.1.1 Introduction

In its 'epoch-making' Climate Decision[39] from March 2021 the Federal Constitutional Court[40] struck down parts of Germany's Federal Climate Protection Act (*Bundesklimaschutzgesetz*). The act aimed at Germany's climate neutrality in 2050, following the Paris Climate Agreement and EU law to limit warming to well below 2 °C and, if possible, to 1.5 °C. Its original provisions provided a target of reducing green house gas emissions by 55% by 2030. However, for the years onwards there was no specific regulation on emission cuts, only an empowerment for the government to enact reduction levels by an executive order.

---

[38] Section 99 (1) Code of Administrative Court Procedure (VwGO).

[39] Order of the First Senate of the Court (Beschluss des Ersten Senats), 24 March 2021, 1 BvR 2656/ 18 (paras 1–270), published on 29 April 2021.

[40] References are made to the (shorter) English version of the Climate Decision; available at https://www.bundesverfassungsgericht.de/e/rs20210324_1bvr265618en.html.

The Constitutional Court found a violation of fundamental rights of the claimants and ordered the German parliament to be more specific on how to achieve the climate protection targets from 2031 onwards. The German parliament reacted quickly. Already in June 2021 it passed an amendment to the Federal Climate Protection Act that provided—inter alia—climate neutrality by 2045.

The complainants were a group of youth, among them the German climate activist Luisa Neubauer. Therefore, the Anglo-American or European name of the case would be 'Neubauer, et al. v. Germany'. In Germany, cases are not cited by the names of the litigants, but by the place and page of publication. Leading cases get their names by their characteristic feature, like 'Climate Decision' (*Klimabeschluss*).

3.1.2 The *Leitmotiv*

The *leitmotif*[41] of the unanimous court ruling, a not easy-to-read composition of 270 paragraphs, can be clearly heard in paragraph 192:

> One generation must not be allowed to consume large portions of the CO2 budget while bearing a relatively minor share of the reduction effort, if this would involve leaving subsequent generations with a drastic reduction burden and expose their lives to serious losses of freedom (…).
>
> (…) since the current provisions on allowed emission amounts have now already established a path to future burdens on freedom, the impacts on future freedom must be proportionate to the standpoint of today – while it is still possible to change course.

Obviously, the Constitutional Court's concern is about a fair burden sharing. To this aim, the eight justices of the first chamber (*Senat*) take a close look at the span of time between the present day and the year of envisioned carbon-neutrality in German (then 2050). The Court expects serious 'losses of freedom' caused by legislation that has to provide the necessary cuts of carbon dioxide emissions and raises the question of intergenerational equity.

3.1.3 Article 20a of the Basic Law

The legal reasoning turns inter alia to Article 20a of the Basic Law that reads:

> Mindful also of its responsibility towards future generations, the state shall protect the natural foundations of life and animals by legislation and, in accordance with law and justice, by executive and judicial action, all within the framework of the constitutional order.

The constitutional provision explicitly names the protection of natural foundations of life and the responsibility towards future generations. The Court derives many legal arguments from it and even combines it with the Paris Climate Agreement. However, it had to decide on constitutional complaints and therefore on individual rights and freedoms put forward by the complainants (liberty, life, property, cf.

[41] A closer analysis is provided by Minnerop (2022).

Article 2 (1) and (2) and Article 14 of the Basic Law). At this point the justices were confronted with a crucial problem of the climate case.

Article 20a goes back to a constitutional amendment of reunified Germany in the year 1994. It was not designed to impose specific and enforceable obligations vis-à-vis the state. The provision is just an 'aim to be pursued by the state' *(Staatszielbestimmung)*.[42] The assignment to this specific category of German constitutional doctrine makes it impossible for individuals to build a court case solely on the two constitutional requirements of Article 20a of the Basic Law. Especially, it does not grant a 'right to a healthy environment'. Without a right, there is no remedy. The German Act on the Federal Constitutional Court is very clear on that issue. A constitutional complaint (*Verfassungsbeschwerde*) may be lodged by persons claiming that their fundamental rights have been violated by an act of German public authority.[43] Therefore, the group of youth that brought the constitutional complaint to challenge the Federal Climate Protection Act, took a big risk of losing their case on the simple ground that they could not invoke any fundamental right.

3.1.4 Groundbraking: New Fundamental Rights for Young People!

In this situation the German Constitutional Court opened a new chapter in the understanding of fundamental rights and ruled in favour of the young complainants.

The new approach makes the court ruling groundbraking. Especially in environmental law, statutes and regulations change even more quickly than the climate they have to protect. The legislator appears to be 'motorised'. In times of freedom and stability, major changes in legal doctrine are quite rare. However, the Constitutional Court is able to discover an 'additional dimension' of fundamental rights of the young generation.[44] The legal focus is put on something that is essential, priceless, though easily overlooked and often running out: time!

It is fair to say that the Court did nothing less than to establish a new fundamental right for young people and their future freedoms. The *telos* of this right is easy to understand: It is meant to secure fundamental freedoms for the time span till climate neutrality (then 2050, now 2045). During the decades to come, young people's' enjoyment of fundamental rights is endangered because of its dependency on the coherence, the time table/amount of carbon cuts provided by the Climate Protection Acts.

---

[42] Härtel (2020).

[43] Section 90 (1) Act on the Federal Constitutional Court.

[44] BVerfG (note 41) para. 182 seq.

3.1.5 Summary of the Court's Findings

The Court's 'Treatise on Climate Protection and Generational Equity' and its postulates are a bit difficult to digest. The author takes the liberty to select 10 findings of the Court and summarize them in his own words:

1. Safeguarding fundamental freedoms over time (*intertemporale Freiheitssicherung*) requires that the legally allowed enjoyment of freedoms by one generation must not lead to unreasonable restrictions for a future generation.[45]
2. Fundamental rights to classical freedoms (life, liberty property) are guaranteed over time which is an additional dimension of these freedoms.[46]
3. Climate Protection Acts with their necessary emission cuts must be carried out in such a way that the resulting burden for the enjoyment of fundamental rights is distributed fairly among the generations. This requirement has a solely objective dimension.[47]
4. The Paris Climate Agreement is to be qualified as a concretisation of Article 20a of the Basic Law on the legal level of statutory law. This follows from the clear commitments to the Agreement made by the German legislator.[48]
5. Regulatory measures to reduce carbon dioxide emissions pursue a legitimate goal established by Article 20a of the Basic Law.[49]
6. Article 20a of the Basic Law does not grant (green) access to justice for NGOs.[50]
7. Article 20a of the Basic Law establishes an (objective) legal requirements in the field of climate protection that may justify legislative restrictions of fundamental rights.[51]
8. The Federal Climate Act, challenged in the case, does not contain any provisions for a carbon reduction path for the period after 2030 in the field of climate protection.[52]
9. Without a carbon reduction path from 2030 onwards, the challenged Federal Climate Act is unconstitutional on the ground that—inter alia—intergenerational equity, as required by Article 20a of the Basic Law, is not secured.[53]
10. The breach of (objective) constitutional requirements under Article 20a Basic Law violates as well (subjective) fundamental rights to life, liberty, and property on the ground that these classical freedoms are guaranteed over time (see above No 1 and 2).[54]

3.1.6 Groundbraking, but also Convincing?

Apart from being groundbraking, is the Climate Decision convincing? Yes and no. The legal reasoning is solid and strong when it explains that the commonly accepted way to climate neutrality by carbon cuts may have an impact on the individual opportunities to enjoy fundamental freedoms over time. Even if the fair burden

[45] BVerfG (note 41) para. 182 et seq.

[46] BVerfG (note 41) para. 183.

[47] BVerfG (note 41) para. 146.

[48] BVerfG (note 41) para. 210.

[49] BVerfG (note 41) para. 205 et seq.

[50] BVerfG (note 41) para. 136.

[51] BVerfG (note 41) para. 197.

[52] BVerfG (note 41) para. 14.

[53] BVerfG (note 41) para. 183 last sentence.

[54] BVerfG (note 41) paras. 189,190.

sharing is a political question, it can become, as the Climate Decision brilliantly elaborates, a case for courts that are called upon to ensure 'freedoms over time' vis-à-vis a manifestly insufficient approach of a legislator towards climate neutrality.

Furthermore, it would be unfair to criticise the Court for using a very academic style in its writing. With a bit of exaggeration it can be said that a groundbreaking German court decision is supposed to be written like the first or second book (*Dissertation/Habilitation*) of a young scholar on his or her long and winding road to a Chair of Public Law. Without comprehensive references to relevant case law and (German) legal scholarship and without this—difficult to read—effort to clarify and establish 'legal concepts' (*Rechtsbegriffe*), a court's legal reasoning with a calling to answer fundamental questions would immediately be criticised as 'not scientific' (*unwissenschaftlich*) and hence not accepted by the German legal community, regardless of its outcome.

The criticism of the author concerns the question on how the Court applied the well established constitutional-law doctrines of 'interference' and 'duty to protect' to the case. The Court comes to the conclusion that the 'not strict enough' [words of the author] Climate Protection Act establishes an interference with the classical freedoms of life, liberty, and property which is not sufficiently 'justified' and therefore constitutes a violation of fundamental rights. This approach, hidden behind the language of legal scholarship, does not even stand a common sense test. Low emission cuts cannot count as a case of state intrusion into life, liberty, and property. They let things go and give even more room to a present-day enjoyment of fundamental rights that is linked with carbon emissions. The real problem is the shortsightedness of such a regulatory approach. Therefore, the relevant question is, whether there is a constitutional duty under which the legislator has to 'protect' young people against such emission cuts that will have to compensate today's too low carbon cuts by too harsh ones in the future in order to reach climate neutrality in time. The Court writes pages to elaborate on the 'duty to protect'[55] in all academic subtlety, not forgetting the important contributions of the European Court of Human Rights.[56] However, the Court lacks the courage to reduce the leeway of the legislator and establish a duty to protect in the given case. That is regrettable, even more so as the Court's approach of 'safeguarding freedoms over time' (*intertemporale Freiheitssicherung*) perfectly fits into the legal context of a 'duty to protect'.

Instead, the Court finds 'special effects' [words of the author] of the challenged Climate Protection Act to establish an interference. The Act challenged is considered to have an *effect* (*Wirkung*) that is interference-like (*eingriffsähnlich*). To avoid a major raising of eyebrows in the legal community the Court adds that its *interference-like effect* has also a *pre-effect* (*Vorwirkung*) that affects young people and their freedoms well in advance and therefore already in the present. The Court's

[55] BVerfG (note 41) para. 143 et seq.

[56] BVerfG (note 41) para. 147.

complete composite goes by the difficult name of 'advance interference-like effect' (*eingriffsähnliche Vorwirkung*).[57]

This new doctrine of the Court is unnecessary, artificial, and dangerous. The clear constitutional-law distinction between state interference on the one side and a required state action on the other side, being of major importance for the protection of fundamental rights, gets blurred all of a sudden. Citizens are conferred a fundamental right that, in the case of judicial enforcement, results in a stricter present-day legislation. This is not how fundamental rights against the state are supposed to work. Why this confusing new doctrine? A possible answer could lie in the Court's own position. Using the 'advance interference-like effect' it can reach the legal category of a classical 'state interference'. As a result, parliamentary legislation on climate protection must—inter alia—stand a strict proportionality test which, of course, widens the Court's own scope of the judicial review significantly.

All in all, the Climate Decision, translated by the Court into English,[58] French[59] and Spanish[60] and hence meant to be the German voice in a worldwide discussion of climate litigation, is certainly a step forward for constitutional law and fundamental rights of young people in times of climate change. However, it will be interesting to see whether the new legal concepts made in Germany will be followed, rejected or ignored by the international legal community.

3.2 Environmental Impact Assessment and Access to Justice

3.2.1 Introduction

The Environmental Impact Assessment (in the following: EIA) is an example on how a concept for sound decision-making in environmental matters came from the US[61] to the EU and then was shaped 'European-style' by a lot of Directives addressed to the Member States and its institutions, among them Germany and its judiciary.

The refreshingly provocative US question, whether Administrative law as such is 'unlawful',[62] has to the knowledge of the author no German equivalent. However, discomfort with a whole body of law is sometimes expressed concerning EU

[57] BVerfG (note 41) para. 183 et seq.

[58] Available at: https://www.bundesverfassungsgericht.de/e/rs20210324_1bvr265618en.html; accessed 2 February 2023.

[59] Available at: https://www.bundesverfassungsgericht.de/SharedDocs/Entscheidungen/FR/2021/03/rs20210324_1bvr265618fr.html; accessed 2 February 2023.

[60] Available at: https://www.bundesverfassungsgericht.de/SharedDocs/Entscheidungen/ES/2021/03/rs20210324_1bvr265618es.html; accessed 2 February 2023.

[61] As early as in 1969 the USA introduced EIA laws and procedures under the National Environmental Policy Act (NEPA) which became effective on January 1, 1970; available at https://en.wikipedia.org/wiki/National_Environmental_Policy_Act; accessed 2 February 2023.

[62] Hamburger (2014).

Directives. This is due to their character as specific normative acts that have to be transposed into domestic law. Therefore, they are 'hidden' and follow a regulatory approach that appears to the citizens as legal bits and pieces, not properly named, but just numbered and impossible to remember. The development history of the EIA regulatory framework may show the problem.

3.2.2 History of the EIA

In 1985 the European Union, at the time still the European Economic Community (EEC), approved a first Directive on EIA[63] that covered only 'projects' (not 'plans and programmes'). In 1997 Directive 97/11/EC[64] brought the Directive in line with the United Nations Economic Commission for Europe (UNECE) Espoo Convention on EIA in a Transboundary Context. The Directive of 1997 widened the scope of the EIA Directive by increasing the types of projects covered, and the number of projects requiring mandatory environmental impact assessment (Annex I). It also provided for new screening arrangements, including new screening criteria (at Annex III) for Annex II projects, and established minimum information requirements.[65]

A new step was reached 1998: The UNECE Convention on Access to Information, Public Participation in Decision-Making and Access to Justice in Environmental Matters[66] (Aarhus Convention) of 25 June 1998 to which the EU and all Member States are parties was signed. The Directive 2001/42/EC[67] (SEA Directive) filled a gap by regulating the (Strategic) Environmental Assessment of 'plans and programmes'. Directive 2003/35/EC[68] sought to align the provisions on public participation with the Aarhus Convention. Directive 2009/31/EC[69] amended the Annexes I

[63] Council Directive 85/337/EEC of 27 June 1985 on the assessment of the effects of certain public and private projects on the environment, OJ L 175, 05/07/1985, p. 40.

[64] Council Directive 97/11/EC of 3 March 1997 amending Directive 85/337/EEC on the assessment of the effects of certain public and private projects on the environment, OJ L 073, 14/03/1997, p. 5.

[65] This summary is taken from: http://ec.europa.eu/environment/eia/eia-legalcontext.htm; accessed 2 February 2023.

[66] The text of the Convention is available at: https://www.unece.org/fileadmin/DAM/env/pp/documents/cep43e.pdf. The Aarhus Implementation Guide is a useful tool for a proper understanding and available at: http://www.unece.org/fileadmin/DAM/env/pp/Publications/Aarhus_Implementation_Guide_interactive_eng.pdf; accessed 2 February 2023.

[67] Directive 2001/42/EC of the European Parliament and of the Council of 27 June 2001 on the assessment of the effects of certain plans and programmes on the environment, OJ L 197, 21.07.2001, p. 30.

[68] Directive 2003/35/EC of the European Parliament and of the Council providing for public participation in respect of the drawing up of certain plans and programmes relating to the environment and amending with regard to public participation and access to justice Council Directives 85/337/EEC and 96/61/EC, OJ L 156, 25.6.2003, p. 17.

[69] Directive 2009/31/EC of the European Parliament and of the Council of 23 April 2009 on the geological storage of carbon dioxide and amending Council Directive 85/337/EEC, European Parliament and Council Directives 2000/60/EC, 2001/80/EC, 2004/35/EC, 2006/12/EC, 2008/1/EC and Regulation (EC) No 1013/2006, OJ L 140, 5.6.2009, p. 114.

and II of the EIA Directive, by adding projects related to the transport, capture and storage of carbon dioxide (CO_2). Directive 2011/92/EU[70] codified the initial EIA Directive of 1985 an its three amendments. In order to improve the EIA's effectiveness and efficiency, Directive 2011/92/EU was amended by Directive 2014/52/EU.[71] This legislation reflects the current legal situation. There is no official consolidated version of the EIA Directive. However, there is an informal one[72] by the EU Commission.

3.2.3 Definition

Against the backdrop of detailed and—meanwhile transposed—EU Directives, the German EIA operates basically the same way as in other Member States.

A legal definition of the EIA process being the core of the EIA procedure is to be found in Article 1(2)g) of Directive 2011/92/EU as amended by Directive 2014/52/EU:

'Environmental impact assessment means a process consisting of:

> (i) the preparation of an environmental impact assessment report by the developer (…);
>
> (ii) the carrying out of consultations (…);
>
> (iii) the examination by the competent authority of the information presented in the environmental impact assessment report and any supplementary information provided, (…) and any relevant information received through the consultations (…);
>
> (iv) the reasoned conclusion by the competent authority on the significant effects of the project on the environment, taking into account the results of the examination referred to in point (iii) and, where appropriate, its own supplementary examination; and
>
> (v) the integration of the competent authority's reasoned conclusion into any of the decisions referred to in Article 8a.' (e.g. by the author: development consent).'

Summing up in one sentence: the objective of the EIA process is to identify, describe, and assess environmental impacts.

3.2.4 Access to Justice

Under the Aarhus Convention and the accordingly adjusted EU law it is common ground that decisions, acts or omissions that are subject to a public participation provisions are reviewable in a court of law. That includes EIA-requirements.

---

[70] Directive 2011/92/EU of the European Parliament and of the Council of 13 December 2011 on the assessment of the effects of certain public and private projects on the environment, OJ L 26, 28.1.2012, p. 1.

[71] Directive 2014/52/EU of the European Parliament and of the Council of 16 April 2014 amending Directive 2011/92/EU on the assessment of the effects of certain public and private projects on the environment, OJ L 124, 25.4.2014, p. 1.

[72] Available at: http://ec.europa.eu/environment/eia/pdf/EIA_Directive_informal.pdf; accessed 2 February 2023.

Access to justice for the public concerned is a fundamental pillar of the concept established by Article 9 of the Aarhus Convention where court proceedings are a means to reach a better enforcement of environmental law. The point of reference for activities that are subject to public participation is Article 9(2) in conjunction with Article 6 of the Aarhus Convention. This guarantee of access to justice has been adopted by Article 11 EIA Directive, other EU Directives[73] and laws of the Member States. Only here, in the legal context of public participation requirements, the EU has reached a harmonisation by pieces of EU secondary law.

German courts have to apply general principals of EU law when—for whatever reason—confronted with a denial of access to justice on the domestic level. Article 19(1) of the Treaty on European Union (TEU) requires that 'Member States shall provide remedies sufficient to ensure effective legal protection in the fields covered by Union law'. In addition, Article 47 of the Charter of Fundamental Rights enshrines in its first paragraph the right to an effective remedy, providing that 'Everyone whose rights and freedoms guaranteed by the law of the Union are violated has the right to an effective remedy before a tribunal in compliance with the conditions laid down in this Article'.[74] This Article corresponds to Article 6 and Article 13 of the European Convention on Human Rights (ECHR) which enshrine, respectively, the right to a fair trial and the right to an effective remedy.

3.2.5 Traditions of Administrative Justice

A 'green actio popularis' is not required, neither by the Aarhus Convention nor by EU law. Article 11(1) of the EIA Directive allows Member States to avoid an actio popularis in environmental matters and to confine the standing of claimants to those who can demonstrate

> a sufficient interest, or alternatively, the impairment of a right.

The two possible restrictions on standing take into account different traditions of administrative justice.[75] Especially in Germany and other countries of Central Europe (eg.: Austria and the Czech Republic) the concept of administrative justice as a redress for the impairment of subjective (public) rights ('recours subjectif') has a long history,[76] only disrupted by totalitarian legal thought.[77] Whereas other

---

[73] Article 25 of the IED Directive, 2010/75/EU; Article 6 (2) of the Access to Environmental Information Directive, 2003/4/EC; Article 13 of the Environmental Liability Directive, 2004/35/EC; Article 23 of the Seveso III Directive, 2012/18/EU.

[74] Article 47 of the Charter of Fundamental Rights corresponds to Article 6 and Article 13 of the European Convention on Human Rights (ECHR) which enshrine, respectively, the right to a fair trial and the right to an effective remedy.

[75] For detailed comparative studies on administrative justice confer: Darpö (2013); Eliantonio et al. (2012); Seerden (2018).

[76] Thorson (2016), p. 111.

[77] Gutmann (2018).

Member States follow more or less the traditional French system[78] where a qualified interest[79] is sufficient to start judicial proceedings and invoke public law ('recours objectif').

Article 11 (3) EIA Directive gives a very clear—but sometimes not very well accepted[80]—answer by following the wording of Article 9 (2) Aarhus Convention:

> What constitutes a sufficient interest and impairment of a right shall be determined by the Member States, consistently with the objective of giving the public concerned wide access to justice.

A German judge, part of a Member State judiciary that belongs to a tradition of 'recours subjectif' has—as a first step—to determine the meaning of what it calls 'impairment of a right' in the national context. The concept of rights has to be revisited and, if necessary, modified in a way to make it consistent with the EIA Directive ('consistent interpretation'). The result to be achieved under the EIA Directive (Article 288 para. 3 TFEU) can be described as follows: Access to justice in cases of non-compliance with the EIA Directive.

3.2.6 Public Interest Litigation (Trianel)

Public interest litigation must be made possible for the public concerned. The possibility to defend private rights (like health or property) linked with an EIA provision is not enough. Admittedly, it is not easy to take this step without new legislation, just by interpretation. Especially in Germany, litigation on the ground of subjective rights is deeply rooted in the legal culture of today as a means of effective protection against State power and to legitimize the (non elected) members of judiciary when they interfere in the activities of the administrative or legislative branch.

If the right to invoke EIA laws is recognized within the domestic legal order, it is clear under Article 11(3) EIA Directive that NGOs shall be holders of those rights as well:

> Such organisations (scil.: NGOs) shall also be deemed to have rights capable of being impaired for the purpose'

Accordingly, the EU Court of Justice (CJEU) ruled in a case, widely known as Trianel case,[81] brought by a German Higher Administrative Court for the State of North Rhine-Westphalia, that the standing of NGOs cannot be limited to private interest litigation. To give a reason it is enough to refer to the golden words of the former Advocate General Sharpston:

---

[78] On the nineteenth century background of the Franco-German divide on administrative justice: Keller (2016).

[79] The qualified interest may be more 'narrow" (e.g.: Hungary) or 'wide" (e.g.: Belgium, England/Wales, France and Sweden) cf. Eliantonio et al. (2012), p. 67.

[80] See the 'Trianel-Case': C-115/09 *Bund für Umwelt und Naturschutz*, EU:C:2011:289.

[81] C-115/09 *Bund für Umwelt und Naturschutz*, EU:C:2011:289.

the fish cannot go to court.

In other words, the environment cannot protect itself if it is threatened or harmed. It is a public good and should be supported by public voice.[82]

The Trianel case shows in a paradigmatic way the above mentioned Europeanisation of the German administrative justice system. Well-established case law[83] of the CJEU accepts the 'procedural autonomy' of the Member States and its limits. National procedural conditions must not

- be less favourable than those relating to domestic actions (principle of equivalence),
- make it impossible in practice to exercise those rights (principle of effectiveness).

While the principle of equivalence is easy to apply and almost goes without saying, the principle of effectiveness requires access to justice and does not allow to let 'EIA rights' without judicial protection. Otherwise domestic law would not be in compliance with the principle of effective judicial protection set out in the EU Treaties and with the requirements enshrined in Article 47 of the Charter of Fundamental Rights of the European Union.

3.3 Air Quality

3.3.1 The Janecek Case

The facts of the most prominent German/EU case on air quality are as follows: Mr. Janecek lived on the Landshuter Allee on Munich's central ring road. The limit value fixed for emissions of particulate matter (PM 10) was exceeded much more than 35 times a year. He brought an action for an order requiring the public entity to draw up a plan for better air quality in order to ensure compliance with the maximum number of instances permitted. The German Federal Administrative Court (Bundesverwaltungsgericht) asked the Court of Justice whether an individual person was entitled to require the national authority to draw up a plan for better air quality.

The ruling: The CJEU answered in the affirmative. It held that the obligation of a public entity to establish or improve the air quality planning in case of exceedances of limit values can be invoked by a person concerned. For the Court this follows from the aim of the provisions to protect human health. Furthermore, the relevant provisions of the Directive are unconditional and sufficiently precise. They

[82] Cf. the title of a speech given by the former European Commissioner for Environment Janez Potočnik, available at: http://europa.eu/rapid/press-release_SPEECH-12-856_en.htm#footnote-7.

[83] Beginning with Case 33/76 Rewe v Landwirtschaftskammer für das Saarland [1976] ECR 1989, paragraph 5.

have a binding effect on the Member State and put a clear obligation on domestic authorities to establish or improve air quality plans when the requirements are met.

The Court's confirmation of an EU individual right to a better air quality plan should come as no surprise for those who are familiar with previous jurisprudence. Already in 1991 in *Commission v Germany*[84] the Court took a clear stand in favour of individual environmental rights. The case concerned the first EU air quality directive which provided limit values for levels of lead in ambient air. The Court pointed out that those limit values were imposed to protect human health and clearly explained in para 19:

> ... persons concerned must be in a position to rely on mandatory rules in order to be able to assert their rights.

Like in many German cases of environmental law the domestic provisions are just a mirror of the relevant EU legal instruments. Here, the 2008 Air Quality Directive.[85] This directive—like its predecessor[86]—aims at preventing or reducing harmful air pollution, partly through fixing limit values and alarm thresholds for ambient air pollution levels. For the protection of human health, its air quality standards cover a variety of atmospheric pollutants. The Air Quality Directive of 2008 was a consolidating and amending measure: It repealed the Air Quality Framework Directive and the first Daughter Directive, but the same limit values, margins of tolerance and deadlines are reproduced in Annex XI. To facilitate the understanding of the Air Quality Directive it might be useful to recall the general structure of EU normative acts. The preamble informs about the legal basis (here: the predecessors of Art. 191, 192 TFEU) and the legislative procedure the EU legislator has chosen. Insofar as legal objections are raised, the preamble may be a good starting point for the discussion. It has to be recalled that domestic courts can answer all questions concerning the validity of EU legal acts in the affirmative. If they want to deny the validity of the EU legal act, all domestic courts (not only the highest one's) are under the obligation to refer the case to the CJEU.[87] This is done by a preliminary reference on a question of validity, cf. Art. 267 (1) b) TFEU.

The recitals of the Air Quality Directive provide a good guidance for a better understanding of the legal act and—for example—detect its various predecessors (Recital 3). Recital 18 gives the reader a first notice of the 'air quality plan' and its important role to guarantee air quality standards.

[84] Case C-59/89 Commission v Germany [1991] ECR-I 2626, at paragraph 22.

[85] Parliament and Council Directive 2008/50/EC of 21 May 2008 on ambient air quality and cleaner air for Europe [2008] OJ L151/2.

[86] Council Directive 96/62/EC of 27 September 1996 on ambient air quality assessment and management [1996] OJ L296/55.

[87] Case 314/85, Foto-Frost v Hauptzollamt Lübeck-Ost (1987).

3.3.2 Air Quality Plans

Article 23 (1) of the Air Quality Directive reads:

> **Air quality plans**
> 1. Where, in given zones or agglomerations, the levels of pollutants in ambient air exceed any limit value (...) Member States shall ensure that air quality plans are established (...) in order to achieve the related limit value (...) specified in Annexes XI and XIV.
> In the event of exceedances of those limit values (…) the air quality plans shall set out appropriate measures, so that the exceedance period can be kept as short as possible.

Very often, claims for clean air rely on situations in zones or agglomerations where ambient air exceeds the limit values of nitrogen dioxide (NOx) and/or particulate matter (PM 10[88]). Both air pollutants are mainly caused by traffic, especially diesel exhaust.[89]

Within the big European cities and urban agglomerations where you find lots of traffic and diesel exhaust the limit values for NOx and PM 10 are not unlikely to be exceeded. If there is an exceedance the air quality plan must be improved to keep the exceedance period, as the above cited Article 23 points out, 'as short as possible'.

3.4 *Dieselgate and the 'Rights' of NGOs*

3.4.1 Facts of the Case

In September 2015 US Environmental Protection Agency detected that Volkswagen, the biggest German car manufacturer, had installed inadmissible defeat devices with 'cheat software' in its diesel vehicles. An illegal device that reduced the nitrogen oxide emissions of the diesel engine to the permissible values on the test bench alone, but not in normal road operation. As a result, the automaker faced a large number of claims for damages worldwide ('Dieselgate'). In the following years, *Volkswagen* developed a new software update that was intended to remedy the previous defects in the exhaust system.

In an product-approval decision—at issue here—the German Federal Motor Transport Authority affirmed vis-à-vis *Volkswagen* that there was no longer any impermissible defeat device after the new software update had been installed. According to the opinion of the Federal Authority the defeat device, operating at lower outside temperatures, was legally compliant as a 'thermal window' under Article 5 (2) of Regulation (EC) No. 715/2007 on type-approvals, which allowed exceptions to the ban on defeat devices.

---

[88] PM 10 stands for particulate matter with a diameter of 10 μm or less.

[89] European Environmental Agency (EEA), EEA Report No 5/2015, Air quality in Europe — 2015 report, p. 32, Box 5.1. Available at: www.eea.europa.eu/publications/air-quality-in-europe-2015; accessed 2 February 2023.

3.4.2 Legal Proceedings

Deutsche Umwelthilfe, an environmental NGO recognized under German law, brought an action against the contested approval decision before the Administrative Court, Schleswig-Holstein, Germany. The court noted that, under German law, the NGO does not have standing to bring legal proceedings against the contested decision. The German Environmental Appeals Act limits public interest litigation in environmental matters to an exhaustive list of administrative action. This list does not cover product-approvals or type-approvals like the one challenged. The Administrative Court launched a preliminary reference[90] to the CJEU under the Article 267 TFEU and asked whether the NGO can derive legal standing directly from EU law. Recently, the CJEU, sitting as Grand Chamber, answered in the affirmative.

3.4.3 The Judgment of the CJEU from November 2022

In its November 2022 judgement[91] the CJEU recalls the far reaching guarantee of access to justice under Article 9 (3) of the Aarhus Convention. This catch-all clause provides that

> each party must ensure that, where they meet the criteria, if any, laid down in its national law, members of the public have access to administrative or judicial procedures to challenge acts and omissions by private persons and public authorities which contravene provisions of its national law relating to the environment.

Furthermore, the CJEU relies on the right to an effective remedy, enshrined in Article 47 of the EU Charter of Fundamental Rights which reads in its paragraph 1:

> Everyone whose rights and freedoms guaranteed by the law of the Union are violated has the right to an effective remedy before a tribunal in compliance with the conditions laid down in this Article.

For the 15 EU justices sitting as the Grand Chamber there was no doubt that Article 9 (3) of the Aarhus Convention, read in conjunction with Article 47 of the Charter, precludes a situation like in Germany where a recognized environmental NGO is unable to challenge a decision granting or amending an EC type-approval which may be contrary to Regulation No 715/2007, the later seen as EU and national law relating to the environment.

[90] Case 3 A 113/18, Decision of 20 November 2019.

[91] Case C-873/19, Judgment of 8 November 2022.

3.4.4 Guidance and an Open Question

The CJEU has no authority to interpret German law on court standing of NGOs. However, guidance is given to the German court on how to take the final decision on the case. The EU Court highlights the possibility of consistent interpretation with its findings on NGO standing. According to that method of interpretation, the German court could interpret national procedural law in a manner consistent with the Aarhus Convention and with the right to an effective remedy, enshrined in Article 47 of the EU Charter of Fundamental Rights, in order to grant court standing for the German NGO.[92] At the beginning of the year 2023, when this contribution was written, matters of the case stood like this.

First of all, it is exciting to see how a foundational question of the justice system, like NGO standing, suddenly emerges from a quite technical and unspectacular context. For the author it is an example of the—sometimes difficult to discover—existence of the 'hidden beauty' of Administrative law. Furthermore, it is not uncommon and perhaps the 'nature of the beast' that judgments, rendered by the CJEU in a preliminary reference procedure, leave important questions open, although an answer is definitely needed to take the final decision on the case. Here, the German court is called upon to reconsider its own system of legal standing and the question whether the established right of the NGO to launch legal proceedings can be considered as a 'right' in the sense of general standing rules in Germany. In the following we will briefly present our affirmative position and raise some points of interest. This is designed to provoke discussion.

3.4.5 Rights-Based Standing in Germany

Starting point has to be the German Constitution. In accordance with Article 19 (4) of the German Basic Law anyone asserting that his or her rights are being infringed on by public authority may have recourse to the courts. Therefore, Section 42 (2) of the Code of Administrative Court Procedure (*Verwaltungsgerichtsordnung*) grants court standing only if claimants are able to show that the challenged act or omission may violate their individual rights. This is the case, for example, if a decision by the administrative agency would impair a person's property or health.

By contrast, environmental NGOs, like the claimant, are engaged in matters of public interest. They are a part of the public concerned under the Aarhus Convention and launch legal proceedings to enforce environmental law. In Germany, an NGO's legal power to bring legal proceedings has not been considered as a 'right' in the sense of Section 42 (2) of the Code of Administrative Court Procedure on general

[92] Alternatively, if a consistent interpretation is not possible, and having regard to the absence of direct effect of Article 9 (3) of the Aarhus Convention, the German court has to take account of the EU Charter of Fundamental Rights. Its Article 47 enshrines a right to an effective remedy which individuals may rely on. If necessary, it will then be for the German court to disapply the national provisions precluding the court standing of the environmental NGO.

standing (*Klagebefugnis*). NGOs have standing for public interest litigation only if statutory law explicitly says so. The Environmental Appeals Act *(Umwelt-Rechtsbehelfsgesetz)* lists important cases. However, challenging 'product-approvals' or 'type-approvals' is not listed. Here, a legal interpretation in favour of NGO-standing would contradict the clear wording of the Environmental Appeals Act making it *contra legem* and therefore inadmissible.

3.4.6 Adjustment of the Traditional Concept of 'Right'?

The author argues that the traditional concept of right in the German administrative justice system should be adjusted by including NGO-rights.

Rights are legally protected interests. The interests of the environmental NGO to have access to justice in environmental maters are protected by the legal order. The protection follows from the Aarhus Convention and EU law as interpreted by the CJEU. Of course, on the level of German law this type of a mere procedural right to enforce public interests is a major exception. It must be well justified. The justification lies in solving a problem of global scale: preventing and redressing damage to the environment. Therefore, judicial proceedings are used to enforce environmental law effectively if, and only if public authorities fail to do so. Germany committed itself to this model of 'environmental law enforcement' by signing and ratifying the Aarhus Convention (*pacta sunt servanda*) and—even more obvious—by being a Member State of the EU and being obliged to sincere cooperation.

It would exceed the scope of this contribution to present issues in detail. Here, we limit ourselves to suggest three points of interest for a discussion to come. They concern the eternal problem of the development of the law (*Papinian*) as first inspiration, the concept of claim (*Windscheid*) and the (forgotten) flexibility of the system of subjective-public rights (*Jellinek*).

3.4.7 Papinian's Definition

Law and doctrine are living instruments. In the course of time they necessitate adjustments for the reason of public utility. It has always been like that. The Roman jurist Papinian[93] (142–212) can still inspire us by his statement on the further development of the *ius civile* by the magistrates' law, the *ius praetorium* (D.1,1,7,1):

> Ius praetorium est, quod praetores introduxerunt adiuvandi, vel supplendi, vel corrigendi iuris civilis gratia propter utilitatem publicam.
>
> *Ius praetorium* is what the magistrates (praetores) introduce (by their edicts) in order to assist, or supplement, or correct the ius civile for the reason of public utility.

[93] Papinian is presented by Knütel (2001).

3.4.8 Windscheid's Claim

Windscheid's claim is a (narrow) concept of claim. His seminal work is more than just an inspiration. It contains the foundational concept of substantial 'claim' or 'right' as opposed to bringing an action which is conceived as a purely procedural matter. The German jurist Bernhard Windscheid[94] (1817–1892) was a member of the Pandectistic school. His scientific programme was coined by a practical impetus. He wanted to adjust the existing Roman law into an operable German private law, fit for the purposes of the German society in the second half of the nineteenth century. Although being controversial in his days, in the end, he was successful like no one else. Already in 1856 Windscheid came to the conclusion that the *actio* under Roman law needed to be 'translated' and understood in his days as a substantial claim or a right (*Anspruch*). Even today the German Civil Code (*Bürgerliches Gesetzbuch—BGB*) relies on Windscheid's idea and uses it to define a claim (*Anspruch*), cf. Section 194 (1) BGB.

At the end of the nineteenth century, different than in France, the German science of public law had just began[95] and was in a dilemma. The state-of-the-art French administrative justice, mostly developed by the Council of State (*Conseil d'Etat*), could neither be ignored nor completely adopted as coming from the biggest rival in continental Europe. Therefore, during the following decades the German concept of administrative justice moved away from the iconic French remedy (*recours pour excès de pouvoir*) to the German private law science and developed administrative justice around 'Windscheid's claim'.[96]

As the German word *Recht* can mean right or law, a composite was coined to describe an individual right: 'subjective right' (*subjektives Recht*), admittedly a pleonasm in English. Rights or claims against the state needed to have a name of its own. Already in the nineteenth century these rights were baptised as follows: 'subjective-public rights' (*subjektiv-öffentliche Rechte*[97]) which appears even like a solecism in English. Till today, the concept of 'subjective-public rights' is foundational. Fundamental rights and freedoms are regarded as such rights.[98]

German judicial review of public acts or omissions by the Federal Constitutional Court or by State Courts is conceptualized as protection of (subjective-public) rights (*Rechtsschutz*) and hence conceived as 'purely legal and completely apolitical'. Before this background, court procedures and its types of claims (Klagearten) are 'not an aim in itself', but serve as a device to enforce subjective-public rights.

One of the unfortunate consequences of this deeply rooted focus on substantial rights is that procedural law and the law on evidence is more or less regarded as a mere 'practical business' and quite often not properly represented in the curriculum

[94] Windscheid is presented by Falk (2001b).

[95] Keller (2016), p.73.

[96] Keller (2016), p.73.

[97] Keller (2016), p.71.

[98] Keller (2016), p.74.

of legal education at German law faculties. By the way, the strict dichotomy of substantial right (*materielles Recht*) and—its helping hand called—procedure (*Verfahren*) does not leave much room for the German jurist to understand and to accept the Anglo-American category of 'remedy'.

3.4.9 Jellinek's System

A nineteenth century public law scholar, Georg Jellinek[99] (1852–1911), refined and opened the categorisation of rights in the German administrative justice system. In his famous book of 1892[100] on the system of subjective-public rights he explains:

> At last, it is (...) a highly important question which formally unprotected interests should be elevated to protected ones.
>
> *Endlich ist es (…) eine höchst wichtige Frage, welche formell schutzlosen Interessen zu geschützten erhoben werden sollen.*

According to Jellinek it is 'a question' whether interests that do not have the protection by a court procedure should have one. A question can be discussed and answered in the negative or in the affirmative. The Aarhus Convention, EU law and the EU Court of Justice give an affirmative answer. In the words of Jellinek: The interests of the environmental NGO to bring an action against the type-approval are 'elevated to protected ones' by international and supranational law.

It remains to be seen, if German judges and scholars will follow the suggested path. If so, the rigorism of nineteenth century legal postulates on the concept of (private) rights would be removed from administrative justice. Green access to justice and its specific requirements would be an addition to the classical one. Environmental public interest litigation could be integrated into a more coherent system of German administrative justice in environmental matters for the twenty-first century.

4 Conclusion

The way how German judges decide environmental matters under public law is to be understood before a multifaceted background. This includes German history, the impact of the Basic Law on Administrative law, Europeanisation by EU Directives and Internationalisation by the Aarhus Convention. Standard rules on interpretation, stemming from the nineteenth century, the court system in a federal state, the special style of legal reasonings and the active role of the German judge in administrative court procedures are also important features.

---

[99] On Jellinek: Jouanjan (2005).

[100] Jellinek (1892).

The Climate Decision of the German Constitutional Court establishes a new fundamental right for young people securing their future freedoms in the time span till climate neutrality. The groundbreaking judgment is convincing. It is questionable only insofar as the Court affirms a state interference with freedoms, instead of relying on a state duty to protect freedoms. EU Directives regulate in detail how environmental impact assessments are to be carried out in Germany. Court cases in this context raised questions of access to justice and standing. The *Trianel* judgment of the CJEU established NGO-standing and public interest litigation in Germany. In the field of air quality it was the *Janecek* judgment of the CJEU that made air quality plans judicially enforceable.

A recent judgment by the CJEU in a not yet finished *Dieselgate* case affirms the right of an environmental NGO to challenge product-approvals before a German court. The author argues for rethinking general standing requirements and adjusting the traditional German concept of subjective-public right by including NGO rights.

References

Bunke H (1997) Willy Hundertmark. Erinnerungen an ein widerständiges Leben. Edition Temmen, Bremen, p 93

Champeil-Desplats V (2014) Méthodologies du droit et des sciences du droit. Dalloz, Paris, p 82

Chapoutot J (2014) Histoire de l'Allemagne (1806 à nos jours). PUF, Paris, p 39

Cover RM (1983) The Supreme Court, 1982 term-foreword: nomos and narrative. Harv Law Rev 97:4

Darpö J (2013) Effective justice? Synthesis report of the study on the Implementation of Articles 9.3 and 9.4 of the Aarhus Convention in the Member States of the European Union. Available at: https://www.unece.org/fileadmin/DAM/env/pp/a.to.j/Analytical_Studies/2013_A2J_Synthesis_report_Final.pdf

Dubey B (2003) Administration indirecte et fédéralisme d'exécution en Europe. Cahiers de droit européen 39:87

Eliantonio M et al (2012) Standing up for your right(s) in Europe, a comparative study on legal standing (Locus Standi) before the EU and Member States' Courts. Intersentia, Cambridge, 2013. Available at: http://www.europarl.europa.eu/RegData/etudes/etudes/join/2012/462478/IPOL-JURI_ET(2012)462478_EN.pdf

Falk U (2001a) Jhering. In: Stolleis M (ed) Juristen, Ein biographisches Lexikon, Von der Antike bis zum 20. Jahrhundert. Beck, p 334

Falk U (2001b) Windscheid. In: Stolleis M (ed) Juristen, Ein biographisches Lexikon, Von der Antike bis zum 20. Jahrhundert. Beck, p 672

Gutmann (2018) Ideologie der Gemeinschaft und die Abschaffung des subjektiven Rechts. Available at: https://www.jura.uni-muenster.de/de/institute/lehrstuhl-fuer-buergerliches-recht-rechtsphilosophie-und-medizinrecht/studieren/recht-und-rechtswissenschaft-im-nationalsozialismus

Hamburger P (2014) Is administrative law unlawful? University of Chicago Press

Härtel I (2020) Klimaschutzverfassungsrecht: Klima-Staatszielbestimmungen im Föderalismus. Natur und Recht 42:577

Jellinek G (1892) System der subjektiven öffentlichen Rechte. Mohr Siebeck, p 66

Jouanjan O (2005) Une histoire de la pensée juridique en Allemagne (1800–1918). PUF, Paris

Keller M (2016) In: Renaudie O (ed) L'intérêt à agir devant le juge administratif, sous la direction de Olivier Renaudie, Berger-Levrault, Collection: Au fil du débat. L'intérêt à agir devant le juge administrative. Berger-Levrault, Paris, p 69
Knütel R (2001) Papinian. In: Stolleis M (ed) Juristen, Ein biographisches Lexikon, Von der Antike bis zum 20. Jahrhundert. Verlag C.H. Beck, München, p 485
Meder S (2021) Rechtsgeschichte, 7th edn, p 449
Minnerop P (2022) The 'advance interference-like effect' of climate targets: fundamental rights, intergenerational equity and the German Federal Constitutional Court. J Environ Law 34:135
Rückert J (2001) Savigny. In: Stolleis M (ed) Juristen, Ein biographisches Lexikon, Von der Antike bis zum 20. Jahrhundert. Beck, p 555
Ruffert M (2012) Vorrang der Verfassung und Eigenständigkeit des Privatrechts: eine verfassungsrechtliche Untersuchung zur Privatrechtsentwicklung des Grundgesetzes, 2001
Seerden R (ed) (2018) Administrative law of the European Union its Member States and the United States, 3rd edn. Intersentia, Cambridge
Thorson B (2016) Individual rights in EU law. Springer
Wahl R (2003) Verfassungsstaat, Europäisierung und Internationalisierung. p 411
Werner F (1959) Verwaltungsrecht als konkretisiertes Verfassungsrecht. DVBl, p 527

Dr Matthias Keller is a Presiding Judge at Verwaltungsgericht Aachen, Germany.

The Court of Justice and Protected Areas in the European Union

Colin T. Reid

Abstract The Court of Justice of the European Union has played a major role in ensuring that the legislation designed to protect habitat important for biodiversity is given priority and that Member States do in fact resist the pressures to allow protected sites to be used for other purposes. Cases have made it clear that States have to take their conservation commitments seriously and are subject to external oversight. The Court has insisted that a precautionary approach is adopted in applying this area of law and that the procedures for analysing the risks and potential acceptability of any proposed project are not by-passed. Cases have also considered the status of sites during the extended designation procedures but shown weaknesses in terms of rapid intervention. The Court has thus been important in ensuring that the legislation offers not just an empty promise but a valuable tool for protecting nature across the European Union.

1 Background

The protection of particular areas of land and sea as being of particular value to wild species of fauna and flora is central to the efforts around the world to tackle the biodiversity crisis. Although a more holistic approach looking after the wider landscape is increasingly being viewed as offering the best option for progress,[1] protected areas continue to play a vital role and are recognized as 'essential for biodiversity conservation' and as 'the cornerstones of virtually all national and international conservation strategies'.[2] Accordingly, the latest Conference of the Parties of the Convention on Biological Diversity has set the target that by 2030 at

[1] CBD-COP 5 (2000), Decision V/6.

[2] Dudley (2008), 2.

C. T. Reid (✉)
Dundee Law School, University of Dundee, Dundee, Scotland, UK
e-mail: c.t.reid@dundee.ac.uk

G. Antonelli et al. (eds.), *Environmental Law Before the Courts*,
https://doi.org/10.1007/978-3-031-41527-2_10

least 30% of the land and sea should be effectively managed within systems of protected areas (or covered by equivalent measures).[3]

The International Union for the Conservation of Nature refers to the "almost infinite array of approaches" to protected sites. This diversity is shown by the fact that in producing its categorization of such sites under six headings it notes that sites that are locally termed "National Parks" may in fact fall under any of the categories proposed, from those strictly set aside for nature to those where the exploitation of natural resources is controlled with a view to sustainability.[4] Whatever the details of the protection provided, such protection must be based on law. This means that there is a vital role for the courts in interpreting and applying that law. The aim of this chapter is to look at experience within the European Union (EU), exploring the notable contribution that the Court of Justice has made to the operation and effectiveness of the protected areas declared under EU law.

Nature conservation within the EU is predominantly a responsibility of the Member States, but the EU has acted to ensure that the most important and endangered species and habitats are provided with a high level of protection across its territory.[5] In pursuing this objective the Court of Justice of the European Union (CJEU; formerly the European Court of Justice) has played a major role in ensuring that environmental considerations are taken seriously and that the States live up to the commitments they have made and it is decisions from this Court that are the focus here. Cases come before the CJEU either as the culmination of action being taken against a Member State alleged not to be meeting its obligations under EU law or when the case is referred to it by a national court which decides that in order to determine the dispute being heard it is necessary to resolve a matter of the interpretation or validity of EU law.[6]

At the heart of the EU's legal intervention on protected areas are two measures, the Birds and the Habitats Directives.[7] As Directives, these do not themselves apply directly within the Member States but require all the States to ensure that they have in place domestic laws that fulfil their requirements. National courts are bound to interpret and apply the domestic law in line with the Directives. Where a State is considered to be failing to live up to the terms of a Directive the European Commission can launch infringement proceedings which can end up with the State being taken before the CJEU and ultimately subjected to substantial fines if it does not comply.[8] Consequently, in every Member State there should be laws in place to ensure that the habitats and species covered by the Directives enjoy the level of

[3] CBD-COP (2022), Final Text, Target 3.

[4] Dudley (2008), vii and 11.

[5] See generally García-Ureta (2020), Jackson (2018), Born et al. (2014).

[6] Treaty on the Functioning of the European Union (2010), arts 258–260, 267 (consolidated version available at [2012] OJ C 326/47).

[7] Council Directive 79/409/EEC on the protection of wild birds, amended and restated as Directive 2009/147/EC; Council Directive 92/43/EEC on the conservation of natural habitats and of wild fauna and flora.

[8] Treaty on the Functioning of the European Union (note 6); see Hedemann-Robinson (2015), Part 1.

protection prescribed, whether as separate measures directly implementing the Directives or because the terms of their more general national laws on this subject already meet or exceed the standards required.

In terms of habitat protection, the focus is on the Natura 2000 network, which aims to be a coordinated network of protected areas across the EU. Comprised of Special Protection Areas (SPAs) under the Birds Directive and Special Areas of Conservation (SACs) under the Habitats Directive, the Natura network has been established on the basis that Member States have been required to designate the most suitable areas for conservation and ensure that these are protected. This does not mean that they are set aside purely for nature—many are the location of major agricultural, industrial and infrastructure activities—but looking after the biodiversity interest must take an element of priority. The Natura network covers roughly 18% of EU land across all 27 EU countries—most of which remains privately owned, controlled, and managed.

The core protection for both SPAs and SACs is set out in article 6 of the Habitats Directive, which imposes a general obligation on Member States to "take appropriate steps to avoid … the deterioration of natural habitats and the habitats of species as well as disturbance of the species for which the areas have been designated, in so far as such disturbance could be significant in relation to the objectives of this Directive."[9] More specifically, under article 6(3) plans or projects[10] likely to have a significant effect on sites must be assessed for their impacts[11] and should be approved only if the authorities have ascertained that the integrity of the site will not be adversely affected. In other words, unless it is clear that the conservation interest will be unaffected, the project should not be allowed to go ahead. There is, however, a "get-out" clause under article 6(4) which allows potentially harmful activities to proceed provided that three conditions are met: that there is no satisfactory alternative, that the plan or project must be carried out "for imperative reasons of overriding public interest, including those of a social or economic nature", and that the Member State takes "all compensatory measures necessary to ensure that the overall coherence of Natura 2000 is protected".

Environmental matters have historically provided one the areas where the CJEU has been most commonly called into action to ensure that Member States live up to their obligations under EU law.[12] The CJEU has played a major role in ensuring that the protection of nature, as proclaimed in the Directives' objectives, is given priority and that States do in fact resist the pressures to use protected sites for other purposes. A selection of leading cases is presented here under several headings. Firstly, a pair of early cases are included that made clear that Member States had to take

[9] Habitats Directive (note 7), art 6(2).

[10] For brevity, the rest of this chapter refers just to "projects".

[11] This assessment may be part of a formal Environmental Impact Assessment, as regulated in detail by further EU measures (Directive 2011/92/EU as amended by 2014/52/EU), but will be required even when those rules do not apply.

[12] See for example European Commission (2012); other areas have now taken over as those generating most activity.

their conservation commitments seriously. Then there is consideration of just two of the very many instances that show how far the Commission has used the Court as part of the oversight of what Member States are doing. Thirdly the crucially important *Waddenzee* case is discussed, establishing the precautionary approach to be adopted in applying this area of law. Then there are cases that confirm the sequential approach to analysing the risks and potential acceptability of any proposed project, and the determination to avoid this being by-passed. Further cases come to terms with the "life-cycle" of sites as formally protected areas and finally the time-scale for action is considered.

2 Taking Conservation Seriously

The significance that the Court has consistently placed on ensuring that the environmental commitments made by the Member States are satisfied can be shown by two of the earlier cases in this area. The first of these, the *Leybucht Dykes* case,[13] predates the Habitats Directive and was based on the Birds Directive as originally enacted (which did not allow for an exception such as set out in article 6(4) of the Habitats Directive). It reveals the pressures in balancing conservation with other interests and the Court's clear view that the former must be taken seriously.

The case concerned proposals for rebuilding coast-protection dykes in the Leybucht, a bay in north-western Germany between the Ems and Weser rivers, and an area designated as an SPA under the Birds Directive. This work, by enlarging the existing dykes which set the boundary of the SPA, would have the effect of moving that boundary and reducing the area protected.

The Court based its conclusions on giving effect to the key obligation under the Directive to avoid the deterioration of protected sites. Although the Member States had discretion in selecting the sites most suitable for classification as SPAs, once designated the State could not claim the same discretion in modifying or reducing the areas covered, because that would amount to giving them a unilateral power to escape the obligation to provide protection for the areas designated. The Court recognised that in the absence of an express power to reduce the area protected such action might still be permissible,[14] but only where there was a general interest superior to the interest represented in the ecological objective of the Directive. Here, so long as the protected area was reduced by the minimum possible, the proposed works could meet that test since they reduced the risk of flooding and because the navigation activity under finished arrangement would remove disturbance from a significant part of the area, with distinct ecological benefits. It was emphasized, though, that it was only because of these benefits that that any economic

[13] C-57/89 *Commission v Germany* [1991] ECR I-883.

[14] See the discussion of the *Cascina Tre Pini* case in Sect. 6 below.

considerations, relating to the continuing access of fishing vessels to a port, could be taken into account.

The result was thus that the work could go ahead, but the reasoning of the Court laid down a very clear marker that it was only in exceptional cases that projects reducing the area or effectiveness of an SPA would be allowed. It was only the importance of protecting sizeable areas from flooding and the ecological benefits resulting from the adjustment to navigation channels, leaving large areas undisturbed, that entitled the socio-economic considerations to be taken into account. Without these considerations conservation would have come first. At the time, this insistence on placing nature first was seen as remarkable. Nature conservation was starting to be seen as more significant, with the concept of biodiversity and its value about to be marked by the signing of the Convention on Biological Diversity,[15] but there was, and still too often is, a large gap between the rhetoric on this subject produced by governments and any real commitment to allowing conservation to override the economic and social interests that have traditionally dominated political calculations.

The realisation that such priority would be given to the need to protect nature caused two reactions. In the first place Member States became unhappy as they realised that they had made such a strict commitment that gave them virtually no room for manoeuvre when conservation clashed with their own socio-economic priorities. Secondly, conservationists feared that, as a consequence of that, States would become very reluctant to designate sites or might even try to rewrite the Birds Directive completely. The solution was to apply to SPAs the terms of the Habitats Directive which was already being developed at that time, incorporating the exception where there are imperative reasons of overriding public interest. Conservation remains, therefore, the top, but not the completely overriding, priority when a designated site may be affected.

This high priority is also shown in the *First Corporate Shipping* case, where again the question was the balance between protecting nature and socio-economic considerations.[16] The case involved the proposal to put forward parts of the Severn Estuary in south-west England as a candidate for becoming an SAC, a step which the port authority for Bristol feared would restrict its opportunities for further development. The fundamental question was whether considerations other than ecological ones could be taken into account in determining which areas should be designated. The Habitats Directive does state that the measures taken should "take account of economic, social and cultural requirements and regional and local characteristics"[17] and this was relied on to argue that these could provide a justification for excluding sites from the list put forward for designation even though they met the scientific criteria for inclusion.

---

[15] Convention on Biological Diversity, signed at the Rio Earth Summit in 1992.

[16] C-371/98 *R v Secretary of State for the Environment, Transport & Regions, ex parte First Corporate Shipping Ltd* [2000] ECR I-9235.

[17] Habitats Directive (note 7), art 2(3).

The Court, however, disagreed. The starting point for its reasoning was the objective of Natura 2000 to create a coherent ecological network of protected sites, with the proposal by a Member State just the first step in the process towards becoming an SAC. Only if all the sites that met the ecological criteria set out in the Directive were put forward at the initial stage would it be possible to establish a network capable of meeting the Directive's objectives of maintaining or restoring the natural habitat concerned at a favourable conservation status, bearing in mind that the natural range of habitats and their species might well cross national boundaries. Since each Member State on its own did not know of the position in other States, it would be wrong to allow any one State, on the basis of local reasons, to ignore sites that met the ecological criteria. If economic and social criteria could shape the selection at national level, the EU Commission could not be sure of having a full list of suitable sites when it came to the further stages of confirming which sites were to become SACs and thus establish the coherent network that was sought. Although in part based on the multi-stage process for designating SACs, which relies on national proposals being reviewed at EU level,[18] the key message is clear—that social and economic considerations would not be allowed to override ecological ones at every step in the process.[19] As will be shown in Sect. 5 below, there is a place for such concerns, but they must not be allowed to provide a way to by-pass the whole structure of protective laws.

3 Oversight of National Action

Membership of the European Union requires acceptance of "a new legal order … for the benefit of which the States have limited their sovereign rights, albeit within limited fields."[20] This means that Member States are under the scrutiny of the European Commission and of other States and can be subject to infringement actions if they do not live up to their commitments. A couple of cases, chosen from a very long list of similar ones, illustrate how this system works and how States can be called to account when they are not providing adequate protection for important sites.

In the *Santoña Marshes* case, the Spanish government was held to have fallen short of its obligations under the Birds Directive by not designating an area as an SPA and not taking adequate measures to avoid deterioration of the habitat.[21] On the first point, it was recognized that although Member States did have some margin of discretion in the choice of SPAs, the ornithological criteria must take precedence, as

[18] See the discussion in Sect. 6 below the *Dragaggi* case.

[19] On designation generally, see Shoukens and Woldendorp (2014).

[20] Case 26/62 *NV Algemene Transport-en Expeditie Onderneming van Gend & Loos v Nederlandse Administratie der Belastingen* [1963] ECR 1, [3].

[21] C-355/90 *Commission v Spain* [1993] ECR I-4221.

noted above. It was common ground that the marshes are of exceptional importance to many species of birds but the measures taken by Spain to designate part of the marshes as a nature reserve did not cover 40,000 square metres of the important habitat and did not include any management plans. Accordingly, the obligation to classify "the most suitable territories in number and size"[22] had not been satisfied.[23] In terms of protection, it was held that the Directive required States to take active measures to ensure conservation, not simply to achieve the outcome that birds were successfully conserved. Action was required immediately, not just waiting until populations started to decline and then considering what, if anything, should be done. More specifically, a number of projects that had been undertaken—road building, industrial and aquaculture developments and waste-water discharges—were damaging to the health of the ecosystem and therefore by allowing these to take place Spain has again fallen short of meeting its obligations. In relation to both implementing conservation policy and allowing specific projects to go ahead, the Spanish government's action was subject to external scrutiny and condemnation.

A similar picture is shown in the *Poitevin Marshes* case where the French government was found wanting.[24] Both French and international assessments of ornithological significance had identified areas of between 78,000 and 58,000 hectares as being important areas for bird conservation, but under 27,000 hectares had been designated by France as an SPA. The Court ruled that this was inadequate. Even within the area designated, the various measures claimed to be providing protection were not sufficient, because they had been introduced too late, were not binding on farmers working in the area or covered only the management of water resources. The fact that a number of areas that should have been designated within the SPA had been destroyed by agricultural activity meant that the France had further failed to meet its obligation to ensure that important sites did not deteriorate.

What both these cases, and many, many others, demonstrate, is that the national governments are under scrutiny by the Commission, and ultimately the Court. The national decisions on which areas to designate for protection, on the means by which protection is to be secured and on allowing specific projects to go ahead are all ones where discretion is limited by the need to ensure compliance with the terms of the Birds (and in other cases Habitats) Directive, terms which allow minimal, if any, scope for economic and social considerations to be taken into account at those stages.

---

[22] Birds Directive (note 7), art 4(2).

[23] The inventory of *Important Bird Areas in Europe* (1989), and now its successors, have repeatedly been recognised as a very important guide on this point, creating a presumption that the listed sites should be designated unless good reasons otherwise are provided.

[24] C-96/98, *Commission v France* [1999] ECR I-8531.

4 Precautionary Approach

A further significant intervention by the Court has been its insistence that a precautionary approach be taken in applying the Birds and Habitats Directives. This was established in the *Waddenzee* case, dealing with the licensing of mechanical cockle fishing on the Dutch coast.[25] The case had two significant issues, on both of which the Court's interpretation favoured nature over other considerations.

The first issue was whether the decision on the annual renewal of licences for cockle fishing amounted to the approval of a "plan or project" falling within article 6(3) of the Habitats Directive and thus required an appropriate assessment and was eligible to be approved only if there were to be no adverse effects on the integrity of the site. Here it was held that the aim of the Directive was "to prevent activities which are likely to damage the environment from being authorised without prior assessment of their impact on the environment."[26] Even though the activity had been going on for years, each annual renewal required consideration of the impact on the site and the process laid down in the Directive had to be followed.

The second issue related to the approach to be taken in carrying out and acting on the outcome of the assessment that was required. The Court looked at the wording of the Directive which states that approval for a project can be given "only after having ascertained that it will not adversely affect the integrity of the site."[27] With reference to the precautionary principle, which is embedded as one of the bases for EU action on the environment,[28] the Court held that an assessment was necessary wherever it could not be excluded, on the basis of objective information, that the project would have a significant effect. Moreover, if that assessment revealed any doubt as to the presence of adverse effects, then approval must be withheld. Only where it had been made certain that there would be no adverse effects, with no reasonable scientific doubt remaining, could approval be given.[29]

This decision confirmed the priority that the Court gives to the conservation of nature. That priority, however, is not their own invention but is based on a clear reading of the words in the Directives and the Treaty, which commits the EU to "aim[ing] at a high level of protection" for the environment.[30] It might be speculated, though, that the Member States are not always happy at the words they agreed to being read in this way. As the earlier cases have shown, the aspiration of protecting nature can come into conflict with the economic and social considerations which governments tend to find more pressing, and the many cases where States have been

[25] C-127/02 *Landelijke Vereniging tot Behoud van de Waddenzee and Nederlandse Vereniging tot Bescherming van Vogels v Staatssecretaris van Landbouw, Natuurbeheer en Visserij* [2004] ECR I-7405.

[26] Ibid at [26].

[27] Habitats Directive (note 7), art 6(3).

[28] Now in art.191(2) of the Treaty on the Functioning of the European Union (note 6).

[29] On further developments in the use of the precautionary approach see Lees (2017).

[30] Ibid.

found wanting provide evidence of how often nature is relegated to the back-seat in the thinking of national and local governments.

5 Sequential Approach

There are, nevertheless, circumstances where a project that may, or will, cause harm can still be approved. The Court, though, has been very alert to ensure that it is only projects which have passed through the multiple stages for such exceptional approval that are authorised. Attempts to by-pass the sequential procedure, by not designating relevant sites or by being too willing to say that there is no risk of adverse effects, have been firmly suppressed. The issue of ensuring that those sites that meet the scientific criteria are designated has been dealt with above and a similarly strict view has been applied to the later stages.

The rule that projects which may have an adverse effect on a site cannot be approved has been reasserted on several occasions, notably in the *Sweetman* case.[31] Here the relevant authority in Ireland had decided that building a new highway that cut into an SAC would not "adversely affect the integrity of the site"[32] as a whole, even though it would involve the loss of a small area of the limestone pavement which had caused the area to be designated as an SAC, a loss that would have a locally significant impact. The Court strongly disagreed, reaffirming the need to take a precautionary approach. In order to fulfil the obligation to prevent the deterioration of natural habitats,[33] sites had to be preserved at a favourable conservation status. The permanent and irreplaceable loss of even a small part of a designated site clearly had an adverse impact and as such would stand in the way of approving any project unless an exception could be justified.

The Court has also been strict on those who try to escape the rigours of the sequential process by arguing that any potential adverse effects are being avoided by mitigating action being taken, so that in fact there is no obstacle to approval.[34] This was discussed in the *People over Wind* case, relating to a wind farm development in Ireland.[35] The question was whether the mitigating steps being proposed by the developer could be taken into account at the initial stage of determining whether or not the project might have an adverse effect on the site. The significance of this is that if it were decided, on the basis of the proposal as qualified by the mitigation, that there would be no such effect, then there would be no need to have a full

[31] C-258/11, *Sweetman v An Bord Pleanála* ECLI:EU:C:2013:220.

[32] Habitats Directive (note 7), art. 6(3).

[33] Ibid., art 6(2).

[34] An important terminological note is that in EU phrasing "mitigation" is limited to steps taken to reduce the actual impact of a development on the site itself, with any offsetting or compensatory measures trying to balance harm with enhancements elsewhere (such as under the US Wetland Mitigation Scheme) referred to as "compensation".

[35] C-323/17, *People Over Wind v Coillte Teoranta*, ECLI:EU:C:2018:244

assessment of the project. It would therefore be determined, at that preliminary stage, that there was no obstacle to approval being given.

The Court was firmly of the view that this was not appropriate. The Directive created a clear sequential process intended to maintain or restore sites to favourable conservation status and required a careful and precautionary approach at each stage. The very fact that measures designed to reduce the harmful effects of a project were being put forward for consideration presupposed that it was likely that the site was going to be significantly affected and therefore a full assessment should be carried out. In this way the scale and effectiveness of the proposed mitigation could be properly reviewed to determine whether or not there was likely to be an adverse effect at the end of the day.

If there was such an adverse effect, then approval had to be refused unless the criteria for an exception were met: that there is no satisfactory alternative, that the plan or project must be carried out "for imperative reasons of overriding public interest, including those of a social or economic nature", and that the Member State takes "all compensatory measures necessary to ensure that the overall coherence of Natura 2000 is protected".[36] At that stage the Court is again vigilant to ensure that the criteria are in fact met, for example in *Commission v Portugal* where it was held that alternatives to the chosen route for a motorway had not been properly considered.[37] It has, however, been argued that Member States and the Commission have been rather too lenient in their assessment of the issue of whether an exceptional case has been established.[38]

The Birds and Habitats Directives aim to protect the Natura network from damaging activities by insisting that any plan or project that is likely to have a significant effect is fully assessed and that any proposals that might have adverse impacts are refused in all but a few, carefully justified circumstances. The Court has been strong in insisting that these procedures and tests are applied and in trying to cut off the routes by which these rigorous standards might be avoided.[39]

6 The "Life-Cycle" of Sites

The next pair of cases to be considered deal with the legal life-cycle of a protected site, examining its status before it is formally classified and its potential declassification. The first point looks at the initial stages of a site becoming protected and is a consequence of the multi-stage, and therefore prolonged, designation process for SACs.[40] It is inevitable that proposals may come forward that affect sites that are in

[36] Habitats Directive (note 7), art. 6(4).

[37] C-239/04, *Commission v Portugal* [2006] ECR I-10,183.

[38] McGillivray (2012).

[39] A less positive assessment is offered by Wandesforde-Smith and Watts (2014).

[40] Habitats Directive (note 7) arts 3–4.

process of being designated, before they are entitled to benefit from the full effect of the protective measures. This issue arose in the *Dragaggi* case, where sediment gathered from dredging work was due to be dumped on reclaimed land at a time when the land in question had been proposed, but not adopted, as an SAC.[41] The Court held that since sites that had been proposed might be excluded from the list approved by the Commission for formal designation ("sites of Community importance"),[42] it was not appropriate to apply the rules protecting sites until a site was identified on that list. Nevertheless, although the full detailed rules did not apply, the Court held that the State was still under an obligation to safeguard the ecological interest of the site in order to fulfil the general purpose of the Habitats Directive.

This was expanded on in the *Bund Naturschutz in Bayern* case, where motorway proposals would affect sites identified as potential SACs.[43] Again it was accepted that the full protective rules could not apply at that stage. The Court noted, though, that in order for the designation process to meet its aim of enabling "the creation of a coherent ecological network of special sites of conservation",[44] the Commission had to have available an exhaustive list of sites of relevant ecological interest, and those sites had to remain in that condition until the decision-making process was completed. Member States were therefore obliged to ensure that the ecological interest of the sites was not seriously compromised before the Commission's decision could be made. The legal framework for doing so was left in the hands of the Member State, since the specific rules in the Directive could not apply until a site became a site of Community importance, but the overall objective of the Directive had to be fulfilled. Again the Court has shown its determination to see that the objective of protecting nature agreed by the States is not undermined by States taking decisions that threaten the survival of valuable habitat.

The opposite side of that coin is shown in the *Cascina Tre Pini* case, where it was argued that development at the Milan-Malpensa airport meant that a designated site had been degraded so as no longer to meet the criteria for recognition as a site of Community importance.[45] The declassification of protected sites is not always provided for in the relevant legal regimes,[46] but the Habitats Directive does envisage the "adaptation" of the list of sites in cases where "natural developments" render that appropriate.[47] The Court held that this meant that declassification was possible. The purposes of the Directive would not be served by maintaining a definitively degraded site on the list, the site would no longer contribute to the Natura network and

---

[41] C-117/03 *Società Italiana Dragaggi SpA v Ministero delle Infrastrutture e dei Trasporti, Regione Autonoma del Friuli-Venezia Giulia* [2005] ECR I-167.

[42] Habitats Directive (note 7), art 4(2).

[43] C-244/05 *Bund Naturschutz in Bayern eV and Others v Freistaat Bayern* [2006] ECR I-8445.

[44] Ibid. [39].

[45] C-301/12 *Cascina Tre Pini Ss v Ministero dell'Ambiente e della Tutela del Territorio e del Mare* ECLI:EU:C:2014:214.

[46] Reid (2019), 841; and see the *Leybucht Dykes* case (note 13).

[47] Habitats Directive (note 7), art 9.

resources might be wasted on managing a site of no use to conservation. The overall objective of the Directive was thus again the leading consideration and in the absence of specific procedural provisions, it was held that the removal of a site from the list should follow the same process as for inclusion on it.

Nevertheless, mindful of the dangers of sites being lost, the Court noted that not all degradation of a site would justify declassification. The Directive requires States to protect sites to avoid their deterioration, and if their failure to do so led to a site being degraded, that should reinforce their duty to take the measures necessary to safeguard the site rather than allowing them to avoid it. Similarly, unlawfully permitting damaging projects to go ahead should not enable a State to escape its obligations. It was therefore only where a site had become irretrievably unsuitable, despite the State complying with its duties to protect its natural features, that a proposal for declassification could be considered. Throughout, the focus is clearly on what the Directive is seeking to achieve, both in terms of producing real benefits for conservation and ensuring that States do take the action necessary to further that goal.

7 Interim Measures

The discussion so far might suggest that the Court has always been a good friend of nature, but there is one significant area where it has been found wanting. This is in relation to providing rapid and effective intervention when sites are under threat. The process by which a case reaches the Court, whether as a result of infringement action by the Commission or as a reference from a national court, is notoriously slow. An example of this can be identified in the high-profile early cases such as *Lappel Bank*, where operations that destroyed the valuable site had been completed before the Court was able finally to rule that the national measures purporting to authorise them were in fact unlawful.[48] Where a case begins in a national court, the initial decision on any interim measures will lie with that court, subject to its own procedures and standards, but these may require the party seeking an interim order to provide security to compensate the other party's losses in the event of the intervention being found to be unjustified, a major hurdle for parties such as conservation NGOs to overcome.[49]

The Court does have available a broad power to impose interim measures,[50] including the power to impose substantial daily fines for non-compliance.[51]

[48] C-44/95 *R v Secretary of State for the Environment, ex parte Royal Society for the Protection of Birds* [1996] ECR 3805.

[49] As in the *Lappel Bank* case itself, as discussed by the House of Lords at [1997] Env LR 431 at 440–441.

[50] Treaty on the Functioning of the European Union, art 279; see Hedemann-Robinson (2015), Part 1.

[51] E.g. C-121/21R *Czech Republic v Poland* ECLI:EU:C:2021:752

However, these measures apply only once a case has reached the Court itself,[52] and have rarely been deployed.[53]

A recent example of these powers being used is in *Commission v Poland*, relating to management measures in part of the Białowieża Forest that were ultimately held not to provide the protection required for designated sites there.[54] The story here, though is far from straightforward.[55] The Commission's request for interim measures was granted on a provisional basis by the Vice-President of the Court in July 2017, but the Court in November 2017 authorised the implementation of some of the contested measures, subject to certain qualifications, before the case was considered under the expedited procedure available to the Court and decided in April 2018. The complexities here, and the wider experience, show that although not always unsympathetic to the need for rapid intervention, the Court does not always deliver a sharp halt to potentially damaging activities, creating a real risk that the objectives of the Birds and Habitats Directives will on occasions be defeated.

8 Conclusion

In environmental matters, there is a tendency for the rhetoric adopted by governments to promise more than they are actually willing to deliver when the steps necessary to protect the environment come into conflict with economic and social pressures. The cases considered here all show that where that rhetoric has been given legal form in the treaties and legislation of the EU, the Court of Justice has played a significant role in ensuring that it is followed through into practice. The commitments in the Treaty to a high level of protection and the use of the precautionary principle,[56] and the processes created in the Birds and Habitats Directives to limit harmful development affecting protected sites, are all phrased so as to give significant, but not absolute, priority to nature over other considerations. This is taken seriously by the Court, which seeks to ensure that Member States do in fact take steps to care for biodiversity, in line with the overall objective of creating the Natura network as a meaningful way of caring for biodiversity across Europe. In the light of increasing recognition of the biodiversity crisis facing the world, some of these decisions may not seem that surprising, but in the early days of the Birds and

[52] This will come only after the initial proceedings in the national court, potentially at several levels, leading up to a reference to the Court of Justice, or after the completion of the various stages required before the Commission commences the formal infringement proceedings (Treaty on the Functioning of the European Union, art 258).

[53] Examples include: C-76/08R *Commission v Malta*; C-573/08R *Commission v Italy*; C-441/17R *Commission v Poland*. See Hedemann-Robinson (2015), 122–146.

[54] C-441/17, ECLI:EU:C:2018:255; see Krämer (2018).

[55] Ibid, [54]–[60].

[56] Treaty on the Functioning of the European Union (note 6), art 191(2); see Hedemann-Robinson (2015), Part 1.

Habitats Directives actually insisting on conservation taking priority over other considerations was a major development. Without the Court insisting that the formal procedures are followed and the conservation of nature is given the dominant role set out for it, it is likely that on occasions only lip-service would be paid to these commitments and that many more sites would have been damaged or lost.

The protection offered is far from perfect,[57] and the absence of reliable rapid intervention is a weakness. Nevertheless, the presence of the Court and its insistence on living up to the objectives enshrined in legislation has meant that the governments have had to pay much more heed to their commitments in this context when compared with the promises they have made internationally to tackling the climate and biodiversity crises. The Court ensures that the legislation is not just an empty promise but a valuable tool for protecting nature across the EU.

References

Born CH, An C, Schoukens H, Misonne D, Van Hoorick G (eds) (2014) The habitats directive in its EU environmental law context: European nature's best hope? Routledge, Abingdon

CBD-COP 15 (2022) Kunming-Montreal Global Biodiversity Framework, agreed at the 15th meeting of the conference of parties to the UN convention on biological diversity (CBD/COP/15/L25)

CBD-COP 5 (2000) Fifth ordinary meeting of the conference of the parties to the convention on biological diversity, 15–26 May 2000—Nairobi, Kenya (CBD-COP 5), Decision V/6

Dudley N (ed) (2008) Guidelines for applying protected area management categories. IUCN, Gland

European Commission (2012) 30th annual report on monitoring the application of EU law 2012. https://commission.europa.eu/system/files/2017-02/Commission_report_complete_2012%255BCOM%25282013%2529726%255D_en_0.pdf. Accessed 5 January 2023

European Commission (2023) Natura 2000. https://ec.europa.eu/environment/nature/natura2000/index_en.htm. Accessed 5 Jan 2023

García-Ureta A (2020) EU Biodiversity Law: Wild Birds and Habitats Directives. Europa Law, Groningen

Hedemann-Robinson M (2015) Enforcement of European Union Environmental Law: legal issues and challenges, 2nd edn. Routledge, Abingdon

Inventory of Important Bird Areas in Europe (1989) In: Grimmett R, Jones TA (eds) International Council for Bird Preservation, Cambridge

Jackson ALR (2018) Conserving Europe's wildlife: law and policy of the Natura 2000 network of protected areas. Routledge, Abingdon

Krämer L (2018) Relief in environmental matters court of justice of the EU. J Eur Environ Plan Law 15:259

Lees E (2017) Concretising the precautionary principle in habitats protection – *Grüne Lige Sachsen v Freistaat Sachsen* and *Orleans v Vlaams Gewest* 19 ELR 126

McGillivray D (2012) Compensating biodiversity loss - the EU Commission's approach to compensation under article 6 of the Habitats Directive. J Environ Law 24:417

Reid CT (2019) Protection of sites. In: Lees E, Viñuales JE (eds) The Oxford handbook of comparative environmental law. Oxford University Press, Oxford

[57] Wandesforde-Smith and Watts (2014).

Shoukens H, Woldendorp HE (2014) Site selection and designation under the habitats and birds directives: a Sisyphean task? In: Born CH, An C, Schoukens H, Misonne D, Van Hoorick G (eds) The habitats directive in its EU environmental law context: European nature's best hope? Routledge, Abingdon

Wandesforde-Smith G, Watts NSJ (2014) Wildlife conservation and protected areas: politics, procedure, and the performance of failure under the EU birds and habitats directives. J Int Wildl Law Policy 17:62

Environmental Protection Before Italian Administrative Courts

Maurizio Santise

Abstract This chapter aims at investigating the multilevel protection of the environment in the Italian legal system. Hence, starting from the Italian legislation the author analyzes the fundamental role played by the Courts in the definition of the environmental legal framework, as well as in its development. A specific focus is dedicated to environmental administrative law and administrative law Courts.

1 Introduction: A Sketch of Multilevel Environmental Protection in the Italian Criminal and Civil Law

The environment has a multilevel protection, scattered between supranational sources and domestic law. In the latter respect, specific forms of protection are found not only in administrative, but also in criminal and civil law.

With particular reference to criminal law, Law No. 68 of 2015 introduced into the Criminal Code new crimes to safeguard the environment, thus changing the previous legal framework that almost exclusively entrusted environmental protection to contraventions and administrative penalties (cfr. Legislative Decree No. 152 of 2006). A new Title VI-bis was, therefore, introduced within the second book of the Criminal Code, entitled *Crimes against the environment.*[1]

[1] The legislature has introduced for six new crimes: environmental pollution; environmental disaster; trafficking and abandonment of highly radioactive material; obstruction of control failure to clean up; inspection of seabed. Unnamed disaster. These are cases codified in the form of crimes of danger, as repression is often not sufficient to protect the good-interest protected by the norm, but it is necessary to take preventive measures. See *Circolare 19 febbraio 2016 - Monitoraggio in tema di reati ambientali a seguito delle novità normative introdotte con la legge n. 68/2015 e la*

M. Santise (✉)
Tar Campania, Naples, Italy
e-mail: m.santise@giustizia-amministrativa.it

G. Antonelli et al. (eds.), *Environmental Law Before the Courts*,
https://doi.org/10.1007/978-3-031-41527-2_11

In civil law, the provision in Article 844[2] is relevant. In a logic that is not only proprietary, the provision prefers protection in a specific form, rather than protection by equivalent.

The rule is implementation of the principle of solidarity and tolerability, as well as the social function of property (Article 41 Const.) and covers indirect immissions.

It is a protection connoted by its private nature, placed to protect the interests of the owner (his right to property).

The rule operates with regard to indirect immissions (not directly produced by the owner, but which are nonetheless the effect of the owner's normal course of business).

Otherwise, direct immissions represent a tort governed by Article 2043 Civil Code. The distinction is not only theoretical: art. 844 Civil Code requires proof of the intolerability of the immission; in art. 2043 Civil Code, on the other hand, proof of the unlawfulness of the conduct is required; in art. 844 Civil Code, however, the subjective element is not relevant, but rather the objective element (i.e., the objective exceeding of the tolerability threshold is relevant, in the absence of any investigation into the subjective element).

According to jurisprudence, Article 844 of the Civil Code can also be invoked to protect the right to health; the exclusively proprietary logic is therefore superseded, with the result that the exercise of this form of protection is also admissible against the tenant.[3]

2 Italian Environmental Law and the Role of Constitutional Court

The Italian legal system began to take an interest in environmental protection in the first half of the last century, although not in a direct and unified way.

Law No. 778 of 1922 (*Protection of Natural Beauties and Properties of Special Historical Interest*) was protecting more than the environment, still not understood as a unitary good, its components, consisting of the landscape and natural beauties.

A more organic regulation, but still with regard to natural beauty, was provided by Law No. 1497 of 1939 (*Protection of Natural Beauty*).

There was still, however, no direct reference to the environment.

This probably explains the silence of the Italian Constitution (which came into force in 1948) on this point.

---

legge n. 6/2014, in www.giustizia.it. Another rule often referred to in criminal case law for the protection of the environment is Article 674 of the Criminal Code. *See* Cassaz. n. 2240, 2016.

[2] Article 844 of the Civil Code provides that the owner of a fund may not prevent the immissions of smoke or heat, exhalations, noises, shaking and similar propagations arising from the fund of the neighbor, if they do not exceed the normal tolerability, having regard also to the condition of the places. *See* Santise (2021b).

[3] *See* Cass. civ. n. 28,893, 2018, in *Foro it.* 2019.

In the 1970s, in parallel with the rise of urban and industrial development and the maturing in opposition to it of a widespread ecological consciousness, that a notion of the environment as an autonomous and unitary legal good, as the object of legal protection against human aggression, was elaborated by doctrine.

As the administrative jurisprudence claims, "*The emergence of this 'new' legal asset stems from the work of reducing to unity the legislation of the time, which was then marked by regulations of a sectorial nature placed to safeguard the constituent elements of the landscape and natural beauty such as soil, air and water, which had already been the subject since the time before the Constitution of protection in other forms and for other purposes, namely through a regime of an essentially conservative nature proper to the protection of landscape and cultural protection (in particular with Law June 9, 1939, no. 1497 - Protection of Natural Beauties; later repealed in 1999). To the typical characteristics of traditional conservation protection, essentially focused on constraints on human activities to protect the value of natural and scenic beauty of the property, there was in that era of industrial growth developing by legislation an action of supervision, prevention and repression of human conduct harmful to the individual constituent elements of the environment on the basis of sectoral regulatory disciplines.*"[4]

According to the Administrative Supreme Court, the first real response of the Italian legal system toward environmental protection came only in the 1980s with Law No. 349 of 1986, which established the Ministry of the Environment and provided rules on environmental damage. The former has the function of promoting, preserving and recovering environmental conditions in accordance with the fundamental interests of the community and the quality of life, as well as the conservation and enhancement of the national natural heritage and the defense of natural resources from pollution. Article 18 stipulates that "*any intentional or negligent act in violation of provisions of law or measures adopted on the basis of law that compromises the environment, causing damage to it, altering it, deteriorating it or destroying it in whole or in part, obliges the perpetrator to pay compensation to the state.*"

It is the beginning of an organic protection towards the environmental good.

The Constitutional Court, called upon to judge the legitimacy of the abovementioned rule, in its historic judgment of Dec. 30, 1987, No. 641,[5] consecrates the environment as a "unitary intangible good although with various components, each of which may also constitute the object of care and protection; but all of which, as a whole, can be traced back to unity."

The Court specifies that, "*The fact that the environment may be usable in various forms and different ways, just as it may be the subject of various norms that ensure the protection of the various profiles in which it is expressed, does not diminish or affect its nature and substance as a unitary good that the legal system takes into consideration. The environment is protected as a determinative element of the quality of life. Its protection does not pursue abstract naturalistic or aestheticizing*

[4] Cons. St., Adunanza plenaria, n. 10 del 2019, in www.giustizia-amministrativa.it.

[5] In www.cortecostituzionale.it.

purposes, but expresses the need to a natural habitat in which man lives and acts and which is necessary to the community and, for it, to citizens, according to widely felt values; it is imposed first and foremost by constitutional precepts (Articles 9 and 32 Const.), so it rises to a primary and absolute value."

The Court, moreover, points out that the environment is "*a legal good in that it is recognized and protected by norms*" *and cannot be appropriated, but, "belonging to the category of so-called free goods, it is usable by the community and individuals.*"

The matured conviction to protect the environment, which had already found several responses in the international arena[6] and then also within some rulings of the Court of Justice,[7] leads to an amendment of the Constitutional Law No. 1 in 2001 and the inclusion of the environment, ecosystem and cultural heritage among the matters devolved to the exclusive competence of the State.

With the Legislative Decree of April 3, 2006, No. 152, it has been provided "Norms on environmental matters," drawing an organic protection of the environment. These norms still represent the reference for environmental protection in the Italian legal system.

Article 1 specifies that the aforementioned legislative decree "has its primary objective in the promotion of the levels of quality of human life, to be achieved through the preservation and improvement of the conditions of the environment and the prudent and rational use of natural resources."

However, Constitutional Law No. 1 of Feb. 11, 2022 introduced for the first time a clear recognition of environmental protection in the Constitution.

On the one hand, and for the first time, the lawmaker intervened on the fundamental principles, supplementing Article 9 with a second paragraph, in which it protects "*the environment, biodiversity and ecosystems, also in the interest of future generations,*" as well as, through state law, to regulate ways and forms of animal protection; on the other hand, it amended Article 41 Const, paragraphs 2 and 3,

---

[6] Consider, to take just one example, the 1992 Rio Declaration on Environment and Development, which solemnly proclaimed (in Article 1) that "Human beings are at the center of concerns for sustainable development. They have the right to a healthy and productive life in harmony with nature." Also of significance in the development of environmental protection is the Kyoto Protocol concluded on December 11, 1997, which is one of the most important international legal instruments aimed at combating climate change. It is the first international agreement that contains commitments by industrialized countries to reduce emissions of certain greenhouse gases responsible for global warming.

[7] As early as 1985, the Court pointed out that the environment was one of the "essential aims of the Community" (EU Court of Justice, Feb. 7, 1985, Case 240/83) and later, in 1988, an "overriding requirement" (EU Court of Justice, Sept. 20, 1988, Case 302/86) that could limit the application of Article 30 TEC (now Article 34 TFEU) on the free movement of goods, provided that the trade-restrictive rule met the criteria of non-discrimination and proportionality. With the entry into force of the Single European Act in 1987, competences in the field of the environment became part of the Treaty, through the inclusion of a specific title (VII), consisting of three articles, and the definition of the main objectives of community action (preservation, protection and improvement of the quality of the environment, protection of human health, prudent and rational utilization of natural resources).

adding the environment and health alongside the original limits on private economic initiative and placing environmental ends alongside the social ones to which public and private economic activity should be directed and coordinated.[8]

The constitutional reform merely crystallizes well-established jurisprudential principles, but nonetheless represents further evidence of the Italian legal system's sensitivity to an increasingly enhanced protection of the environment.[9]

3 The Environment Before Administrative Law Courts

Administrative jurisprudence has recognized environmental protection as a unitary good since the 1970s.[10]

In particular, it has been pointed out, thanks also to the elaboration of the doctrine, that "*the normative qualification of an environmental asset arises from the ascertainment of its objective material characteristics, so that the legal act (law or administrative measure) that qualifies it and institutes the relative protection regime has a declaratory nature, of ascertainment of a quality immanent to it; furthermore, with respect to the unitary consideration of the property for the purposes of environmental protection, the aspects linked to its material composition (i.e. whether the property is in turn composed of a collection of individual material things) and its dominical, public, collective or private regime to which it is subject are recessive,*

[8] Riva (2023).

[9] The jurisprudence of the Constitutional Court: (a) has denied the possibility of identifying a "matter" in the technical sense that can be qualified as "environmental protection," since it does not seem configurable as a strictly circumscribed and delimited sphere of state competence, and, on the contrary, inextricably invests and intertwines with other interests and competences" (Cort. cost n. 150, 2018; n. 407, 2002); (b) recognized the exclusive legislative power of the state over determinations that meet needs deserving of uniform regulation throughout the country. In particular, the discipline of the EIA, pertaining to the "protection of the environment and the ecosystem," as an exclusive legislative matter placed in the hands of the State, contains within it provisions capable of transversally affecting the competences of the Regions in order to provide throughout the national territory a uniform discipline. (Cort. cost. n. 198, 2018; n. 77, 2017; n. 249, 2009; n. 407, 2002); (c) affirmed that Administrations and businesses must always necessarily and loyally cooperate to seek the solution most conducive to safeguarding environmental values, so that the maximum achievable improvement in the environmental performance of the establishment is identified, a true "balance point" so that a "tyrant right" does not emerge (Cort. cost, n. 58, 2018); (d) has held that state regulation constitutes a limit for regulatory interventions by the Regions and Autonomous Provinces that, while pertaining to matters within their competence, nevertheless present profiles of interference with said environmental protection requirements (Cort. cost. n. 180, 2015; n. 58, 2015; n. 67, 2014; n. 314, 2009), *See Rassegna monotematica di giurisprudenza redatta dall'Ufficio del massimario della Giustizia amministrativa* (2023), *Le valutazioni ambientali*, in www.giustizia-amministrativa.it.

[10] *See* Aru and Betzu (2016); Scognamiglio (2013); Comporti (2013); Fimiani (2011); Bolognini (2009).

since the elevation of the property to the status of environmental property in any case determines a functionalisation of the relative faculties."[11]

3.1 Council of State, Plenary Assembly No. 10, 2019

One of the most problematic issues that reached the Plenary Council concerned the environmental tort and the need to consider this figure as existing well before its recognition by law in 1986 (Law No. 349).

To the Plenary Council was, therefore, referred the question of whether an order for the reclamation of polluted sites pursuant to art. 244 of the so-called environmental code (legislative decree 3 April 2006, n. 152) a company not responsible for pollution prior to the time in which the reclamation of polluted sites was required by law, but which took over from the responsible one as a result of merger by incorporation in the regime prior to the company law reform.

Plenary Assembly No. 10 of 2019 clarified that even before reclamation was introduced with Article 17 of Legislative Decree No. 22 of 1997, environmental damage constituted a tort, provided for in Article 2043 of the Civil Code.

The Plenary Council distinguished reintegrative measures from restorative measures of the damaged environmental good.

Both the measures introduced by Legislative Decree No. 22 of 1997, later transfused into the currently applicable Environmental Code, and the remedy of compensation for damages already recognized on the basis of Art. 2043 of the Civil Code, and then by Law No. 349 of 1986, have, according to the supreme court, the same (*"restorative-reintegrative"*) function of protecting the environment. The former aim in particular not to limit protection to the monetary equivalent of the damage produced, as was the case in the past, but to prevent its occurrence and, if not, to place the removal and related charges on the responsible party. The prevention function is, moreover, consubstantial to the general action of public authorities to protect the environment. It emerges from the legislative evolution on the subject, which was realized on the impetus of European law and the pivotal *"polluter pays"* principle (now enshrined at the supranational level by Art. 191 of the Treaty on the Functioning of the European Union and Art. 3-ter of the Environment Code under d.lgs. no. 152 of 2006), by which the aim is to make the costs of damage caused to the environment fall on the responsible parties rather than on the community and repaired with public money, or on blameless parties who nonetheless find themselves in some material or legal relationship with the polluted site. In fact, only through preventive and protective action it is possible to prevent damage to the environment and, therefore, having ascertained the relevant responsibilities, all the necessary procedures must be activated to remove the harmful situation, with the risk of their non-implementation and the impossibility of full reparation by equivalent of the costs of

[11] Cons. St., Adunanza plenaria, n. 10, 2019.

restoration. In addition to the preventive function, European legislation - in particular Directive 2004/35/EC of April 21, 2004 *on environmental liability with regard to the prevention and remedying of environmental damage*") - have thus emphasized the need to ensure environmental restoration, based on the observation, expressed in Recital No. 13, that the civil liability provided by national legal systems is not always "*a suitable instrument for dealing with pollution of a diffuse and general character in cases where it is impossible to link negative environmental effects to acts or omissions of certain individual persons*." The amendments to the Environmental Code, and in particular to Article 311, on "*Compensatory action in a specific form*", introduced by the 2013 European Law (Law No. 97 of August 6, 2013), centered as a whole on giving it priority status and instead subordinate to protection by equivalent in order to protect the environment in a more effective way.

It can therefore be considered unequivocal, according to the Plenary Assembly, that the measures introduced in 1997, and now governed by Articles 239 of Legislative Decree No. 152 of 2006, have as a whole a purpose of safeguarding the good of the environment against any event of danger or damage, in which any matrix of sanction with respect to the relevant author is absent. Further, such measures do not belong to the "*lato sensu punitive law*," although an ascertainment of responsibility is indispensable (see Cons. St., Ad. plen., ord. Nov. 13, 2013, no. 21 and 25).

These measures take the form of obligations, under the supervision of the competent public administration (through the approval of the reclamation project) with the sole purpose of restoring the factual situation prior to the environmental pollution and removing its effects. Reclamation constitutes, in a nutshell, "*a publicistic instrument aimed not at monetizing the decrease in the relative value (in this substantiating protection by equivalence), but at enabling its material recovery at the care and expense of the person responsible for the contamination*." Reclamation has, therefore, the function of reinstatement of the legal asset "injured by the wrongful act" proper to civil liability and echoes the remedy of reinstatement in specific form pursuant to Art. 2058 of the Civil Code, already expressly provided for environmental damage by Art. 18, paragraph 8, of Law No. 349 of 1986 in the form of "restoration of the state of the places at the expense of the person responsible," as previously set out. Before, therefore, the latter provision was introduced, the general case of reinstatement in a specific form pursuant to the same Article 2058 of the Civil Code was in any case to be considered applicable by virtue of the alternating relationship with the remedy of monetary equivalent provided for in the case of a tort by Article 2043 of the Civil Code. Moreover, similarly to what happened at the time of the establishment of the Ministry of the Environment and the case of tort relating to that asset, even when the reclamation of polluted sites was introduced into the legal system, with the aforementioned Legislative Decree No. 22 of 1997, the limit of excessive onerousness (provided for by paragraph 2 of Article 2058) was not provided. This is, in the opinion of the Plenary Assembly, a difference that, far from marking the incompatibility between the remedy of the reclamation of polluted sites with respect to the institution of tort liability, is instead explained in the light of the preeminent value assigned by the Constitution to the environment in the hierarchy of legal goods, on the basis of the aforementioned Articles 9 and 32 of the Fundamental Charter and the above-mentioned collective dimension of the damage to that good.

The Plenary Assembly reiterates that in the case of environmental damage, with the introduction of reclamation obligations by Art. 17 of Legislative Decree No. 22 of 1997, the area of illegality was not extended with respect to the conduct previously considered compliant with the law, but the remedies were broadened with respect to acts of aggression against the environment already considered damaging to a legal asset worthy of protection.

Article 18 of Law No. 349 of 1986 did not, therefore, have an innovative scope in terms of considering the environment as a protected legal asset. Instead, the genetic source of its protection was identified "*directly in the Constitution, considered dynamically, as current and living law, through the combined provision of those provisions (such as Articles 2,3,9,41 and 42) that concern the individual and the community in its economic, social, and environmental habitat.*"

On this point, moreover, see Cass. civ. III, June 19, 1996, No. 5650, confirming the sentence of compensation for damages suffered by the municipalities involved in the Vajont disaster, which identified as the only new element the division of competencies in environmental protection between the state, local entities and environmental protection associations.

The Plenary Assembly added that the tort thus typified enshrined on the most general and systematic level the collective and super-individual dimension of environmental damage, including all social costs resulting from assaults on the natural habitat, consisting of external diseconomies of production and emerging, among other things:

- by the reference made by Paragraph 5 of Article 18 l. No. 349 of 1986 to the "cost necessary for restoration" for the determination of the damage, in the case of the impossibility of its "precise quantification."
- by the provision contained in Paragraph 8, whereby in case of conviction it is ordered "where possible, the restoration of the state of the place at the expense of the responsible party."

The aforementioned Constitutional Court of December 31, 1987, no. 641, affirmed that the damage that can be compensated for the environmental tort "*is certainly patrimonial, although it is freed from an arithmetic-accounting conception and is rather concretized in the economic relevance that the destruction or deterioration or alteration or, in general, the impairment of the good has in and of itself and that is reflected on the community which is burdened with economic burdens*" and it is pointed out that the environment, although not "*an appropriatable good,*" nevertheless "*lends itself to being valued in economic terms and a price can be attributed to it.*"

According to the Constitutional Court, the latter corresponds to the costs of public action for preservation and protection "*including policing that regularizes the activity of subjects and creates oversight of compliance with constraints; management of the property in an economic sense with the aim of maximizing the enjoyment and usability of the community and individuals and developing environmental resources.*" On this basis, the Court continues, "*it is therefore possible to compare the benefits with the alterations; the estimation and planning of preservation, improvement and recovery interventions can be carried out; the costs of damage can be assessed. And for all this, the environmental impact can be put back into monetary terms. All of this makes it possible to give the environment and thus environmental damage an economic value.*"

The Plenary Assembly specifies, therefore, that the restorative function of environmental tort is not limited only to the difference in value of the injured asset compared to what it had before the damage but it is extended to all the costs necessary to restore the overall damage inflicted on the natural ecosystem. Under this perspective, the damage to the environment compensable under Article 18 l. no. 349 of 1986 seems to be qualified in specific form under Article 2058 of the Civil Code (see Cass. civ., I, July 3, 1997, no. 5993; III, ord. November 21, 2017, no. 27546). However, this form of restoration does not encounter the limit of excessive onerousness, so that in this respect the protection of the environment is strengthened compared to ordinary tort instruments.

3.2 *Protection of "diffuse and collective interests"*

Always discussed in administrative jurisprudence is the legitimacy of associations placed to protect the environment to challenge administrative measures that harm the interests protected by those associations.

The issue, in Italian law, recalls the distinction between diffuse and collective interests[12] and goes back to an affair involving precisely environmental protection.

In particular, the Council of State, in its landmark ruling No. 253 of 1973,[13] deemed admissible a judicial action brought to protect a public interest (in this case, safeguarding the beauty of Lake Tovel), by a private association (Italia Nostra) that provided in its bylaws as its own purpose the protection of this interest. The final recognition of the protection of the collective interests later came through the Plenary Assembly's ruling No. 24/1979, which admitted the judicial protection of collective interests in the presence of an entification of diffuse interests with a gradual approximation of the same to the legitimate interest.

[12] Diffuse interests are non-personalized interests of an adhesively ad-hoc character referring indistinctly to a collectivity of subjects (e.g., consumer, saver). The diffuse character of the interest derives from the characteristic of the good that is not su-ject to individual appropriation or enjoyment (environment, landscape, artistic heritage). Where, then, these interests referable to a community of individuals are, also, characterized by the fact that this community has organized itself through the establishment of a body responsible for their protection, they are defined and identified as collective interests. See on this point, for an in-depth study, Santise (2021a).

[13] Cons. St. n. 253, 1973, in *Foro It.*, 1974.

From that time onward, therefore, there has been an awareness of the admissibility of judicial protection of collective interests as long as an exponential body of the same can be identified; the latter can, therefore, be of the most varied nature, but it must be characterized by an organization functionalized for the protection of class interests.

The prevailing jurisprudence affirms the principle that "*the diffuse interest is transformed into a collective interest, and becomes, therefore, a legitimate interest protectable in court, when, regardless of the subsistence of legal personality, the entity demonstrates its representativeness with respect to the interest it intends to protect. Representativeness that must be inferred from a series of indices elaborated, not without contrasts, by jurisprudence over the past thirty years.*"[14]

In more details, the criteria developed by case law to identify the legitimacy of entities can be summarized as follows: "*it is necessary first of all to point out that it must be an entity whose bylaws provide as an institutional purpose the protection of a given asset for collective use, that is, of a given widespread or collective interest. Secondly, the entity must be able, by its organization and structure, to achieve its purposes and be endowed with stability, in the sense that it must carry out its activities externally on an ongoing basis. The action therefore must assume connotations such as to create in the entity a substantive situation deserving of protection, in order to exclude the legitimacy of so-called associations of convenience, whose activities do not reflect actual collective needs. Finally, the collective body must be the bearer of a localized interest, that is, there must be a stable territorial connection between the area of afference of the body's activity and the area in which the asset for collective use that is allegedly injured is located* (so-called *vicinitas*)."

The recognition of the judicial protection of collective interests was first confirmed by article 18 par. 5 which provided that "*Associations identified on the basis of Article 13 of this law may intervene in judgments for environmental damage and appeal to administrative jurisdiction for the annulment of illegitimate acts.*" Article 13 provides that "*environmental protection associations of a national character and those with a presence in at least five regions are identified by decree of the Minister of the Environment on the basis of the programmatic purposes and internal democratic order provided for in the statute, as well as the continuity of action and its external relevance, after receiving the opinion of the National Council for the Environment to be expressed within ninety days of the request [...].*"

The legislative provision that anchors the legal standing of associations to prior typification has raised the question of whether other associations not identified by ministerial decree can sue.

Thus, two opposing approaches became widespread.

According to the minority, only the legislative could identify the collective bodies entitled to challenge the administrative measure.

The opposite theory instead claimed that associations not singled out by the legislative should also be allowed to have legal standing.

---

[14] Cons. St. n. 3507, 2008, in www.giustizia-amministrativa.it.

The latter interpretive perspective was the basis for the February 20, 2020 Council of State Plenary Assembly ruling No. 6, which adhered to the dual-track theory.

In particular, the administrative judges pointed out that diffuse interest "*is a substantial interest that exceeds the sphere of individuals to assume a shared and non-exclusive connotation, as the interest of 'all' in relation to a good from whose individual enjoyment no one can be excluded, and whose enjoyment does not exclude that of all others*". In this sense, the diffuse interest "*does not rise to a substantial 'personal' situation susceptible to judicial protection (i.e., it is not protected by a right or legitimate interest) since the system cannot offer legal protection to a substantial individual interest that is not wholly or partially exclusive or susceptible to individual appropriation.*"

Further, "it is only projected into the collective dimension that the interest becomes susceptible to protection, as a synthesis and not summation of the interest of all the members of the community or category, and that therefore is endowed with the protection proper to the legitimate interest, so that [...] it must recisely exclude that associations, in requesting in their own name judicial protection, activate a "right" of others. The legal situation acted upon is their own. It relates to widespread interests in the community or in the category, which vi-vono devoid of protection until a collective subject, structured and representative, embodies them. Not by virtue of a *fictio* but by virtue of a judgment of identification and selection of the interests to be protected, as well as the rigorous verification of the representativeness of the collective subject that promotes their protection" (points 6.1 and 6.2 of the judgment).

The Plenary Assembly observes that the circumstance that *"the care of the general public interest is entrusted to the administration does not detract, however, from the fact that it is subjectively referable, albeit indiscriminately, to social formations, and that the latter, in their associated dimension, represent the actual users of the common good whose care is in question. The situations are in fact different and heterogeneous: the administration has a duty to take care of the public interest and therefore enjoys a legal situation capable of affecting communities and categories (power); the associations representing the community or categories, on the other hand, embody the substantial interest, they are its beneficiaries, and therefore the legal situation of which they are holders is that proper to the legitimate interest."*

On this basis, therefore, the Plenary Assembly adheres to the dual-track theory, allowing associations, even those not pre-identified by the legislature, to take legal action as long as they comply with the criteria established over time by case law.

This seems the solution to adhere with because it is in line with the principle of effectiveness of protection and consistent with Articles 2 and 118 of the Constitution, and not in violation of Article 81 of the Code of Criminal Procedure, since associations assert their own interest and not the interest of others.

Subsequent jurisprudence has not only embraced the principles distilled by the Plenary Council, but has extended environmental protection.

In particular, Cons. St., April 19, 2021, no. 3170 starting from the assumption that the concept of protection of the environmental good must be understood in a broad sense, in order to include any situation capable of causing harm to the environment, albeit directly aimed at the protection of interests of a more circumscribed or different nature, held that, also with reference to objections addressed to acts of a town-planning nature, it is possible to recognize the legitimacy to act of environmental associations provided for by art. 18, paragraph 5, of Law No. 349/1986, whenever it is inferred that such acts are likely to endanger the environment.

For several years now, administrative jurisprudence has been consistent[15] in recognizing the *ex lege* legitimacy of environmental associations not only in the case of acts pertaining to environmental matters "*strictu sensu*", but also for those that, in a broad sense, "*affect the quality of life in a given territory*" In particular, it was clearly stated that "*the legitimacy to act must be recognized to environmental associations, to which Articles 13 and 18 paragraph 5 of Law 349/1986, both in relation to the protection of environmental interests in the strict sense summarized by the presence of a specific constraint, and for environmental interests in the broad sense, including precisely the preservation and enhancement of cultural heritage, the environment in the broad sense, the urban, rural and natural landscape, monuments and historic centers and the quality of life*" (Cons. St. Apr. 14, 2011, no. 2329, and Council of St. Oct. 9, 2002, no. 5365; further, *ex multis*, also T.A.R. Campania, Naples, May 3, 2018, no. 2964; T.A.R. Abruzzo, L'Aquila, Nov. 20, 2001, no. 679, which had already held the existence of the legitimacy of environmental associations in cases concerning construction and urban planning matters against acts of variation of a land-use plan).

According to administrative jurisprudence, the legitimacy of standing recognized to environmental associations is not, therefore, limited to cases in which the object of the complaints is the violation of norms directly and exclusively placed to safeguard the environment. The protection of environmental interests can, in fact, also be pursued through the challenge of general administrative acts to the extent that they negatively affect environmental issues (see T.A.R. Lombardia, October 22,

---

[15] Cons. St. n. 2095, 2013, on the point specified, first of all, that the Associations identified by the Minister of the Environment pursuant to Articles 13 and 18 of Law No. 349/1986 and Articles 309 and 310 of Legislative Decree No. 152 of April 3, 2006 are legitimized, but that: "the criteria, long developed by case law, based on the actual and not occasional commitment to the protection of certain widespread or super-individual interests, on the existence of a statutory provision qualifying said protection as an institutional task of the association and its territorial articulations as well as on the spatial proximity of the source of the feared harm to the protected legal interests to the main center of the association's activity or its specific peripheral structure, continue to apply to all associations lacking the aforementioned legal standing." However, although in the first instance it had been recognized as an additional requirement that of the "representativeness of the local community of reference," the Fifth Section pointed out that, "according to the most recent and well-established jurisprudence of this Council of State, it is not sufficient, it is also necessary to prove a specific and concrete injury resulting from the work opposed, even in terms of mere depreciation of neighboring properties, as a result of the challenged act, otherwise only an abstract legitimacy "ad causam" could be said to be demonstrated, but not also that concrete and current injury that justifies the existence of the interest to appeal."

2013, no. 2336). In such cases, in fact, it can be held that the consistency of the (environmental) interests brought before the court remains unchanged.[16]

3.3 Remediation Obligations of the Bankruptcy Trustee

The Council of State, Plenary Meeting No. 3/20211,[17] dealt with the case of whether it is possible to require the bankruptcy receiver, who has taken possession of the property, to clean up the area affected by pollution.

Thus, in one case, the remediation order concerned the issue regarding the transmissibility of the remediation obligation in the event of a corporate merger (Ad. pl. No. 10/2019); here, however, the remediation obligation is at issue for the bankruptcy trustee, who has no connection with the company that caused the environmental damage.

In fact, the bankruptcy trustee is appointed only to manage the bankruptcy estate.

Plenary Assembly No. 3/2021 believes that the obligation to clean up can also be imposed on the bankruptcy trustee, consistent with the rationale of ensuring maximum protection for the environmental good and in line with the regulations of the Environmental Code, which provides for the "polluter pays" principle also recognized in the supranational sphere.

The Plenary Assembly states, therefore, that obligated is not only the person responsible for the pollution, but also the person who has some relationship with the property. The prerequisite for the obligation to clean up to be imposed is, therefore, that the person is in possession of the property.

However, the legal notion of "possession" and the distinction with "detention," a distinction peculiar to our legal system but unknown in EU law, is not relevant for this purpose. What matters for the purpose of imposing the obligation is that there is contact with the polluted property and that the person has at least possession of it.

The receiver has possession of the property, being able to dispose of it, albeit with the specific purpose of liquidation.

Therefore, as stated by the P.A., the receiver is required to reclaim the area.

The reconstruction now illustrated, in addition to being more in line with environmental protection, is consistent with bearing the economic burden on the company's assets (i.e., it falls on the company itself). And this is so even if at the time of the receiver's takeover there are no funds to carry out the reclamation obligation, which the receiver must bear. The bankruptcy receiver is not a successor, but is a temporary manager of the assets, functionalized for the purpose of satisfying creditors.

[16] Cons. St. n. 3711, 2015; Cons. St. n. 36, 2014; Cons. St. n. 839, 2015, in www.giustizia-amministrativa.it.

[17] *See* News n. 14, 2021, *Ufficio studi, massimario e formazione della Giustizia amministrativa*, in www.giustizia-amministrativa.it.

4 Remediation Obligations of the Owner/Manager of the Polluted Site

Another issue discussed was whether the owner/manager of the polluted site can be required to provide measures to clean up the area.

The Supreme Court, in Judgment No. 3077 of Feb. 1, 2023,[18] pointed out that an obligation cannot be imposed on those who have not directly caused the pollution. The only measures that can be deployed are limited to what is provided by Article 253 of the Environment Code.[19] To the owner who did not cause the pollution are, likewise, inapplicable the criteria for imputation of liability under Articles 2050 and 2051 of the Civil Code,[20] since the rules defined in Part IV of the Environment Code for the remediation of contaminated sites are special within the legal system.

In the above mentioned ruling, the Supreme Court essentially drew the perimeter of the obligations of the owner not responsible for the pollution, limiting the application of the multiple measures provided by the Environmental Code to what is expressly provided for in Article 253.

The principle provided by the Supreme Court, according to which the owner who is 'not responsible' for the pollution is obliged, pursuant to Article 245, paragraph 2, of the same Legislative Decree No. 152 of 2006, to take the preventive measures referred to in Article 240, paragraph 1, lett. (i), but not the emergency safety and remediation measures referred to lett. (m) and (p), is an application, on the European Union side, of Directive no. 2004/35/EC, which has resulted in the application in domestic legal systems of the "*polluter pays*" principle, stipulating that the operator whose activity has caused environmental damage or the imminent threat of such damage will be held financially liable so as to induce operators to take measures and

---

[18] In www.cortedicassazione.it.

[19] The provision reads as follows: 1. The interventions referred to in this title constitute a real charge on contaminated sites when carried out ex officio by the competent authority pursuant to Articles 250 and 252, paragraph 5. The real charge shall be entered in the land registers kept by the offices of the Land Agency upon approval of the reclamation project and shall be indicated in the land use certificate. 2. The expenses incurred for the interventions referred to in paragraph 1 shall be assisted by a special real estate lien on the said areas, pursuant to and in accordance with Article 2748, second paragraph, of the Civil Code. Said lien may also be exercised to the detriment of the rights acquired by third parties on the property. 3. The privilege and the repayment of expenses may be exercised, against the owner of the site blameless for the pollution or the danger of pollution, only following a reasoned order of the competent authority justifying, inter alia, t h e impossibility of ascertaining the identity of the person responsible or justifying the impossibility of exercising recourse actions against the same person or their unfruitfulness. 4. In any case, the owner not responsible for the pollution may be required to reimburse, on the basis of a reasoned measure and in compliance with the provisions of Law No. 241 of August 7, 1990, the costs of the interventions adopted by the competent authority only within the limits of the market value of the site determined as a result of the execution of such interventions. In the event that the owner not responsible for the pollution has voluntarily remediated the polluted site, he shall have the right of recourse against the polluter for the expenses incurred and any greater damage suffered.

[20] These are forms of aggravated extra-contractual liability, or, by some considered to be objective in nature, because it disregards the subjective element of the injurer.

develop practices to minimize the risks of environmental damage. Supranational legislation has clearly privileged the goal of eliminating environmental damage over the compensatory (equivalent) perspective, and it has enshrined the rule that it is the operator who causes environmental damage or an imminent threat of such damage who should in principle bear the cost of the necessary preventive or remedial measures, while the cost of the competent authority's remedial action should be borne by the operator (the person who performs or controls a professional activity or to whom decisive economic power over the technical operation of that activity is delegated) by including the cost of assessing the damage and assessing the imminent threat.

The text of Legislative Decree No. 152 of 2006 has aligned with the aforementioned directive, with the result that the reactions to environmental damage are represented by effective restoration (primary remediation) or complementary and compensatory remediation in accordance with the principles of precaution, preventive action, correction, as a priority at the source, of the damage caused to the environment, as well as the "*polluter pays*" principle (Art. 3-ter).

The Supreme Court has clarified the following criteria:

1. primary repair aims to restore damaged natural resources and/or services to their original condition;
2. the complementary one, where they do not return to their original condition, tends to compensate for the failure to fully restore damaged natural resources and/or services;
3. compensatory repair equalizes the temporary loss of resources from the date of occurrence of the damage to the date when the primary repair did not have full effect.

Article 311 Legislative Decree No. 152 of 2006 establishes the liability of those who carry out specific professional activities.

Article 308 excludes the costs of precautionary, preventive and restoration actions taken in accordance with the provisions of Part VI from being borne by the operator if he can prove environmental damage or imminent threat of such damage:

1. was caused by a third party and occurred despite the existence of abstractly suitable safety measures;
2. is a consequence of compliance with a mandatory order or instruction issued by a public authority, other than those issued as a result of an issue or incident attributable to the operator (Paragraph 4);

In addition, the manager shall not be required to bear the costs of actions under paragraph 5 taken in accordance with the provisions of Part VI if he proves that he is not attributable to wilful or negligent conduct and that preventive action to protect the environment was caused by:

1. an issue or event expressly permitted by a permit in full compliance with the conditions therein;

2. an emission or activity or any other way of using a product in the course of an activity that the operator demonstrates was not considered likely to cause environmental damage according to the state of scientific and technical knowledge at the time the emission was released or the activity was carried out.

The exegetical approach of the Supreme Court seems to be in partial contrast to the guidance expressed by Cons. State, Ad. plen, Oct. 22, 2019, No. 10, according to which "*The reclamation of the polluted site can also be ordered against a company that is not responsible for the pollution, but which has succeeded it as a result of merger by incorporation, in the regime prior to the reform of corporate law, and for conduct prior to when the reclamation was introduced into the legal system, the harmful effects of which persist at the time of the adoption of the measure.*" Thus, the Plenary Assembly, affirms that the damage to the environment can be framed in the general scheme of civil tort under Art. 2043 of the Civil Code.

5 The Jurisdiction

The circumstance that principles on environmental law are distilled by both the Supreme Court and the administrative law courts demonstrates the competing jurisdictions on this issue and the related need to define the allocation of jurisdiction.

The administrative judge is the judge called upon to verify whether the public administration has legitimately applied the rules dictated to protect the environment whenever a significant transformation of the territory, likely to adversely affect environmental values, comes to the fore.[21] In the Italian legal system, environmental protection is administered through two types of jurisdictions. Supreme Court 23 April 2020, no. 8092 specified that the administrative law courts have jurisdiction on disputes arising from the challenge of administrative measures adopted by the Ministry of the Environment for the precaution, prevention and restoration of the environment, under Art. 310 d.lg. no. 152 of 2006. Instead, civil law courts have jurisdiction with regard to compensation or injunction lawsuits brought by individuals to whom the fact producing environmental damage has caused damage to health or property, according to the provisions of Article 313, paragraph 7. The fact that the harmful activity is carried out in accordance with authorization measures of the public authority does not affect the jurisdiction (since the aforementioned measures cannot be recognized as having the effect of weakening fundamental rights of third parties) but exclusively on the powers of the civil law judge, who, in the event that the harmful activity derives from material conduct that does not comply with the administrative measures that make its exercise possible, will provide for sanctioning, by inhibiting it or bringing it back into compliance, the activity revealed to be harmful because it does not comply with the administrative regulation, while, in the hypothesis in which such compliance turns out to be the case, it will have to

[21] Valletta (2022).

disapply the aforementioned regulation and impose the cessation or adjustment of the activity in order to eliminate its harmful consequences.[22]

6 "Silent Consent" Between Public Administration and Private Actors

The jurisprudence of the Council of State has generally affirmed that in accordance with the EU legislation and the principles expressed by the Court of Justice of the European Union, the issuance and revocation of authorizations must result from an explicit measure and follow precise procedural rules, and that they are, as a rule, preventive in nature.

The Italian legislative, well aware that the silence of consent may conflict with the protection of special interests, meaning those pertaining to acts and procedures concerning cultural and landscape heritage, the environment, the protection of hydrogeological risk, national defense, public security and immigration, health and public safety, has provided, in Article 20, Paragraph IV, Law 241 of 1990 the exclusion of silence consent.

According to Article 13 Paragraph I of Law No. 394 of December 6, 1991 (framework law on protected areas) after the expiration of 60 days from the request for authorization for concessions or authorizations related to interventions, the authorization is deemed to be issued.

[22] See: Cass., Sez. Un. n. 2338, 2018, *in Giust. civ. Mass.*, 2018, where it is specified that the injunctive and compensatory protection of damages in a specific form have a priority and proper character with respect to the general compensatory action under Article 2043 Civil Code, and this is with reference both to the right to health of constitutional rank and to the right to property with regard to the protection granted to the holder of the right in rem by Article 844 Civil Code. Therefore, it belongs to ordinary jurisdiction the claim of the private individual who complains about the concrete way in which the production cycle is carried out, assuming its dangerousness to health or other fundamental personal rights and requesting the adoption of the necessary measures to eliminate the actual and potential damage and intolerable immissions; see also Cass., Sez. Un. n. 11,142, 2017, in *Giust. civ. Mass.*, 2017, where it is stated that: "it belongs to ordinary jurisdiction the claim of the private individual who complains about the concrete methods of exercise of the relevant production cycle, assuming the danger to health or other fundamental rights of the person and requesting the adoption of the necessary measures to eliminate the actual and potential damage and intolerable immissions, given that the challenged conduct integrates the material extrinsication of an ordinary business activity, when no particular executive rules or technical application are dictated directly in administrative measures, so that the public power does not result in any way involved ". See, also, Cass., Sez. Un. n. 2052, 2016, in *Foro amm.*, 2016, 2, 266 ss, according to which jurisdiction belongs to the ordinary court when the action is brought by subjects to whom the injury-producing fact has caused personal damage to health or property; and that in the conflict between the right of the enterprise to the exercise of the activity authorized by the public administration and the right to health, the latter prevails; Cass., Sez. Un. n. 23,536, 2019, in *Foro amm.*, 2020; Cass., Sez. Un. n. 20,571, 2013, in *Giust. civ. Mass.*, 2013; Cass., Sez. Un. n. 16,848, 2012, in *Foro amm.*, 2012; Cass., Sez. un. n. 15,660, 2005, in *Foro it.*, 2006.

The norm introduces the mechanism of silence consent in relation to the authorization that must be issued by the Park Authority with regard to building interventions that are to be carried out in the territory under its control; these are, therefore, interventions that could potentially affect the landscape, but which, nevertheless, are subject to the mechanism of silence consent. The Park Authority's authorizing control, therefore, concerns any intervention that is intended to be carried out on the area inside the perimeter in order to verify whether it complies with the Park Plan in Article 12 and the Park Regulations, in Article 11.

Thus, a potential contrast emerges between Art. 20 Paragraph IV co. of Law No. 241/1990, which, as seen, excludes silence consent in relation to proceedings concerning special interests, and Art. 13 Paragraph I of Law No. 394 of December 6, 1991, which adopts an opposite solution.

Administrative jurisprudence has, therefore, questioned whether or not the 1991 law was tacitly repealed following the rewriting of Article 20 Law No. 241/1990 by Law No. 80 of May 14, 2005. The Plenary Assembly of the Council of State, called upon to settle the jurisprudential contrast, in ruling No. 17 of July 27, 2016, found Article 13 Law No. 394/1991 to be in force and, therefore, the silent consent in relation to special interests to be operational.

This is for both formal and substantive reasons.

Under the first profile, according to the Supreme Administrative Court, there is no tacit or implicit repeal of Art. 20 Paragraph IV Law No. 241/1990, because Art. 13 Law No. 394/1991 provides a particular structure of a specific and special procedure, however, capable of ensuring the full protection of the protected interest.

According to the Council of State, there is no incompatibility between silent consent and the protection of special interests.

This assumption is also in line with the case law of the Constitutional Court and the Court of Justice.

In fact, the Plenary Assembly recalled several rulings in which the Constitutional Court[23] declared illegitimate regional laws that introduced silence-consent in environmental matters only because they were productive of the effect of determining lower levels of environmental protection than those provided by state law; likewise, it affirmed the constitutional illegitimacy of regional provisions that reduced the terms of silence-consent provided by the corresponding prescriptions of state laws; according to the principle that the state is responsible for determining the regulation throughout the national territory, with the consequence that regional intervention is possible only for achieving an extension of the levels of protection. The Constitutional Court has, therefore, never declared illegitimate state regulations introducing silence consent in relation to special interests.

The case law of the Court of Justice also does not preclude this possibility.

---

[23] Cort. Cost. n. 302, 1988; n. 437, 1992; n. 194, 1993; n. 26, 1996; n. 404, 1997; n. 315, 2009; n. 209, 2014.

The administrative judges usually refers to judgment of the Court of Justice Feb. 28, 1991, Case C-360/87, which held that tacit termination of proceedings is incompatible to EU law, only when an express administrative assessment such as a technical assessment or verification is necessary to ensure effectiveness of the protected interests (e.g. health protection).

7 "Silent Consent" Between Public Administrations

The generalization of silent consent came in force with the implementation of Article 17 bis Law No. 241/1990, introduced by Article 3 Law No. 124 of August 7, 2015, as recently amended.

This is the extension of the principle regulating the relations between public administration and private parties by Article 20 with respect to the relations among public administrations.

This general rule, unlike Article 20, also applies to cases where the Administrations are in charge of environmental, landscape-territorial, cultural heritage and public health protection interests.

The only difference from ordinary cases is that the time limit for providing their assessment is 90 days.[24] This new legislation refers to the generality of proceedings.

8 The Review of Environmental Assessments

The protection of the environment is guaranteed both by a series of assessments of the administrative authority, in charge of guaranteeing the environmental interest, and by an "intrinsic" review by the administrative judge of the same environmental assessments.

In the first respect, Legislative Decree No. 152 of April 3, 2006 (the so-called Environmental Code) provides in Title I of Part II the general and defining lines of the institutes of EIA (environmental impact assessment), SEA (strategic environmental assessment) and AIA (integrated environmental authorization), and describes their procedure analytically in the following provisions. The AUA (single environmental authorization) was established by Presidential Decree No. 59 of 2013, "*Regulations governing the single environmental authorization and the*

[24] The Council of State in its advisory opinion n. 1640 of July 13, 2016, pointed out that Article 17 bis stands as a new paradigm in horizontal relations (between public administrations) in all cases in which the procedure is destined to conclude with a multi-structured decision, because it introduces the mechanism of silence of consent between public administrations. That will allow the proceeding administration to conclude the procedure with the adoption of the final act, regardless of the inert behavior of the competent public administration.

simplification of administrative requirements in environmental matters incumbent on small and medium-sized enterprises and facilities not subject to integrated environmental authorization, pursuant to Article 23 of Decree-Law No. 5, converted, with amendments, by Law No. 35 of April 4, 2012," and it replaces a basic core of seven permits provided for by Legislative Decree No. 152 of 2006, as well as additional permits provided for by the regulations of each region. The AUA complements the prevention system, consisting of SEA, EIA and AIA, and it is aimed at facilitating small and medium-sized enterprises, including "micro-enterprises" as defined in Ministerial Decree April 18, 2005. VINCA or IA (environmental impact assessment) is an institution under European Union law designed to ascertain in advance whether certain projects are likely to have a significant impact on Sites of Community Importance (SCI), Special Areas of Conservation (SAC) and Special Protection Areas (SPA). Introduced by Article 6(3) of the Habitats Directive 92/43/EEC with the aim of safeguarding the integrity of sites through the examination of the interference of plans and projects not directly related to the conservation of the habitats and species for which they have been identified, but capable of affecting their environmental balance, it is regulated at the national level by Article 5 of Presidential Decree No. September 8, 1997. No. 597, as replaced by Article 6 of Presidential Decree No. 120 of March 12, 2003. Pursuant to Article 10, paragraph 3, of Legislative Decree No. 152 of 2006, the aforementioned assessment is integrated in EIA and SEA procedures. In cases of integrated EIA-INCA, SEA-INCA procedures, the outcome of the impact assessment is binding for the purpose of expressing the reasoned SEA opinion or EIA measure, which can be favorable only if there is certainty regarding the absence of significant negative impact on Natura 2000 sites.[25]

Administrative jurisprudence has pointed out that the administration is not called upon to ascertain the existence of possible environmental impacts of the work, but is required to seek, in the comparative weighing of potentially conflicting instances, a balance between the interests pursued with the realization of the opus and the opposing requirements of preservation of the environmental context *lato sensu* intended. He went on to specify that environmental assessments are not a mere technical act of management or administration in the strict sense, since they are rather measures by which a real function of political—administrative direction is exercised with particular reference to the proper use of the land, in the broad sense, through the care and balancing of the multiplicity of conflicting public interests (urban planning, nature, landscape, as well as economic—social development) and private interests.

It has also made it clear in case law that although the principle of cumulative application of environmental assessment disciplines applies—as a result of which each of the regulations must be applied in full, without prejudice to the application of the others—the same is tempered by principles and criteria of efficiency and

---

[25] Ufficio del massimario della Giustizia amministrativa (2023), Rassegna monotematica di giurisprudenza, Le valutazioni ambientali, in www.giustizia-amministrativa.it.

simplification of the decision-making process, aimed at reducing the bureaucratic burden on stakeholders.

With specific regard to the procedural profiles, the jurisprudence of the Council of State, regardless of the specific environmental assessment from time to time examined for the purposes of dispute resolution, has consistently affirmed that the judicial evaluation of legitimacy, excluding absolutely the substitutive character of the same, must show the existence of flaws detectable *ictu oculi*, due to their abnormality, unreasonableness, contradictoriness and superficiality in view of the intense profiles of administrative discretion on the level of appreciation of the public interests involved and their weighing against the interest in the execution of the work.[26] It is not, therefore, permissible for the administrative judge to substitute his own assessment for that reserved for the discretion of the administration, constituting hypothesis of prohibited encroachment of the jurisdiction of legitimacy in the sphere reserved for the public administration[27]; the judicial review must be strictly maintained on the level of verification of the non-pretextuality of the assessment of the factual elements acquired and cannot make use of criteria aimed at highlighting the mere non-sharing of the assessment itself. This is because the review of technical evaluations finds its limit in the evaluations reserved by special rules to the administration and cannot go so far as to verify the accuracy and agreeability of the administration's technical appreciations in environmental matters. It follows that the administrative judge cannot order c.t.u. or verification in order to exercise the most penetrating controls, with particular regard to the ascertainable profiles, and the technical expertise cannot entail an inadmissible substitute judgment for a technical discretionary assessment carried out by the Administration and entrusted to specific Bodies in possession to the highest degree of the necessary technical and scientific expertise—and as such "infungible," except in cases of macroscopic abnormality or illogicality.[28]

9 Conclusion

Environmental law is a system in which the integration of sources has enabled the birth of law itself. Starting from few general principles under international law, environmental law has then developed in a multiple sources of domestic law (constitutional, civil, criminal, administrative law).

Within this multilevel framework, Italian Courts have been playing a pivotal role over the years.

---

[26] Cons. St. sez. IV, n. 1240, 2018; sez. IV, n. 1392, 2017.

[27] Cass. civ. sez. un., n. 2312 and 2313, 2012; Cort. cost., n. 175, 2011; Cons. St., sez. VI, n. 871, 2011.

[28] Cons. Stato, sez. IV, n. 1240, 2018.

References

Aru S, Betzu M (2016) Il risarcimento del danno ambientale tra esigenze unitarie e interessi territoriali. Giur Cost

Bolognini S (2009) La natura (anche) non patrimoniale del danno ambientale e il problema della legittimazione attiva. Resp Civ Prev

Comporti G (2013) Il danno ambientale e l'operazione rimediale. Dir Amm

Fimiani P (2011) Nota a corte di cassazione, 22 marzo 2011, sez. II, n. 655. Giust Civ

Riva C (2023) L'ambiente in Costituzione. Cambia qualcosa per il penalista? www.sistemapenale.it

Santise M (2021a) Coordinate ermeneutiche di diritto amministrativo. G Giappichelli Editore, Torino

Santise M (2021b) Coordinate ermeneutiche di diritto civile. G Giappichelli Editore, Torino

Scognamiglio C (2013) Danno ambientale e funzioni della responsabilità civile. Giur Cost

Valletta A (2022) La tutela dell'ambiente, lo sviluppo sostenibile e il giudice amministrativo. Rivista della Corte dei Conti (www.rivistacorteconti.it)

Maurizio Santise is Judge at the Regional Administrative Law Court (T.A.R. Campania) and Adjunct Professor of Administrative law at Suor Orsola Benincasa University.

Biodiversity Law Before the Courts

Maria Vittoria Ferroni and Tiziana Bandini

Abstract The starting point of this research is the assumption that the rate of biodiversity loss, together with climate change, is looming on the planet. Human-induced changes to ecosystems and the extinction of species have been more rapid in the past 50 years than at any time in human history. Several scientific studies show that humanity has already exceeded some of the planetary boundaries, putting itself in danger.

This chapter, in its first part, focuses on the definition of biodiversity as a representation of life, as it includes from the smallest and most basic gene up to the most complex ecosystem, and about the legal multilevel framework for the protection of biodiversity.

The second part discusses how the legal system and Courts respond to biodiversity's decline. The caselaw mentioned in this chapter tend to highlight the multifaceted nature of the protection accorded to biodiversity in the world. The different anthropocentric or ecocentric approach to nature changes the type of rights and of protection recognized.

Maria Vittoria Ferron drafted the par. 1, 2, 3, 3.1, 6, 6.1, 6.2; Tiziana Bandini drafted the par. 4, 5, 5.1, 5.2, 5.3, 5.4, 6.3, 6.4. Both Authors drafted the conclusions 7.

M. V. Ferroni (✉)
Department of Political Sciences, Sapienza University of Rome, Rome, Italy
e-mail: mariavittoria.ferroni@uniroma1.it

T. Bandini
Department of Environmental Science and Policy, La Statale University of Milan, Milan, Italy

G. Antonelli et al. (eds.), *Environmental Law Before the Courts*,
https://doi.org/10.1007/978-3-031-41527-2_12

1 Introduction. Going at the Heart of the Biodiversity Threat: Biodiversity Loss

It has been estimated that 8.7 million species of plants, animals and micro-organisms are existing in the world.[1] Out of these, about 1.4 million species have been identified so far. Each species is adapted to live in specific environment, from mountain peaks to the depth of seas, from polar ice caps to tropical rain forests and deserts. All this diversity of life is confined to only about 1 kg thick layer of lithosphere hydrosphere and atmosphere which form biosphere.

The starting point of this research is the assumption that the rate of biodiversity loss, together with climate change, is looming on the planet, since the two phenomena are intrinsically linked and correlated. A scientific study showed that human societies might have already exceeded some of the planetary limits[2] by placing themselves in a danger zone, could lead, if not quickly reversed, to the sixth largest extinction event in the history of life on earth, caused by the impacts of human activities on the planet.

Human domination of the earth and its natural resources has increased dramatically over the past century, substantially altering natural ecological processes over three-quarters of the earth's surface. This domination of the biosphere has contributed to increased human well-being, but the downside for humans and the environment is increasingly apparent.

Human-induced changes to ecosystems and the extinction of species have been more rapid in the past 50 years than at any time in human history. Today, around one million species of animals and plants species are already threatened with extinction. Furthermore, the global rate[3] of species extinction is already at least tens to hundreds of times higher than the average rate over the past ten million years and is accelerating. According to the Worldwide Fund for Nature (WWF) population sizes of vertebrate species, for example, have declined by an average of 68% over the last five decades.[4] In a two-way process, climate change is one of the main drivers of

[1] The latest data remains Sweetlove (2011).

[2] The Planetary boundaries' theory, in its latest update to 2022, has shown that, with respect to nine macro-processes (1) climate change; (2) change in the integrity of the biosphere and loss of biodiversity; (3) interference with the biogeochemical cycles of nitrogen and phosphorus; (4) depletion of the ozone layer; (5) ocean acidification; (6) water use; (7) land use and deforestation; (8) chemical pollution; (9) the spread of aerosols and dust in the atmosphere), six of the nine planetary limits have already been exceeded by humankind.

The theory of planetary boundaries (PB) and safe operating space (SOS) was formulated with the publication in Nature in 2009 of a study by a group of 29 eminent scientists led by Johan Rockstrom, see Rockstrom et al. (2009). The study was updated in 2015 by W. Steffen, and, lately in 2022 by Wang-Erlandsson et al. (2022), https://doi.org/10.1038/s43017-022-00287-8.

[3] IPBES' Report available https://www.un.org/sustainabledevelopment/blog/2019/05/nature-decline-unprecedented-report/.

[4] WWF's Living Planet Report reveals two-thirds decline in wildlife populations on average since 1970: https://wwf.panda.org/wwf_news/?793831/WWF-LPR%2D%2Dreveals-two-thirds-decline-in-wildlife-populations-on-average-since-1970/.

biodiversity loss, but destruction of ecosystems undermines nature's ability to regulate greenhouse gas (GHG) emissions and protect against extreme weather, thus accelerating climate change and increasing vulnerability to it. This explains why the two crises must be tackled together with holistic policies that address both issues simultaneously and not individually. The results of the assessment of land degradation and the opportunities we have to regenerate it, presented by the Intergovernmental Science-Policy Group on Biodiversity and Ecosystem Services (IPBES) Global Assessment, are alarming.

Lastly, on 2021 November 16, 141 leaders of the countries of the world, during the negotiations of Cop26 held in Glasgow, have reaffirmed and recognized this important battle: " *We[..] emphasise the critical and interdependent roles of forests of all types, biodiversity and sustainable land use in enabling the world to meet its sustainable development goals; to help achieve a balance between anthropogenic greenhouse gas emissions and removal by sinks; to adapt to climate change; and to maintain other ecosystem services. Reaffirm our respective commitments, collective and individual, to the UN Framework Convention on Climate Change and the Paris Agreement, the Convention on Biological Diversity, the UN Convention to Combat Desertification, the Sustainable Development Goals; and other relevant initiatives. Reaffirm our respective commitments to sustainable land use, and to the conservation, protection, sustainable management and restoration of forests, and other terrestrial ecosystems. Recognise that to meet our land use, climate, biodiversity and sustainable development goals, both globally and nationally, will require transformative further action in the interconnected areas of sustainable production and consumption; infrastructure development; trade; finance and investment; and support for smallholders, Indigenous Peoples, and local communities, who depend on forests for their livelihoods and have a key role in their stewardship. Highlight the areas of strong progress in recent years and the opportunities before us to accelerate action. We therefore commit to working collectively to halt and reverse forest loss and land degradation by 2030 while delivering sustainable development and promoting an inclusive rural transformation"*. [5]

2 Definition of Biodiversity

Before starting an excursus regarding how Courts treat the biodiversity category, it is necessary to identify a definition of biodiversity that allows us to understand how and why the decisions of Courts around the globe are so far apart and different.

The term "biodiversity" was coined in 1986 by W.G. Rosen, who merged the Anglo-Saxon expression "biological diversity" with reference to the diversity and variability of living organisms into a single word.[6] Later, this neologism appeared in

[5] https://ukcop26.org/glasgow-leaders-declaration-on-forests-and-land-use/.

[6] Segre et al. (1996), par. 2.4.

1988 in a publication by the sociobiologist E. O. Wilson,[7] whereby biodiversity encompasses "*the variety of living forms*", and is a concept aimed at "*preserving the world as we know it*".

In the legal context the most important definition is that contained in the 1992 Convention on Biological Diversity,[8] also known as the *Rio Convention* or *CBD*, signed at the United Nations Conference on 14 June 1992 and approved by the Federal Assembly on 28 September 1994. The Article 2 defines biodiversity as "*means the variability among living organisms from all sources including, inter alia, terrestrial, marine and other aquatic ecosystems and the ecological complexes of which they are part; this includes diversity within species, between species and of ecosystems*". So, biodiversity is represented as biological organization divided in three levels: genetic diversity, which refers to the genetic pool of a species that differentiates it from others; species diversity, which refers to the variety among species of living organisms in a given ecosystem or habitat; and diversity at the level of ecosystems, communities and habitats. In other words, we can define biodiversity as the richness of life on earth: the millions of plants, animals and microorganisms, the genes they contain, the complex ecosystems that form all the biosphere.

Biodiversity is a broader concept than nature's protection because it also concerns those areas that interact with it. It includes the state of fauna and flora, forests, but also, in addition to natural and semi-natural environments, urban and peri-urban green spaces, agricultural areas (which play an important role in biodiversity) and other environmental components, whereas a large number of species have adapted to live in a agricultural secondary environments (with specific reference to agricultural areas, it should also be pointed out that the process of specialization and intensification of agriculture, through intensive monocultures, sometimes genetically modified, has, in other respects, produced a serious loss of biodiversity).

Ecological diversity represents an intricate network of different species present in local ecosystems and the dynamic interaction among them. The ecological diversity is of great significance that has developed and evolved over millions of years through interactions among the various species within an ecosystem. As all the organisms in an ecosystem are interlinked and interdependent, the value of biodiversity in the life of all the organisms including humans is enormous. Besides its ecological and environmental value, biodiversity has significant socio-economic values as well, because it is a source for man of indispensable goods, resources and services (ecosystem services) that have a key role for the economy of countries. Just consider that the natural environment is responsible for the production of oxygen, maintenance of water-cycle and other biogeochemical cycles. The more a region is rich in terms of biodiversity, the better are the different cycles regulated. For example, forests regulate the amount of carbon dioxide in the air by releasing oxygen as

---

[7] Wilson (1988).

[8] Rio Convention, signed in Rio de Janeiro on 5 June 1992, ratified by Law No. 124 of 14 February 1994. Ratification and Enforcement of the Convention on Biodiversity with Annexes, made in Rio de Janeiro on 5 June 1992. The Text of CBD is available on the Official Website: https://www.cbd.int/doc/legal/cbd-en.pdf.

a by-product during photosynthesis, and control rainfall and soil erosion. As well known, deforestation would further increase carbon dioxide in the earth's atmosphere leading to greenhouse effect and global warming. This will cause irreparable damage to all organisms including mankind. Ecosystems depend on the health and vitality of the individual organisms that compose them. As all the organisms in an ecosystem are interdependent, removing just one species can prevent the ecosystem from operating normally.

Perhaps only in recent decades, also because of the huge loss of biodiversity and the simultaneous effects caused by climate change, we are beginning to understand the importance of these basic ecosystem services that make life possible on earth.[9] The protection of ecosystems is indeed an increasingly recognized ally in the fight against climate change, and environmental protection, management and restoration are precisely the basis of what are called Nature-based Solutions (NbS), as defined by the International Union for Conservation of Nature (IUCN). Working with nature instead of against it is thus one of the foundations of NbS. The concept of NbS thus refers to all those approaches that, through the sustainable management of ecosystems, mitigate the negative impacts of climate change by generating adaptive capacity and resilience. Such approaches include conservation and restoration actions, ecosystem management and protection, specific ecosystem-based interventions and infrastructure-based actions. Furthermore, the recognition of natural capital, i.e. the economic value and conservation of biodiversity and ecosystem services, has led to new global recognition. For example, the World Economic Forum's report *The Future of Nature and Business*[10] calculates the total value of various nature-positive business models and claims that nature-based solutions could be worth up to USD 10.1 trillion and be capable of generating 395 million jobs by 2030. Outlined in this way, NbS appear as solutions that, with minimal costs, which can provide immense and synergetic benefits. In other respects, the future challenge will be to use NbS effectively and not as mere *greenwashing*, or of an elitist nature, especially at the urban level (e.g. by allocating new green areas to the wealthiest parts of the city). On the contrary, they should play an added role in terms of social integration and participation, creating well-being and health.

[9] In recent years, economists have directed research on the economic value of biodiversity in several studies including the «*Millenium Ecosystem Assesment*» (MEA) and the «*The economics of Ecosystems and biodiversity*» (TEEB). This study documents how environmental capital should be considered the basis of the economic systems and indicates how it is necessary to consider nature and biodiversity and their irreplaceable economic value unfortunately often overlooked (for example, in the study the issue of bees and their extinction has been monetized in terms of worldwide economic losses). See Granato (2021).

[10] New Nature Economy Report II: The Future of Nature and Business, p. 10 available https://www3.weforum.org/docs/WEF_The_Future_Of_Nature_And_Business_2020.pdf.

3 The International Convention on Biological Diversity

The protection of biodiversity, with reference to its impact on the entire planet, has been regulated in the legal context by a multilevel framework divided in an international and a European framework. At the international level, several conventions have followed one another over the years, including: the 1971 [11] Ramsar Convention (Iran) on Wetlands; the 1973[12] Washington Convention on International Trade in Endangered Species of Wild Fauna and Flora; the 1979;[13] Bonn Convention on Migratory Species; and the 1979 Bern Convention,[14] all of which aim, in various respects, to protect biodiversity but with the limitation of being sectoral. The Convention considered most important for the protection of biodiversity is the Convention on Biological Diversity (CBD), which was concluded in Rio de Janeiro from 3 to 4 June 1992.[15] The Convention entered into force on 29 December 1993 and has been ratified by most States. The purpose of this Convention was to extend and focus international cooperation on environmental issues of a global nature such as climate change, biodiversity loss and deforestation. [16] The aim was to create an *umbrella convention* that could bring together the efforts of sectoral Conventions. This is a framework convention to enable further agreements to be concluded. Once global objectives have been set, the Parties as provided for in Article 6 shall adopt strategies, plans and programmes to conserve and use biodiversity sustainably. In other words, the provisions of the Convention are characterized by their open nature, because they allow Parties to identify the most appropriate measures to fulfil their general obligations.

[11] *The Ramsar's Convention on Wetlands of international interest signed in Ramsar (Iran) on February 2, 1971 and enforced by Italy adopting the Presidential Decree no. 448 of March 13, 1976 and with the following Decree no. 184 of February 11, 1987. It was considered one of the most important international treaties on biodiversity and the a*ims are to halt the worldwide loss of wetlands and to conserve, through wise use and management, those that remain.

[12] The C.I.T.E.S., an acronym derived from the initials of *Convention on International Trade of Endangered Species* (more precisely Convention on International Trade in Endangered Species of Wild Fauna and Flora) also known as the *Washington Treaty* signed in Washington on March 3 1973 and ratified by Italy with the Act. No. 874 of 19 December 1975.

[13] Convention on the Conservation of Migratory Species of Wild Animals, signed in Bonn on 23 June 1979, ratified in Italy by Ratification Law No. 42 of 25 January 1983, OJ No. 48 of 18 February 1983. It is considered a framework convention on the protection of migratory species, which provides for the support of concerted research activities and the monitoring of endangered migratory species.

[14] Bern Convention, Convention on the Conservation of Wildlife and Natural Habitats in Europe of 19 September 1979, ratified in Italy by Law no. 503 of 5 August 1981.

[15] Convention on Biological Diversity (Convention on Biological Diversity) signed in Rio de Janeiro on 5 June 1992, within the framework of the United Nations Conference on Environment and Development (UNCED).

[16] Marchisio (2008), p. 10.

3.1 The Conference of the Parties: Kuming's Agreement—Montreal

The executive body of the Convention on Biological Diversity is the Conference of the Parties (COP).[17] This is the most relevant body of the Convention, consisting of all governments that have ratified the Treaty, which meets every 2 years to review progress, set priorities and commit to work plans, and which has set targets over the years for global biodiversity protection. In particular, in 2010, during the Nagoya Conference, the Aichi Biodiversity Targets were established, which were not achieved for the most part (as verified in 2020). On October 15, 2021, the first part of COP15 (which was to be held in Kunming in China but held, due to the pandemic, at a distance) was concluded. After the failures of the last decades (just think of the previous 20 Aichi Targets that had to be achieved by 2020, and which only some have been partially achieved) a fast and effective response was requested because humanity is dangerously approaching the "*point of no return*". The main achievement during this first part of the conference was the *Kunming Declaration*, in which countries committed to negotiate an effective global framework for biodiversity post-2020 in 2022. On 22 December 2022, the second part of COP 15 was held in Montreal, during which the Kunming-Montreal Agreement, the global biodiversity framework for this decade, was adopted. In the agreement, recalling the reports of the *Intergovernmental Science-Policy Platform on Biodiversity and Ecosystem Services* (IPBES) it was noted that the rate of biodiversity extinction is accelerating at a rate of ten to one hundred times that of the last ten million years, and about one million species are at risk of extinction. The goal is to reduce the rate of extinction and the risk of extinction tenfold by 2050 (however, the European Union argued that one billion species are at risk of extinction and that the targets for halting human-induced extinctions should be 25% by 2030 and 50% by 2050).

In the final agreement, the conservation target of 30% of terrestrial and marine areas by 2030 was included, although the numerical target of one billion hectares was removed. The target (referred to as 30 × 30) envisages doubling the conservation of terrestrial areas under the Aichi Targets (17%) and tripling that of marine areas (10%), and there are those who believe it will become the guiding target for biodiversity (similar to the Paris Agreement's goal, that aims to limit global warming to 1.5 °C). Regarding the objective of the protection of 30% of the areas, it is not clear whether the 30% should be calculated on a global scale or concerning each State party to the Convention on Biodiversity. In Europe, it was referred to every State. Furthermore, questions are being raised about the type of areas to be included in the 30%. If it certainly includes protected areas to reach 30%, it is believed that areas of organic agriculture, urban forests, or (as in Canada) military easements could also be included. Scientists believe that achieving an effective 30% in good

[17] The purpose of the Conference of the Parties is to coordinate and implement the proper functioning of the Convention and it is vested with various powers: to adopt protocols, amendments, establish subsidiary bodies, working groups, and examine technical and scientific opinions.

conservation status globally will bring 90% of habitats back to balance. However, what emerges from the Agreement is that if the role of protected areas is very important, they are not the solution alone. In addition, it was also emphasized on the one hand that the expansion of protected areas must take place while respecting the rights of indigenous peoples and local communities, and on the other hand that the expansion of protected areas will not be sufficient as it will also have to take action on over-consumption (although the target to reduce the consumption and production footprint by 50% by 2030 was removed, leaving a reference to the need for reduction but without a numerical target. The reference to sustainable diets was also removed). The agreement recognized the interconnections between biodiversity and health, considering the one-health[18] approach. The financial profile is a sensitive issue because the lack of adequate financial funds was one of the main causes of the failure of the targets (Aichi targets) for the 2010–2020 decade. Some African states (Congo, Cameroon, Uganda, Namibia) have expressed disappointment with this agreement as it was not accompanied by the financial means necessary for the effective implementation of the planned measures. Consequently, approximately 20 billion dollars a year (until 2025) and 30 billion dollars, by 2030 will have to be invested to help the less developed countries but still rich in biodiversity. For this reason, there has been a commitment to create a new Global biodiversity fund. The delicate point of this agreement seems to foresee only procedural obligations through the reporting systems required by the Global Diversity Framework. However, there do not appear to be binding obligations and redress measures in the event of non-compliance with the substantive objectives established in the agreement. With regards to the effective biodiversity protection, a crucial point seems to be the new tools research to define the effectiveness and the efficiency of the protection. This is in order to influence the political will of international, European, national and local administrators so that they can eventually deal with the issue effectively, organically and above all urgently. The support of the judges could help to establish more binding and uniform rules for the protection of biodiversity between the different states.

[18] The agreement sets out to eliminate or reform the nature-damaging subsidies of around USD 500 billion per year by 2030, even if there are no references to specific industrial sectors and it is not expected to happen this year. It has also introduced the indication of reducing by 50% the risks arising from pesticides by 2030 and a reference to the prevention and reduction of plastic. With regard to climate change, it is not foreseen that nature will contribute to global mitigation efforts by 2030. The implementation of the measures provided for in the agreement must be accompanied by spatial planning that takes into account biodiversity, reaching by 2030 almost zero loss of areas of high importance for biodiversity.

4 Exploring Litigation: Lawsuits' Qualification Related to the Protection of Biodiversity and the Asymmetry on Standing's Requirements

To date, there is not yet a universally shared definition of litigation on biodiversity or rights of nature ("RoN"), probably due to the strong cultural, ideological and political difference that characterizes some countries compared to others. The term biodiversity litigation may include any regional, national or international legal dispute concerning conservation, sustainable use, access to and sharing of the benefits of animal and plant species, of ecosystems and their relationships and related sectors (as defined in the previous paragraphs).

This definition covers both the cases with the strategic objective of achieving systemic change or improving nature's protection or preventing damage, and those with the specific purpose of protecting particular species or habitats. In recent years, there has been an increase in claims relating to biodiversity.

The different anthropocentric or ecocentric approach towards nature changes the type of rights that are recognized and also the type of protection recognized by the judges.

According to the legal system in force in almost all countries, nature is considered property.

Well, something's property gives to the "owner" the right to damage or destroy it. In this perspective, those who own land, forests, agricultural areas are entitled to use it how they want, even if this may include the destruction of nature's health and well-being.

This is the so-called anthropocentric perspective of nature.

Dealing with the issue of the rights of nature, we need to start from the fundamental assumption that ecosystems and natural resources are not mere properties that can be owned but possess an independent and inalienable right to exist and prosper. In this perspective, laws that recognize the rights of nature recognize ecosystems and natural resources as rights-bearing entities.

Supporters of this school of thought[19] have argued that an ecosystem is entitled to the legal personhood and, as such, has the right to defend itself in court claiming compensation for the damage suffered (for example cases of environmental degradation caused by new project). Laws that recognize RoN expressly provide that an ecosystem has the right to exist, thrive, regenerate its life cycles and naturally evolve. Moreover, when an ecosystem has legal personhood, it is entitled to legal representation by a guardian who will act in its name and on its behalf. This role is

[19] As noted in a recent European study requested by the EU JURI committee, it should be follow two basic lines of reasoning "*First, since the recognition of human rights is in part based on the philosophical belief that those rights emanate from humanity's own existence, then logically, so do inherent rights of the natural world. A second and more pragmatic argument asserts that humanity's own survival depends on healthy ecosystems, and so, protection of nature's rights in turn advances human rights and well-being*", (2021).

often played by individuals or groups of people who are experienced in the field or who live closely as indigenous communities.

The aim of recognizing RoN is to ensure a high level of environmental protection.

On the contrary, in the anthropocentric perspective it is the human being who enjoys the rights of nature, as the only subject of law. Especially within the European Union, rights of nature are very often associated with human rights, in particular the right to a clean and healthy environment.

The Doctrine on the RoN in recent years has spread again thanks to an increase in climate change litigation, with almost 2000 lawsuits underway or concluded[20] against States and companies, together with a growing scientific understanding of the close relationship and interdependence between climate change and biodiversity loss.

Such a push has certainly influenced the worldwide tendency on one hand to recognize nature as rights-bearing entities ("T*he first countries to adopt RoN laws did so in quick succession (United States in 2006; Ecuador in 2008; Bolivia in 2010; New Zealand in 2014). Ecuador's 2008 Constitution recognizes rights for all of Nature. New Zealand has two national laws—the Te Urewera Act (2014) and the Te Awa Tupua Act (2017)—each of which grants rights to a particular ecosystem. In the US, subnational governments have adopted laws recognizing RoN in their jurisdictions, including municipal ordinances, home rule charters, and state constitutional amendments"*[21]) on the other hand to record legal developments that require states to adapt and create environmental resilience and companies' greater due diligence in their operations and above all greater control and rigor over the entire supply chain. Within the context of the climate crisis, the "Rights of Nature" represents one legal theory that can help elevate the urgency of protecting biodiversity in the fight against climate change.

One of the main characteristics of this type of litigation, also defined as "strategic" is that: (a) it presupposes the need for a dialogue between the legal system—as a formal structure—and the administered community—that is, the social structure, (b) the aim is to develop the legal system beyond the specific case being examined by the court.

The fil rouge is to raise awareness on the urgency of taking action against climate change. One of the points in common between climate litigation and RoN litigation is that both having to measure with the "legal conditions" of the claim that is, the interest in bringing proceedings that derives from having a certain result from the case and the legitimacy to file a case that is acquired when the right of action is a right of its own.

Courts in most jurisdictions must consider whether the parties attempting to bring a legal action have "standing" to bring their claim. "Standing" refers to the set of requirements that a plaintiff must meet to demonstrate that they are entitled to

[20] The Sabin Center for Climate Change Law realized a chart on climate-related litigation http://climatecasechart.com/.

[21] See Kauffman Craig et al. (2018), pp. 44 ss.

bring a claim before the court. For example, plaintiffs in the civil law system must show that they were injured, that their injury was caused by defendants' actions, that the dispute is one a court can resolve, and that the court has some authority to order a remedy that would help the plaintiffs. Legal definitions of who has standing change across jurisdictions of different States, but generally include requirements that the parties bringing a claim have a genuine and current stake in the outcome. It should be noted that, according to the UNEP's Litigation Report 2020, the issue of standing has become central, especially in the United States,[22] whereas it has proved less relevant outside the United States,[23] especially in the developing world courts, as the cases in Pakistan,[24] India,[25] Nigeria,[26] Columbia[27] and the Philippines[28] show.

In particular, with regard to cases on biodiversity, which will be analysed below, the opening of extra-EU legal systems towards an ecocentric vision[29] of the right to nature is highlighted, while there remains an anthropocentric approach to RoN within the EU. This means that in the EU context the RoN is a kind of duplication of the right to health that can be undermined by rising temperatures and therefore by an ineffective policy. In fact, Articles 2 and 8 of the European Convention on Human Rights (ECHR) are often mentioned. Article 3 on the prohibition of torture, since the European Court of Human Rights has held that people forced to live in a

[22] See 2017 Litigation Report at p. 28–29. The key cases are, on one hand, Massachusetts v. EPA (2007), where the U.S. Supreme Court concluded that the State of Massachusetts had standing—relying on the State's status as quasi-sovereign within the federal system—to challenge federal government's failure to regulate GHG emissions from new motor vehicles. On the other hand, in Comer v. Murphy Oil USA (2009), the Fifth Circuit Court of Appeals found that landowners harmed by Hurricane Katrina did not have standing to bring a civil conspiracy claim against fossil fuel and chemical companies, because the landowners' injuries were not traceable to the companies' conduct.

[23] In the Urgenda's case, the foundation was permitted to maintain an action against the government, but the individual plaintiffs were not granted standing separate from that of the organization. In Haughton v. Minister for Planning and Macquarie Generation (2011), a plaintiff could pursue a climate case where the plaintiff had a "special interest" in the alleged harms, and not merely "intellectual or emotional concern.

[24] Leghari v. Pakistan (2015).

[25] State of Himachal Pradesh, M.A. Nos. (2014–2016).

[26] Gbemre v. Shell PDC Nigeria Limited, (2005).

[27] Constitutional Court, 2016, Decision C-035/16.

[28] Re Greenpeace Southeast Asia and Philippine Rural Reconstruction Movement, Petition to the Commission on Human Rights (2015). Philippines requesting for investigation of the responsibility of the Carbon Majors for human rights violations resulting from the impacts of climate change.

[29] The legal systems that contemplate the RoNs treat Nature as the bearer of legal personhood. They conceptualize at the level of ecosystem rather than single flora and fauna and at least implicitly recognize that humans are part of this ecosystem. The preamble to the Constitution of Ecuador defines Nature as the indigenous deity "Pachamama" [translated as Mother Earth] "*where life reproduces and occurs*" (Republic of Ecuador 2008). No other definition is offered, deliberately leaving the definition devoid further details. While, for example, New Zealand laws focus on the protection of particular ecosystems, such as the Whanganui River (Te Awa Tupua Act) and the Te Urewera Forest (Te Urewera Act).

polluted area are being treated by the State which is close to torture.[30] In other words, while not assuming an autonomous importance, environmental protection has established itself as a new value that can contribute to a balance between the exercise of human rights—expressly recognized by the Convention—and the general principle of respect for the individual, to which the entire ECHR guarantee system is oriented.

5 Litigation in Brazil, Colombia, Australia, Philippines, EU, Belgium and Finland Between Ecocentric and Anthropocentric Approach

In the light of the reconstruction carried out so far, it is important to focus on some significant cases present in the international and European context in order to represent the two different caselaw approaches. An ecocentric one through which Nature is recognized as a subject of rights and an anthropocentric one in which Nature is protected "by reflex" or as a human right or through specific regulation.

5.1 *Brazil: Landowners' Lialibity for Using Illegally Deforested Land*[31]

A first case is represented by the litigation established in Brazil, where today the Deforestation has increased significantly in recent years in the Brazilian Amazon, with over 5000 km2 of rainforest lost in 2020 alone. The painfully effective scheme to be able to recycle illegally deforested land has been to sell it to a company that then starts farming or ranching, claiming to be unaware of the previous illegal deforestation that took place.

Interestingly, it was Brazilian Federal Prosecutors who took concrete action against deforestation in their jurisdictions, taking several actions related to different areas of the Amazon rainforest, covering states, municipalities, national parks, conservation areas and indigenous lands.

Each action, which covered deforested areas, followed two specific steps: (1) giving to deforester the opportunity to sign a civil agreement regarding (i) reforestation

[30] This is the case of the European Court of Human Rights in Strasbourg, which has given priority to the case filed by six Portuguese children and young people, who, supported by the NGO "Global Legal Action Network", have appealed against 33 Member States of the Council of Europe—including Italy—accusing them of violating their rights, not respecting the commitments made with the signing of the 2015 Paris agreement, at the COP21.

[31] The decision is available on http://climatecasechart.com/climate-change-litigation/wp-content/uploads/sites/16/non-us-case-documents/2020/20200521_Acao-Civil-Publica-no-1007104-63.2020.4.01.3200_decision.pdf.

of the area; (ii) paying compensation with a discount calculated according to specific terms previously disclosed; (2) filing an environmental lawsuit seeking reforestation and compensation calculated according to specific, previously disclosed terms.

A landmark case came to court in late 2020. The Federal Prosecutor's Office and IBAMA filed a civil action against "an uncertain and unlocatable person, owner of the embargoed area" due to the illegal deforestation of sixty-seven hectares of forest. The request was to conduct reforestation of the degraded area and to pay compensation for material and moral environmental damages.

The Judge ruled that the *erga omnes* effect, which is usually associated with human's rights, can also be applied when it comes to environmental protection.

A key role played by this case was to use the argument of the *propter rem obligation*, which refers to the environmental responsibility of the owner. In such vast and remote areas of the Amazon rainforest, deforestation is very simple given the absence of people for thousands of miles. Satellite photos of these areas clearly document the ongoing deforestation. Usually, in fact, land reclamation was used as a strategy to obtain legal registration through notaries, to apply for a mortgage to purchase land, to raise livestock or develop agricultural projects through private or public bank loans.

As a result of this judgment, it is stated for the first time that the burden of reforesting a certain area and implementing compensatory actions is borne by the buyer as owner of the land.

This decision is undoubtedly very important because it has somehow interrupted what could be called a chain of tortuous events linking the change of forest boundaries, deforestation and subsequently the regularization of deforested areas, given the translational effect of the property.

5.2 Colombia: The Atrato Ruling and the[32] Deforestation in the Amazon[33] Violations of Fundamental Rights

Against the background of the devastating environmental and social impacts caused by illegal mining in the Atrato region and the failure of the State to address them, the Colombian Constitutional Court ruled in favour of the claimant communities in 2016. This ruling constitutes a landmark decision in the country's constitutional system, given that the Court not only protected the claimant's rights to life, health, water, food, territory, culture and a healthy environment, enshrined in the country's progressive 1991 Constitution, but also pushed the boundaries of constitutional law by recognising, for the first time, a non-human natural entity as a

[32] Center for Social Justice Studies et al. v. Presidency of the Republic et al. Judgment T-622/16.

[33] https://climate-laws.org/geographies/colombia/litigation_cases/future-generations-v-ministry-of-the-environment-and-others.

subject of rights and by introducing the concept of biocultural rights of indigenous and afro-descendant communities.

Reasoning that the protection of the environment is '*a fundamental objective*' and '*transversal element of the Colombian constitutional order*', that the relationship between the constitution and the environment is '*dynamic and in permanent evolution*' and that 'the greatest challenge contemporary constitutionalism faces consists in safeguarding and effectively protecting the environment', the Court adopted, for the first time, an explicit ecocentric approach to environmental protection, granting legal personhood to the Atrato River.

In January 2015, several communities, Afro-descendant and indigenous organizations living along the Atrato River filed a complaint to stop the large-scale and intensive use of illegal mining and logging methods in the Atrato River.

These practices, according to the claimants, have harmful and irreversible consequences on the environment, affecting the fundamental rights of local communities and the natural balance of the territories where they live. In summary, the claimants ask the court of protection to protect the fundamental rights to life, health, water, food security, healthy environment, culture and territory of the ethnic communities and, consequently, to order a series of measures to articulate structural solutions to the serious health, socio-environmental, ecological and humanitarian crisis concentrated in the basin of the Atrato river, its tributaries and surrounding territories.

The applicants, following their unsuccessful appeals before the courts of first and second instance, submitted the case to the Constitutional Court for review. The latter, in upholding the appeal, declared the Atrato River, its basin and tributary entities "subject to rights to protection, conservation, maintenance and restoration by the State and local communities". This case represents a milestone in the legal landscape as it recognized, for the first time, a non-human natural entity as a subject of rights.

The Court adopted, for the first time, an explicit ecocentric approach to environmental protection, recognizing legal personhood to the Atrato River. The ecocentric approach, according to the Court "*starts from the basic premise that the earth does not belong to humans and, on the contrary, assumes that humans belong to earth, as any other species*"[34] According to this perspective, humanity does not own nature, which is considered by the Colombian Court as a legal entity with own rights.

In this context, the Court introduces the concept of biocultural rights: "In *their simplest definition, [biocultural rights] refer to the rights of ethnic communities to autonomously administer and protect their territories – in accordance with their own laws and customs – as well as the natural resources that constitute their habitat, where their culture, traditions and way of life are developed based on their special relationship with the environment and biodiversity. In effect, these rights result from the recognition of the profound and intrinsic connection that exists*

[34] Corte Constitucional (n 11) 161, Original text available https://www.corteconstitucional.gov.co/relatoria/2016/t-622-16.htm#_ftn86, Sect. 5.9. "*Finalmente, el enfoque ecocéntrico parte de una premisa básica según la cual la tierra no pertenece al hombre y, por el contrario, asume que el hombre es quien pertenece a la tierra, como cualquier otra especie*".

between nature, its resources and the culture of ethnic communities, which are interdependent and cannot be understood in isolation".[35]

Some Scholars reported that "*the most distinctive feature of the Atrato decision is that it couples the river's legal recognition with a comprehensive set of procedural orders to the respondent state entities regarding the formulation of public policies to protect the rights of the river as well as an interinstitutional monitoring mechanism to follow up on their implementation. This approach constitutes a peculiarity of Colombian constitutional law*". [36]

Another case related to Colombia is represented by *Future Generations vs. Ministry of Environment and Others*.[37] In 2018, twenty-five individuals between the ages of 7 and 25 filed a civil lawsuit based on the principle of *neminem laedere* before a Colombian District Court. It is interesting to note that the plaintiffs argue that their right to a healthy environment and life will surely be violated in the future, as well as it is known that in the coming years climate change will significantly impact the quality of life in their cities.

According to the plaintiffs' claim, the Colombian government has been complicit in affecting climate by failing to control deforestation in the Amazon region of the country. The government has allowed it to increase, over 40% per year since 2015. However, the district court dismissed the case. The plaintiffs appealed to the Colombian Supreme Court, which instead upheld the appeal, ruling that the Colombian state's failure to protect the Amazon rainforest affected fundamental rights, not just of the plaintiffs mind you, but of all Colombian citizens.

The Supreme Court ordered the formation of the *Intergenerational Pact for the Life* [38] of the Colombian Amazon and instructed the stakeholders and governments to immediately formulate plans to stop deforestation.

Finally, in a landmark ruling, the Supreme Court recognized legal personhood to the Colombian Amazon.

It is noted that the Supreme Court appealed the Paris Agreement, holding that the Colombian government had a legal obligation to reduce the annual rate of deforestation. This is a high caliber legal judgment, especially in the part where rules that the fundamental rights of life, health, minimum subsistence, liberty and human dignity are fundamentally linked and are determined by the environment and the ecosystem. Applying the principle of precaution, intergenerational equity and solidarity, the Court held that by failing to prevent deforestation has been committed a threat to the fundamental rights of future generations and a violation of constitutionally protected rights.

---

[35] Corte Constitucional (n 11) 161, see Sect. 5.11.

[36] Wesche (2021), p. 540.

[37] Available http://climatecasechart.com/non-us-case/future-generation-v-ministry-environment-others/.

[38] See Sect. 14, p. 46 of the Decision.

5.3 Australia: Local Community and NGO Stop Coal Mining Expansion

In 2013, the citizens of Bulga, a small village in the state of New South Wales, won a major legal case[39] against a mining company and local government, thus blocking the proposed expansion of a coal mine. Coal mining in Bulga is a very important industrial activity that dates back almost two centuries. In 2010, when the exploitation of coal fields was no longer profitable, it became economically viable due to coal prices, the Warkworth Mining Company applied for permission to extend its mining activities around Bulga. The Company not fulfilling its obligations when it started operating in the village, has requested to start extracting coal in areas originally intended not to be contaminated by any anthropic activity due to the presence of unique habitats and biodiversity. Moreover, the new mining operations would have been much closer to the village than the agreed limits.

The Court ruled that the proposed expansion of the coal mine would have an unacceptable impact on local biodiversity, as well as creating noise pollution and negative consequences for the population of Bulga. As a result, Warkworth Mining's claim was rejected. This case represents an important victory in the landscape of biodiversity litigation since, not only has it prevented a "direct damage" to biodiversity but the pronouncement focuses heavily on arguing that mitigate the effects of climate change, By limiting the amount of emissions from its activities, it is a key factor in combating biodiversity loss.

In rejecting Warkworth's request for an extension project, the Court relied on the following important aspects: impact on biological diversity; impact on noise and dust; and social and economic impacts.

Reaching a conclusion, this case represents an important precedent since the Court established a "standard" for decisions on administrative measures for the approval of projects that could affect ecosystems and local communities.

5.4 The Philippines: Oil Exploration in the Tañon Strait Stopped

This case, compared to the others just discussed, presents unique peculiarities in terms of legal standing and legal interest: here, the plaintiffs are resident marine mammals, including whales, dolphins, porpoises, and other cetacean species that inhabit the Tañon Strait.

In November 2007, the oil exploration company JAPEX began drilling exploratory wells in the Tañon Strait, Philippines. The Tañon Strait is considered a habitat

[39] Warkworth Mining Ltd. v. Bulga Milbrodale Progress Ass'n Inc. (2013).

and migratory route for whales and dolphins, but the number of marine mammals found there is steadily declining.

Local lawyers and an NGO sued JAPEX,[40] on behalf of whales and dolphins as well as local fishermen, who are directly affected by the mining company's activities. The case arrived to the Supreme Court of the Philippines, which ruled that oil exploration in the Tañon Strait must be stopped. In court, JAPEX argued that a presidential decree would allow it to extract oil in the Tañon Strait. However, the Court argued that local legislation considers the Strait a protected area, and therefore special laws on protected areas must be considered superordinate to a presidential decree. This case is important not only because it concerns a key site for marine biodiversity, but because the Court ruled that the executive power must respect special legislation that safeguards the environment and biodiversity.

This decision is also important from the legal standing point of view. Indeed, the Court in deciding on the admissibility of the plaintiffs' claim, adopted the so-called Rules of Procedure for Environmental Cases, which state that "*any Filipino citizen representing others, including minors or future generations, may bring an action to enforce rights or obligations under environmental laws*". [41] In the Annotations to these rules, the Supreme Court commented that "*To further encourage the protection of the environment, the Rules enable litigants enforcing environmental rights to file their cases as citizen suits*".

Therefore, it is not necessary to have a direct personal interest, as strictly required by Italian and European courts.

6 Court of Justice Decisions on Nature Protection and Biodiversity Conservation in Europe: the *Natura 2000* Network

In Europe, the new Biodiversity Strategy (approved in May 2020 by the EU Commission—together with the Farm to Fork Plan) in view of the failures of previous interventions, aims to achieve a share of at least 30% of Europe's protected land and marine areas, by leveraging existing *Natura 2000* sites (which under the Habitats Directive and the Birds Directive are the main instrument of biodiversity protection in Europe) while ensuring stricter protection of areas particularly rich in biodiversity and with a high climatic value (e.g. forests). Efforts also appear to be aimed at improving the level and effectiveness of protection in other policy areas related to protected areas (e.g. agricultural areas) of which knowledge and adequate access to data must be implemented; as well as at adopting a series of concrete actions to rehabilitate, recover and restore degraded ecosystems in the European

[40] Resident Marine Mammals of the Protected Seascape Tañon Strait v. Secretary Angelo Reyes (2015) available on https://elaw.org/system/files/180771_0.pdf.

[41] See 40, p. 16.

Union by 2030. The protection and structural and functional restoration of ecosystems requires, according to the Strategy, a change in the European legislative and socio-economic environment through the introduction of a new European framework for biodiversity governance, including clear definition of obligations and timelines and a system for monitoring commitments and ensuring that Member States integrate them into national policies. This approach could lead to a different binding nature to the European Biodiversity Strategy and to those in force in the internal regulations. An important milestone is the very recent "Nature Restoration Law", which has been approved by the European Parliament the 12th July 2023 and that provides binding obligations for all Member States, as well a support for the fight against the loss of biodiversity. This aims to restore within 2030 at least 20% of the terrestrial and aquatic surfaces, with an extension (within 2050) to all the ecosystems in need of recovery. The aforementioned proposal regards not only the protected areas but the ecosystems on the whole, including farmlands and urban areas. Just before that, the protection of biodiversity was only ensured through the Birds Directive and the Habitats Directive (as transposed into internal regulations) and through the sectorial laws regarding the protected areas. These are Directive 79/409/EEC of 2 April 1979 on the conservation of wild birds (recodified by Directive 2009/147/EC [42]) and Directive 92/43/EEC of 21 May 1992 (on the conservation of natural habitats and of wild fauna and flora [43]). More recently, Regulation (EU) No 1143/2014 was adopted, laying down provisions to prevent and manage the introduction and spread of exotic species invasive [44] as well as for the protection of the marine environment, Directive 2008/56/EC (Marine Strategy Framework Directive

---

[42] The Wild Birds Directive in its original version dates to 1979 and was therefore a precursor to the 1992 Rio Convention. Directive 79/409/EEC of 2 April 1979 was replaced by Directive 2009/147/EC of 30 November 2009. The 1979 directive was taken over in 2009 with an update of the lists of protected species and some slight modifications. Migratory birds represent a common heritage of the European peoples and their effective protection is a transnational priority, to be carried out on the basis of a coherent network of sites and initiatives, in particular by protecting sites of international importance for the conservation of migratory birds. The scope of the Directive is not only birds, their nests and eggs, but also their habitats, as the European legislator is well aware of the close and inescapable link between the effective protection of birds and that of their habitats. In order to guarantee the effective protection of the animals covered by the Directive. Article 5 provides for a protection regime with a general prohibition of killing and capture. Exceptions to this general prohibition are made for the cases expressly provided for by the hunting ban (art. 7), which can, however, be provided for and implemented in proportion to the size of the population, and the so-called "derogations" which are subject to a series of peremptory conditions set out in Article 9.

[43] Ferroni (2015), p. 459 ss.

[44] Regulation (EU) No 1143/2014 of the European Parliament and of the Council of 22 October 2014 laying down rules to prevent and manage the introduction and spread of invasive alien species (OJ L 317 4.11.2014, pp. 35–55) sets out the rules to prevent and manage the introduction and spread of invasive alien species in the EU. Invasive alien species (IAS) are defined as: plants or animals that have been transported outside their natural ecological range by humans (intentionally or unintentionally). These species, according to the regulation, cause considerable ecological and economic damage due to their invasiveness (Art. 3). The European legislation aims to minimise and mitigate the negative effects of invasive alien species on EU biodiversity and ecosystems (Art. 1).

MSFD), which provides for an integrated approach and includes all other instruments and directives that act on the marine environment. Another protection instrument for biodiversity is EU CITES Regulation 338/97, which implemented the Convention on International Trade in Endangered Species of Wild Fauna and Flora within the European Union.

The protection of biodiversity through the above-mentioned directives (Habitats and Birds) is based on the conservation of hot spots, *i.e.* identified and protected sites. More specifically, the 'Habitats' Directive 92/43/EEC of 21 May 1992 (on the protection of natural habitats and of wild fauna and flora) defines an overall framework aimed at ensuring the maintenance or restoration, at a favourable conservation status, of natural habitats and species of wild fauna and flora of Community interest through the establishment of a coherent ecological network of special areas of conservation (SACs) called *Natura 2000.* The European ecological network *Natura 2000′* has been until now the main instrument for the protection of biodiversity in Europe, with the aim of maintaining the habitats and species identified in the annexes to the directives in a satisfactory state. That is, in order to ensure the maintenance, or if necessary the restoration, at a satisfactory state of conservation of natural habitat types and the habitats of protected species, the directive constitutes a coherent European ecological network of "special areas of conservation", i.e. areas that the directive requires the member states to designate by their own measures, called *'Natura 2000'*, formed by the sites where the natural habitat types listed in Annex I and the habitats of the species listed in Annex II are found, which also includes the special protection areas provided for in the Birds Directive (Art. 3, para. 1). In addition, the *Natura 2000* network also includes special protection areas classified under the Birds Directive (SPAs) that are designated directly by the Member States.[45] In 2014, the European Commission launched a review of the adequacy of the rules to protect biodiversity, later published on 16 December 2016. [46] In general terms, the review established that the effective achievement of the objectives of the European regulatory framework is closely linked to the substantial improvement of its implementation by the Member States. [47]

---

[45] Article 4 of the Directive regulates the procedure by which Member States identify and notify 'Sites of Community Importance' to the Commission, which in tum draws up a list of them, following which the Member States are called upon within 6 years to designate these sites as Special Areas of Conservation, establishing priorities according to the importance of the sites for the maintenance or restoration, at a favorable conservation status, of one or more natural habitat types in Annex I or one or more species in Annex II and for the coherence of Natura 2000 (Art 4 paragraph 4).

[46] Commission staff working document fitness check the EU Nature Legislation (Birds and Habitats Directives) Directive 2009 1147/EC of the European Parliament and of the Council of 30 November 2009 on the conservation of wild birds and Council Directive 92,143/EEC of 21 May 1992 on the conservation of natural habitats and of wild fauna and flora.

[47] *Commission Staff Working Document SWD (2016) 472 final, 16 December 2016, Monitoring the Adequacy of EU Nature Legislation* (Birds and Habitats Directives) http://ec.europa.eu/environment/naturellegislation/fitness_check/index_en.htm.

Therefore, according to the Commission, its effectiveness and efficiency must be improved. In more detail, according to the audit, the main factors behind the shortcomings in implementation include limited resources, poor implementation, insufficient integration of nature-related objectives into other policy areas (e.g. agriculture), inadequate knowledge and access to data, communication issues and low stakeholder involvement.

6.1 *Italy: Court of Justice, Cascina Tre Pini v. Ministry of the Environment and o., Case C-301/12, 3 April 2014*

Confirming the difficulty of implementing biodiversity protection objectives by means of Sites of Community Interest (SCI), the Court of Justice in a 2014 judgment dealt with a case brought by the Italian courts in relation to a request for the downgrading of a restricted SCI area owned by a private individual who aimed that, despite the existence of that protection area, the expansion of the nearby Milan-Malpensa airport had devastated the site from an environmental point of view, no longer justifying the restriction of his property rights. [48] The judgement stems from an appeal submitted to the Italian courts by Cascina Tre Pini, a company landowner in the *"Brughiera del Dosso"*, a short distance from Milan-Malpensa airport, and focuses on the relationship between property rights and the protection of biodiversity. The above-mentioned company had requested the downgrading of the area it owned on the grounds that, although it had been included among the Special Areas of Conservation, the construction of the nearby Malpensa Airport had led to its severe degradation. The Court of Justice's ruling stated that a Member State cannot request such a downgrading before it has both applied the necessary conservation measures and subjected plans and projects of significant impact on the site to an assessment, otherwise it would be liable to initiate infringement proceedings. The restrictive interpretation provided by the Court of Justice reveals a contraction of the right to property before the interest in the protection of biodiversity. In particular, where the restriction of the right to property is a consequence of the designation of a Special Protection Area or a Special Area of Conservation, a presumption of compliance with the principle of proportionality has been affirmed where justified by the pursuit of general interest objectives (protection of biodiversity). An infringement procedure (No. 2028) was also opened against Italy in 2021 for failure to complete the designation of Natura 2000 network sites.

[48] See De Vido (2014), p. 803 ss.

6.2 *Italy, Court of Justice, Section III, Judgment of 15 July 2010, C-573/08*

Litigation before the EU Court of Justice has developed in relation to the transposition of Article 9 of Directive 79/409/EEC concerning the possibility of derogation from the ban on the killing of wild birds, which must be justified in detail and must provide for the species that may be hunted and the quantities, the circumstances and place in which the derogation may be applied, the means of killing authorised and the authority responsible for controls. The transposition rules in Italy have largely delegated to the regions the task of protecting biodiversity on their territory, and this division of competence [49] has posed problems in the area of the effectiveness of protection [50] such as to lead to EU infringement proceedings that resulted in Italy being condemned by the Court of Justice, which held: "*Since the legislation transposing into Italian law Council Directive 79/409/EEC of 2 April 1979 on the conservation of wild birds does not fully comply with that directive and the system for transposing Article 9 thereof does not ensure that the derogations adopted by the competent Italian authorities comply with the conditions and requirements laid down in that provision, the Italian Republic has failed to fulfil its obligations under Articles 2,7,9, 11,13 and 18 of that Directive*".

6.3 *Finland and Romania: Two Decisions on Wolf's Protection (Canis lupus)*

In March 2020, the Finnish Supreme Administrative Court made it illegal to hunt wolves in the country, ruling that wolf hunting permits issued to Finnish hunters were against the law. The lawsuit was started by Tapiola, an NGO founded by three citizens specifically to protect wolves. Wolves used to be hunted to near extinction in Western Europe, but their numbers are increasing and the species is rapidly recolonizing the territory. In Finland, however, wolves are greeted with hostility and fear by people. Every year, the Finnish government issues a fixed number of licenses to cull so-called "wolves' issue", classified as animals that have killed dogs or livestock. This loophole has allowed wolf hunting in Finland to continue despite the open violation of European law: wolves are a protected species within the EU, of which Finland is a part. In fact, the wolf is listed in Annex IV of the EU Habitats Directive (Council Directive 92/42/EEC of 1992) as a strictly protected species, which means that their killing is prohibited except for a very limited number of reasons. However, Finland has negotiated an exception for wolves in certain parts of

[49] See Brambilla (2000), p. 301 ss.

[50] See Rizzi (2015), p. 524 ss.

the country, which is included in Annex V and imposes fewer restrictions, so that hunting permits from the Finnish authorities are permitted.

In response to this situation, the European Commission initiated infringement proceedings against Finland in 1997, the result of which was stricter national regulations in Finland. However, permits for the culling of wolves were still permitted under Article 16(l)(e): under strictly controlled conditions, on a selective basis and to a limited extent', without specifying a clear purpose and leaving room for the discretion of the authorities. As a result, wolf populations continue to decline. Worse still, despite objections from the public consultation, a plan to reintroduce managed wolf hunting was announced in 2014.

Although environmental organizations are generally authorised to bring public interest cases, by virtue of the right of access to justice under the Aarhus Convention, appeals relating to hunting permits within Finland are nevertheless regulated by the Hunting Act. This means that only local and regional associations are entitled to act. For this reason, three locals decided to set up and register a small NGO named Tapiola, which covered most of the Finnish territory in order to initiate litigation in different regions of Finland by asking the courts to (1) issue an injunction against the permits, and (2) refer the case to the CJEU because the Finnish law violated EU law. However, almost all regional courts rejected these requests for lack of standing. To meet these requirements, Tapiola therefore changed its litigation strategy, dividing the NGOs into six regional organizations. However, the claims were still rejected with regard to both legal standing and merit. Despite this, one of the appeals went to the Supreme Administrative Court, which referred the case back to the CJEU, asking whether and under what circumstances wolf hunting was allowed and whether Finland was violating EU Law. In 2019, the CJEU ruled in favour of the plaintiffs on all claims, imposing very strict restrictions on wolf hunting.

Emphasizing the main objective of the Habitats Directive, namely "*to contribute towards ensuring biodiversity through the conservation of natural habitats and of wild fauna and flora*", [51] the Court ruled that the objective of the permit was not clearly and precisely defined in Finnish law, and the authorities did not demonstrate that the culling of wolves was appropriate to achieve this objective, which in any case should have been supported by rigorous scientific data; the authorities did not consider satisfactory alternatives in their decision-making process; the authorities did not ensure that hunting permits did not harm wolf populations in a favourable conservation status in their natural place an impact assessment of the conservation status of wolves was not carried out when hunting permits were issued. Not all the conditions set out in Article 16(l)(e) are met, compliance with which must be established in particular by reference to the level of the population, its conservation status and its biological characteristics. Therefore, although the CJEU did not have to rule on questions of fact, since this was a preliminary ruling, it conduced that the permits

[51] CJEU's (Second Chamber) judgement, ECLI:EU:C:2019:851, par. 25. Case C-674/17 *Luonnonsuojeluyhdistys Tapiola Pohjois-Savo — Kainuu ry contro Risto Mustonen e a.*

in question did not appear to comply with EU law and lacked supporting scientific reasons as well as a proper assessment of possible alternatives and impacts.

Following the CJEU ruling, the Supreme Administrative Court of Finland ruled accordingly and declared wolf hunting permits illegal.

A year later, in 2020, another EU Court of Justice ruling dealt with the capture and transport of a wolf in inadequate conditions. On November 2016, employees of the DMPA and UN, as a veterinary surgeon, made their way to Şimon a village in Braşov, with the intention of capturing and relocating a wolf which for several days had been loitering around the home of a resident, playing and eating with that resident's dogs.

The DPMA employees coordinated the transport of the captured wolf to the Libearty Natural bear reserve in Zărneşti (Romania).

During the journey, the wolf broke the cage in which it was held and fled into the surrounding woods.

On May 2017, the *Alianța pentru combaterea abuzurilor filed* a criminal complaint against TM, UN and the DMPA, as well as against other people working for the DPMA, concerning offences related to the capture and transport of a wolf in poor conditions. It is clear from that complaint that authorisation to capture and transport the wolf had not been sought.

The Court essentially states that:

1. the protection provided for in Article 12.1 of the Habitats Directive has no limits or borders and therefore "*the view cannot be taken that a wild specimen of a protected animal species located in the vicinity of or within areas of human settlement, crossing such areas or feeding on resources produced by humans, is an animal which has left its 'natural range', or that that range is incompatible with human settlements or man-made developments*";[52]
2. "*the system of protection provided for in Article 12 of the Habitats Directive must therefore be capable of effectively preventing harm to protected animal species*";[53]
3. consequently, "*the capture and transport of a specimen of a protected animal species covered by the prohibitions provided for in Article 12, paragraph 1, of the Habitats Directive» can be only justified if they are the subject of a derogation adopted by the competent national authority pursuant to Article 16, paragraph 1, points b) and c), of that Directive, based, inter alia, on public safety grounds*". [54]

[52] CJEU's (Second Chamber) judgement, ECLI:EU:C:2020:458 par. 39. Case C-88/19 *Alianța pentru combaterea abuzurilor.*

[53] See note 53, par. 47.

[54] See note 53, par. 56.

6.4 Belgium: Decision on traders in Protected Bird Species

The illegal international wildlife trade is one of the main drivers of global biodiversity loss. However, this trade is often not considered a priority by law enforcement agencies and is operated by global criminal networks that are difficult to identify and therefore to prosecute.

This case shows that, with international cooperation and political will, these criminal organizations can be dismantled. In fact, the case involved several states: Belgium, the United Kingdom, Spain, France, Germany, Austria and the Netherlands. Well, the Belgian Cri1ninal Court of First Instance of Eastern Flanders (Ghent Division) [55] emphasized that the defendants committed a direct and irreversible assault on biodiversity, pointing out that the international trade in endangered animal and plant species has approached a lucrative scale comparable to the international drug and arms trade.

The basis of this case was EU-CITES Regulation 338/97,[56] which implements the Convention on International Trade in Endangered Species of Wild Fauna and Flora. The regulation lists species that are protected and endangered and trade in these species is prohibited. The defendants were accused of prosecuting the illegal trade in various bird species by falsifying breeders' declarations and CITES certificates, i.e. for endangered species as specified in Annex A of the Regulation.

Both at first instance and subsequently, the Court of Appeal in Ghent (Belgium) condemned the traffickers and awarded compensation to the NGOs that had filed a civil suit.

7 Conclusions

In conclusion, the decisions reported in this paper seem to highlight the multifaceted nature of biodiversity protection in the world, oscillating between recognition of nature as a subject of right and recognition as a human right.

So far, attempts to defend the rights of nature by seeking judicial protection have yielded limited, if positive results. States that have adopted an ecocentric approach, such as Colombia, Bolivia, but also the Philippines and India, are playing a pioneering role in the recognition of the rights of nature, although even this approach has proved insufficient to slow down the relentless decline of ecosystems.

As rights of nature are emerging as a new legal approach in various countries, more research is needed to assess their actual impacts on environmental protection and to develop good practices with respect to their legal framing and the design of guardianship bodies. In all probability, the implications of this new approach will differ in accordance with these parameters, but also and more importantly with the

[55] CITES crimes—Court of Appeal, Ghent, 7 May 2015.

[56] Available on https://eur-lex.europa.eu/legal-content/EN/ALL/?uri=celex%3A31997R0338.

legal, political, economic and social contexts of different jurisdictions. While the rights of nature approach may have strong impacts in such a context, most countries that have thus far recognised rights of natural entities are quite different.

The recognition of nature as a legal subject has resulted in some important advances with respect to the formulation of environmental policies in different countries. Yet, after the adoption of some rulings, these advances are not yet reflected in concrete improvements in the environmental conditions. Perhaps, there will be improvements in the years to come that result from the implementation of the public policies formulated under the rulings. However, this involves vast challenges. In fact, the Atrato case suggests that rights of nature advocates should be humble in their expectations with respect to the impacts of this approach on environmental protection in weak governance settings. While rights of nature decisions can create an important impetus for change, actually improving environmental protection in such settings requires much more than awarding legal rights to a natural entity in a courtroom. On the contrary, in the European context we can see that the protection of biodiversity has so far been the subject of a sectoral approach (mainly declined as nature protection) rather than a general discipline. In this regard, in order to strengthen the effectiveness, efficiency and, above all, the effectiveness of protection, particular care must be taken to distinguish effectiveness between general rules and sectoral provisions, otherwise we run the same risk as the first effective application of Community law in the individual member states in the 1990s when it was convinced that the only binding Community law was that of regulations and directives (which instead represent the 'secondary' law emanating from the original one). In order to achieve an effective protection and restoration of biodiversity on a global scale, it would be desirable both to create a global Organism for the political direction and administrative management of interventions on the matter, and an International Court endowed with sanctioning powers against the defaulting States. Europe has decided to modify the regulatory paradigm with the "Nature Restoration Law". The adoption of a European regulation directly applicable to the Member States, could produce a shift in the paradigm for the protection and restoration of biodiversity which, at least in Europe, will be left to hard law acts, juridically binding and of which the non-compliance can be contested. The law for the restoration of nature has been approved by the European Parliament the 12th July 2023. In conclusion, it is believed that effective protection of biodiversity requires a decisive change of pace and social economic development at a global level by establishing more binding and sanctionable rules to induce states to seriously fulfil their commitments.

References

Brambilla P (2000) La ripartizione di competenze tra Stato e Regioni nell'attuazione del diritto comunitario in materia di specie cacciabili. Rivista giuridica ambiente:301

De Vido S (2014) Tutela della biodiversità e rispetto dei diritti umani. Le sentenze CGUE nei casi Cascina Tre Pini e deviazione del fiume Archelo. Rivista giuridica ambiente 6:803 ff
Ferroni MV (2015) La protezione degli animali nell'ambiente. In: Dell'Anno P, Picozza E (eds) Tutele parallele norme procedurali, Trattato di diritto dell'ambiente, pp 447–496
Granato N (2021) Profili di valorizzazione della natura. Servizi ecosistemici e capitale naturale. www.giustamm.it
Kauffman Craig M et al (2018) Constructing rights of nature norms in the US, Ecuador, and New Zealand. Glob Environ Politics 18:4
Marchisio S (2008) Il diritto internazionale dell'Ambiente. In: Cordini G, Fois P, Marchisio S (eds) Diritto ambientale. Profili internazionali, europei e comparati, Torino, p 10
Rizzi RM (2015) Tutela dell'ambiente, tutela della caccia e della pesca. In: Dell'Anno P, Picozza E (eds) Tutele parallele. Norme Processuali. Trattato di diritto dell'ambiente. Cedam, Padova, p 524
Rockström J, Steffen W, Noone K, Persson Å, Chapin FS III, Lambin E, Lenton TM, Scheffer M, Folke C, Schellnhuber HJ, Nykvist B (2009) Planetary boundaries: exploring the safe operating space for humanity. Ecol Soc 14(2)
Segre A et al (1996) Politiche per l'ambiente. Utet, Torino
Sweetlove L (2011) Number of species on Earth tagged at 8.7 million. Nature 23:1
Steffen W et al (2015) Sustainability. Planetary boundaries: guiding human development on a changing planet. Science 347(6223):736. https://doi.org/10.1126/science.1259855
Wang-Erlandsson L et al (2022) A planetary boundary for green water. Nat Rev Earth Environ 3:380–392. https://doi.org/10.1038/s43017-022-00287-8
Wesche P (2021) Rights of nature in practice: a case study on the impacts of the Colombian Atrato River decision. J Environ Law 33(3):531–555. https://doi.org/10.1093/jel/eqab021
Wilson EO (1988) Biodiversity, Washington, DC

Maria Vittoria Ferroni is associate professor of Administrative Law (IUS 10) at the Department of Political Sciences, Sapienza University of Rome, where she teaches administrative law and environmental law. Since 2013, member of the Doctoral Board of Public, Comparative and International Law, curriculum environmental administrative law. She is author of several articles in environmental matters: La perdita di biodiversità, gli strumenti di tutela ed il codice dell'ambiente (2022); Diritto all'informazione ambientale nell'ordinamento nazionale (2015); La protezione degli animali nell'ambiente (2015).

Tiziana Bandini is graduated in Law at the University of Milan and then completed a Post Graduate Master at the same University as a Climate Expert. She is an attorney at law, since 2021 registered to the Milan's Bar Association. She is EU Climate Pact Ambassador 2022. She is a climate activist and involved in many environmental NGOs.

Adjudicating Environmental Matters Before Investor-State Dispute Settlement Tribunals: A Question of Legitimacy

Matteo Fermeglia

Abstract This contribution aims to flesh out the role of Investor-State Dispute Settlement (ISDS) tribunals as adjudicators of environmental disputes. ISDS enables private foreign investors to file compensation claims to challenge host States' domestic policies and regulations affecting their investments. Over the last three decades, a staggering number of environmental matters have been deferred to ISDS panels as a result a wide array of measures adopted by sovereign States to pursue environmental protection. However, ISDS bestows inherent features in terms of accountability that depart from those of a traditional environmental adjudicatory body. Nonetheless, decisions taken by ISDS tribunals entail similar implications as national courts while displaying several remarkable legitimacy gaps.

1 Introduction

Investor-State Dispute Settlement (ISDS) is a trans-national system for dispute resolution running in parallel with domestic courts. It equips private entities (both individuals and companies) with the option to sue sovereign States in case of conducts that might affect their Foreign Direct Investments (FDIs) in the States' territory. As a dispute resolution system, ISDS is grounded on the multitude of International Investment Agreements (IIAs), which in turn comprise more than 3000 Bilateral Investment Treaties (BITs), multilateral investment treaties (e.g., the Energy Charter Treaty, ECT) or investment protection chapters included in Free Trade Agreements (FTAs). In addition, ISDS can be opted in and triggered through bilateral contractual agreements between foreign investors and host States. Originally, the ISDS system draws from the international protection of aliens (Stone Sweet and Grisel 2017; Salacuse 2015; Miles 2013). Thus, resorting to international arbitral panels plays out against investors' fears over fear political influence, incompetence, or

M. Fermeglia (✉)
Hasselt University, Faculty of Law, Hasselt, Belgium
e-mail: matteo.fermeglia@uhasselt.be

G. Antonelli et al. (eds.), *Environmental Law Before the Courts*,
https://doi.org/10.1007/978-3-031-41527-2_13

"hometown justice" in domestic court systems (Park and Aguilar 2003). Given the focus of ISDS on the protection of large-scale investments related to infrastructures, environmental issues are often and increasingly been raised as a subject matter before ISDS tribunals. Out of the more than 1110 ISDS cases filed thus far, more than 30% relate to investments in the energy sector (UNCTAD 2022); whereas a large bulk of other cases relate to investments entailing environmental impacts or environmentally-related matters (UNCTAD 2022). Given the potentially relevant impact that ISDS might have by adjudicating cases related to environmental protection, the ISDS system has been hit by severe criticism in view of numerous flaws in the international arbitration system, which include its largely *ad hoc* structure, reliance on private firms and lawyers, and overall departure from basic elements of the Rule of Law (Stoll 2018). The interplay between ISDS and domestic environmental and climate change policies has been clearly spelled out by the Intergovernmental Panel on Climate Change (IPCC) Working Group III in its sixth Assessment Report. Accordingly:

> A large number of bilateral and multilateral agreements, including the 1994 Energy Charter Treaty, include provisions for using a system of investor-state dispute settlement (ISDS) designed to protect the interests of investors in energy projects from national policies that could lead their assets to be stranded. Numerous scholars have pointed to ISDS being able to be used by fossil-fuel companies to block national legislation aimed at phasing out the use of their assets [...] (IPCC 2022).

Concerns over ISDS as an avenue for adjudication of environmental disputes has been raised also in other intergovernmental arenas. OECD launched in 2022 a public consultation on international investment treaties and climate change to receive contributions and inputs from civil society organisations, industry and Academia about ways to reconcile what apparently are two legal regime pursuing opposing, if not mutually excluding objectives. A prominent drawback of the ISDS system relates to its failure to embed public policy considerations and adequately weight them against the long-established standards of protection of investments established in IIAs (Tienhaara 2013). Notwithstanding all the above, ISDS is increasingly addressing environmental and climate change matters at an accelerating pace (UNCTAD 2022). Yet while adjudicating environmental and climate change matters, being an international treaty-based source of adjudication, ISDS shares peculiar features that puts it at odds with a fully-fledged environmental adjudicatory function, which notably include sizeable shortcomings in terms their overall accountability and legitimacy. This contribution therefore navigates such elements while picturing the past and current evolutionary trends of ISDS as an environmental means of adjudication.

This Chapter is structured as follows. Section 2 charts the evolutionary trend of ISDS with regard to environmental matters. Relevant, ISDS tribunals deal with a broad spectrum of domestic environmental policies and regulation while looking exclusively at their concrete impact on foreign investments as opposed to the public policy underlying certain domestic measures. Section 3 depicts the main shortcomings of the ISDS system in terms of its independence, transparency and justiciability. Section 4 concludes.

2 ISDS and the Environment (and Climate Change): A Review of Existing Trends

The burgeoning body of literature exploring the intersection between the FDIs protection regime and the environment over the last decade clearly testifies the ramping attention given to the issue (Levashova et al. 2015; Vinuales 2012; Di Benedetto 2013; Cordonier-Segger et al. 2010). More generally, the tension between foreign investments protection and environmental protection comes as a result of fragmentation of international law (Miles 2013). The environment-investment interplay has become particularly visible in the late 1990s in the context of investor-state arbitration, when environment-related investment disputes began to be filed with international tribunals. This goes hand in hand with the overall trend of proliferation of IIAs, which rose significantly from a few hundred in 1990 to a total of 3196 in 2012, as well as ISDS cases, which rose spectacularly over the early 2000s up to more than 1190 in 2021 (UNCTAD 2021).

According to a recent account of the United Nations Commission on Trade and Development (UNCTAD) out of the total 1190 ISDS cases, 175 ISDS cases (15%) have been filed in relation to domestic measures adopted for environmental protection, whereas 192 cases have been filed by foreign investors in the fossil fuels sector and 80 cases have been filed by investors in the renewable energy sector (UNCTAD 2022).

Out of the above 175 environment-related ISDS cases, 40% were decided in favour of host States (either on the grounds of jurisdiction or merits), where 38% have been decided in favour of investors—thus awarding damages thereto. The remainder was either discontinued, settled or decided in favour of investors yet without damages being awarded. The overall trend in annually filed environmental-related ISDS cases shows a spectacular increase from 1 case in 1992 to 19 in 2013 (Vinuales 2019).

It is not the purpose of this contribution to appraise all ISDS cases related to environmental matters. However, for the purpose of this Chapter major trends can be highlighted, which unfold the pattern of decision-making that characterise ISDS tribunal as environmental adjudicators.

More specifically, three main trends can be underscored as outlined in this Section. All the trends below share the key commonality of being grounded on challenges to host States' domestic measures pursuing environmental protection purposes. I shall corroborate these categories with concrete examples from landmark ISDS decisions.

The first trend relates to market-oriented measures. Within the range of domestic policies and regulations that can be challenged before ISDS tribunals, a notable role is played by any measures addressing potential remuneration for energy generation, as well as other economic activities. Measures falling under this category include any regulatory approach ultimately affecting investments by curbing or annulling expected profits and rate of returns. For example, tax levies on energy production and hydrocarbons extraction, introduction of cap-and-trade schemes addressing highly greenhouse gases emitting sectors, reduction or revocation of subsidies fall under this category.

The key underlying element behind these cases relates to the asserted violations of investors legitimate expectations to being granted a certain regulatory regime as existing at the time the investment was carried out. For, the very existence and permanence of such regimes was the key factor grounding the decision of undertaking the investment in the first place—based on the expected return on investment, long-term agreement conditions, etc. also weighed against host Governments' original representations to foreign investors.

The key elements to be addressed by ISDS tribunals in this strand of cases can thus be summarised in: (a) the actual impact of the challenged measure adopted by the host State on investors' patterns of profitability but for the introduction of the challenged measure or but for the challenged host State's conduct, and (b) the abrupt and arbitrary character of the host State's conduct in terms of depriving the investors of their property and frustrating their expectations in terms of the context surrounding the carrying out of their investments.

A recent striking example of these category relates to the so-called "renewable energy saga", which hit foremost Spain and Italy as well as other European countries (e.g., Czech Republic, Bulgaria and Poland) with regard to the roll-back of the domestic legal regime for financial support to investments in renewable energy generation. The total number of cases filed since 2015 currently exceeds 70 ISDS disputes. While the decisions adopted by ISDS tribunals may follow different approaches despite the almost uniform factual and legal setting, in *Masdar v. Spain*, an ICSID tribunal acknowledged the violation of Article 10(1) of the ECT on Fair and Equitable Treatment (FET).[1] The tribunal ascertained investor's legitimate expectations for the host State's legal framework in place at the time its investment was made not to be amended in any way that would substantially affect its expected return. This representation was further substantiated on previous specific commitments undertaken individually by the Spanish Government towards the investor. Yet in so doing, the *Masdar v. Spain* tribunal stressed the general remark that the protection of foreign investments as secured by international standards of treatment:

> [...] shall not be viewed as implying the stability of the legal and business framework. Economic and legal life is by nature evolutionary. [...] economic stability cannot be a legitimate expectation of any economic actor.[2]

In this category of cases, therefore, the key element to be appraised by ISDS tribunals refers to the asserted loss of profitability of foreign investments due to the host State's intervention as opposed to the underlying public policy drivers of action by host States.

The second category of cases relate to challenges brought directly against decisions taken by host States' authorities at all levels within the context of permitting and administrative procedures for environmental matters. This array of cases covers

[1] *Masdar Solar v. Spain*, ICSID Case no. ARB/14/1, Final Award, 16 May 2018.

[2] Ibid., para. 485.

a wide array of infrastructures ranging from energy generation to waste management facilities to touristic resorts, etc. Moreover, these cases widely range from permitting procedures challenged on the mere grounds of the environmental impact of a specific project to other considerations, including the protection of indigenous communities and the protection of cultural heritage.[3]

The subject of the challenges brought by foreign investors here is mostly centred around host States authorities' conduct throughout the exercise of their administrative powers. Such conduct is often assessed by investments tribunals both from an external perspective—i.e., against the previous representations granted by host States' governments in terms of future uptake of the investment, for example through public statements, bilateral communications, contractual arrangements, etc.—and from an internal perspective—i.e., looking at the actual core elements of the permitting procedures according to due process. These cases relate to the adequacy and legality of host States' administrative action within the context of investors' reliance thereupon for the successful development of their investment.

Among those cases, one of the most striking examples that also made relevant public outreach (and outrage) was the decision in *Eco Oro v. Colombia*.[4] In *Eco Oro*, the subject to the dispute was the failure of the Colombian Government to enact a series of measures to delineate the protected area around a very delicate ecosystem, the *Páramo de Santurbán*. The delay and failure of the Government allegedly came at the detriment of Eco Oro's mining activities in the Angostura mine in the region, which was previously agreed upon and entrusted to Eco Oro through a Concession agreement. While recognising the good faith of the government and the legitimate public purpose served by the same government's measure to protect the ecosystem, the ISDS panel concluded that a violation of the relevant IIA was due in light of Colombia's failure to delimitate the protected area of the *Páramo de Santurbán*. According to the majority of the tribunal:

> Colombia's actions, including its failure finally to delimit the Santurbán Páramo in circumstances where Eco Oro was advised that no environmental licenses could be issued for mining projects in the vicinity of the Santurbán Páramo […] failed to provide Eco Oro with a stable and predictable regulatory environment.[5]

Therefore, the tribunal characterised Colombia's conduct as:

> […] grossly inconsistent and given rise to considerable confusion and uncertainty as to (i) what activities may and may not be undertaken within the páramo as currently delimited; (ii) what the final boundaries will comprise; and (iii) when the final delimitation will be announced […] The Tribunal therefore asks itself whether Colombia's conduct amounts to

[3] *Compania del Desarrollo de Santa Elena S.A. v. Republic of Costa Rica*, ICSID Case No ARB/96/1; *Tecnicas Medioambientales Tecmed S.A. v. United Mexican States*, ICSID Case No ARB(AF)/00/2; *Gold Reserve Inc. v. Bolivarian Republic of Venezuela*, ICSID Case No ARB(AF)/09/1.

[4] *Eco Oro Minerals Corp. v. Republic of Colombia*, ICSID Case No. ARB/16/41, Decision on Jurisdiction, Liability and Directions on Quantum, 9 September 2021.

[5] Ibid., para. 805.

> gross unfairness or manifest arbitrariness falling below acceptable standards. The majority of the Tribunal answers this in the affirmative.[6]

In this case, as well as in other relevant decisions concerning, among others, Environmental Impact Assessment procedures and spatial planning measures, the tribunal's focus lies on the arbitrariness of the host State's conduct, also with regard to the implementation of overarching obligations stemming from international and domestic environmental law.[7]

The third, though not less important category relates to ISDS cases filed as a result of broader measures adopted by host States at the national and sub-national level, which bear far-reaching impacts on different kinds of investments. The obvious example of domestic measures falling under this category is the introduction of general bans or moratoria on economic activities deemed detrimental to the environment. Similarly, other command-and-control measures can be enlisted including restrictions or prohibitions on the production and commercialisation of certain products (e.g., pesticides), introduction of higher standards of environmental performances, introduction of or amendments to liability regimes for environmental damages, etc.[8]

One recent exemplary case clearly falling under this category is *Rockhopper v. Italy*.[9] In *Rockhopper*, the tribunal unanimously found that the decision issued by the Italian government to prohibit exploration of hydrocarbons within 12 nautical miles offshore amounts to a deprivation of the value of the investment in the form of the enjoyment of a concession agreement for the extraction of hydrocarbons. In the tribunal's words:

> As of that moment [i.e., the passing of the oil production ban] the Claimants' right to be granted a production concession was taken away from it. Thereafter the Claimants had neither the right to be granted the production concession, much less the production concession itself. The Claimants went, in one fell swoop, from a position where they had rights to a valuable production concession which would actually lead, under Italian law, to such production concession, to essentially nothing at all. No lengthy elaboration is required to arrive at this conclusion. There was, factually speaking, an immediate and complete deprivation of the Claimants' investment.[10]

[6] Ibid., para. 820.

[7] *Emilio Agustin Maffezini v. Kingdom of Spain*, ICSID Case No. ARB/97/7; *Metalclad Corp. v United Mexican States*, ICSID Case No. ARB(AF)/97/1; *Vattenfall AB, Vattenfal Europe AG, Vattenfall Europe Generation AG v. Federal Republic of Germany*, ICSID Case No. ARB/09/6; *William Ralph Clayton, William Richard Clayton, Douglas Clayton, Daniel Clayton, and Bilcon of Delaware, Inc. v. Government of Canada*, NAFTA Arbitration under UNCITRAL Arbitration rules.

[8] *Methanex Corporation v. United States of America*, NAFTA Arbitration under UNCITRAL Arbitration rules; *Chemtura Corporation v. Government of Canada*, NAFTA Arbitration under UNCITRAL Arbitration rules.

[9] *Rockhopper Italia S.p.A, Rockhopper Mediterranean Ltd and Rockhopper Exploration Plc v. Italy*, ICSID Case No. ARB/17/14, Final Award, 23 August 2022.

[10] Ibid., para. 194.

In these cases, the environmental purposes of a given host State measures is assessed against the concrete magnitude and abruptness of its impact on the acquired and secured property and rights of investors.

Importantly, although from different standpoints and with a different degree of normativity, all the above examples point to ISDS tribunals as deploying an adjudicatory function on environmental matters insofar as tribunals concretely assess the legality of host State's regulatory and administrative conducts, as well as—albeit mostly indirectly—the underlying environmental purposes thereof.

3 Is ISDS a Proper Environmental Adjudicatory Body? A Rule of Law Perspective

As shown in Sect. 2, ISDS tribunals have been increasingly called upon to address normative conflicts between protection of foreign investments and domestic environmental and climate policies. This certainly qualifies ISDS as one of the existing—yet peculiar—environmental adjudicatory fora. Against this backdrop, the general backlash against the ISDS system entails three major criticisms. First, ISDS provisions in IIAs and investment contracts arguably attribute greater power to foreign investors than nations, by permitting investors to file claims against governments. As mentioned above, this conception is ingrained in the view that host States have given reciprocal previous consent to disputes under ISDS by concluding the relevant underlying IIA. This assumption is however contrasted by recent findings stipulating that the ISDS system as such would not be instrumental either in stimulating FDI inflows or channelling investment strategic decisions (Brada et al. 2020; Mehranvar and Sasmal 2022).

Second, ISDS would be undermined by internal structural failures. Recent accounts unfold asserted biases within the ISDS arbitration system while also highlighting a trend of law firms pushing corporations to bring disputes under ISDS provisions (Eberhardt and Olivet 2012). ISDS disputes are normally brought before international arbitration forums. The largest forum referred to by far is the International Centre for Settlement of Investment Disputes (ICSID), a non-profit arm of the World Bank. The second most used forum is the UN Commission on International Trade Law (UNCITRAL). Other forums can be the International Chamber of Commerce (ICC), the Permanent Court of Arbitration (PCA) and the Stockholm Chamber of Commerce (SCC). all the above arbitral forums rely on their own procedural rules.

Third, and stemming from the two arguments above, concerns are growing around the ultimate consequences of ISDS provisions and ISDS adjudication both on the fundamental rights of individuals and collective groups and the environment. By devising a system whereby corporations can demand compensation to sovereign States through taxpayers' money related to their domestic regulatory choices or public policies, ISDS is deemed to trigger regulatory chill (Tienhaara 2018).

Accordingly, the filing of ISDS disputes from private entities (or the mere threat thereof) would lead host States watering down their efforts in protecting vulnerable groups (e.g., indigenous communities) and the environment, as well as pursuing more ambitious climate action (Tienhaara 2018). The concrete magnitude of such phenomenon is grounded on context-specific institutional, social and economic elements (Janeba 2019). Yet the manifestation of regulatory chill has been clearly observed in several instances, especially with regard to cases related to oil & gas and mining infrastructures in the Global South (Perrone 2022). Moreover, and within the ISDS system, the potential major monetary impacts of ISDS liability claims on public budget provides a bold incentive for states to settle the disputes even when faced with high winning odds given the often-clear mismatch in terms of financial and staff resources to handle a lengthy and costly arbitral proceeding.

All these concerns about ISDS however revolve around a set of key elements that characterize the ISDS system as an adjudicatory mechanism. All such elements contribute to a general lack of legitimacy and accountability of the system, which renders it arguably ill-suited to address environmental matters as compared to domestic or other adjudicatory mechanisms based on the principles of the rule of law.

In sum, as voiced by one of the starkest critics of the ISDS system:

> Investment treaty arbitration is often promoted as a fair, rules-based system and, in this respect, as something that advances the rule of law. This claim is undermined, however, by procedural and institutional aspects of the system that suggest it will tend to favour claimants and, more specifically, those states and other actors that wield power over appointing authorities or the system as a whole. On the other hand, other states and investors (especially those that bring claims against a powerful state) can expect to be disadvantaged (Van Harten 2010).

In the following sub-Sections, I will hence zoom in on three relevant elements in this connection, which relate in turn to the ISDS system's *independence*, *transparency* and *justiciability*.

3.1 Independence in ISDS

Independence and impartiality of decision-makers are essential elements of any adjudicatory mechanism based on the principles of the Rule of Law. As such, they constitute part of the general principles of law within the meaning of Article 38(1)(c) of the Statute of the International Court of Justice. Independence and impartiality help safeguard the objectivity and fairness of legal proceedings. Concerns have been widely raised as to the overall independence of ISDS tribunals. The issue of ISDS panels' independence has been noted, among others, by UNCITRAL Working Group III within the context of the ongoing process of ISDS reform:

> Independence and impartiality were described as key elements of any system of justice, including arbitration. The concerns relating to the possible lack of independence and impartiality of decision makers, or of the perception thereof, were said to be particularly acute in

> the field of ISDS, as ISDS cases usually involved public policy issues and involved a State. It was re-affirmed that, in order to be considered effective, the ISDS framework should not only ensure actual impartiality and independence of decision makers, but also the appearance thereof. Therefore, it was said that any reform in that respect should aim at addressing both actual and perceived lack of independence and impartiality (UNCITRAL 2018).

Concerns in this respect have been pointed to both individual and systemic issues within investment arbitration. Indeed, criticism of the present ISDS mechanism is directed against the central feature of party appointment, as well as the propriety of connections between arbitrators and parties, the issues of multiple appointments, double-hatting, issue conflict, and implicit pro-investor bias—as well as, for ICSID arbitrations, role-mixing between arbitrators and members of Annulment Committees. According to empirical accounts, double-hatting (where arbitrators in one proceeding also serve as counsels or experts in other arbitrations) alone could affect more than half of investment arbitrations (Langford et al. 2017). The draft Code of Conduct for investment arbitrators developed by ICSID and UNCITRAL mandates arbitrators to ensure independence by "avoiding any conflict of interest or external control" and impartiality by not having biases or predisposition "towards a disputing party or issues raised in the proceeding". In particular, the draft Code of Conduct stresses "loyalty" as a potential external source of impairment of arbitrators' independence and impartiality although not regulating loyalty itself. New generation IIAs such as the Investment Chapter in the Comprehensive Economic and Trade Agreement between Canada and the European Union (CETA) ban arbitrators' double-hatting altogether. More generally, however, research explains that arbitrators' decisions affect their success and focus on the ways they seek to keep the customers satisfied, in order to gain repeat business (Dezalay and Garth 1996). At the same time, arbitrators' appointment by the parties is a key driver behind investment arbitration over domestic courts. This should not sound surprising as it relates to the very existence of the ISDS system and its traditional self-referential character, which allows for bolder parties' agency to exercise power over the adjudication process.

Yet in the ISDS context, the notion of judicial independence is appraised in a fairly different fashion as it is normally detached from the doctrine of separation of powers. Hence the discussion lies on whether arbitrators should be independent and impartial not from the legislative and the executive, but from private business and corporations (Arato 2015). In this respect, the absence of security of tenure for arbitrators in investment arbitration raises further credible concerns about the lack of procedural fairness in the system. ISDS claims may only be launched by foreign investors. To have a one-way, private-fuelled system of adjudication therefore speaks for the risk of explicit or implicit biases of the service-providers towards the service-consumers (Van Harten 2018). This does not hold true only for investment arbitrators, but also for a whole range of professionals revolving around the ISDS litigation business, such as councils and experts. This carries an additional burden in terms of unbalancing the relationship between richer and poorer country since ISDS is more commonly relied upon by companies based in the West against countries in the Rest, particularly encroaching upon already poorer countries (Schultz and Dupont 2015).

Counterarguments to those critics point to the self-restrain and professionalism of arbitrators and the self-policing nature of the ISDS system. Accordingly, arbitrators rendering biased judgements and awards would not be further appointed as they would lose influence due to their jeopardised reputation (Behn et al. 2020).

Furthermore, yet importantly, the current standards for disqualification of investment arbitrators provides for a highly cumbersome threshold. A party wishing to disqualify an arbitrator under ICSID rules shall prove a manifest lack of impartiality pursuant to Article 14 ICSID Convention. Where the IBA Guidelines on Conflicts of Interest may guide the final appraisal on disqualification, as such they are not binding. In ICSID arbitrations, the final decision on the disqualification of the arbitrator is taken by the other two arbitral panel members (Article 58 ICSID Convention). As a matter of fact, only in one instance an arbitrator was disqualified for apparent issue conflict by repeatedly and consistently upholding specific opinions on a subject matter under dispute before ISDS tribunals both during the proceeding and in public through academic publications, conference presentations, etc.[11]

Last, and relevant for the following discussion, no clear rules exist as to the distinction between arbitrators sitting in the ICSID panels and those sitting in the ICSID Annulment Committees.

3.2 *Transparency in ISDS*

Transparency encompasses allowing broad access proceedings and documents to the public, holding adjudicatory bodies accountable for their decisions and providing avenues for redress. In all the above respects, investment arbitration is characterised by a lack of transparency at almost every step of the arbitral process, throughout the filing of ISDS cases down to the decision-making process and subsequent stages.

Advocates of greater transparency claim that greater participation and access to arbitral proceedings might generate more informed decision-making by arbitrators while enhancing principles associated with the Rule of Law, such as access to justice.

Yet the vast majority of past ISDS proceedings and a fair deal of current proceedings remain fully confidential. In fact, confidentiality is another key driver behind the recourse to ISDS for private entities. As noted by investment lawyers, confidentiality is instrumental to keep the relationship between the parties, secure commercial and industrial secrets and avoid disclosures designed to affect the outcome of the case (Born 2014). Oddly enough, confidentiality might not be disliked by host States either. In fact, the fully confidential character of the proceeding would allow

[11] *Caratube International Oil Company LLP v Republic of Kazakhstan*, ICSID Case No ARB/13/13.

governments to make statements about public policy they would otherwise not disclaim in writing. While these arguments might stand from a mere commercial perspective, the same shall not be from a public policy perspective.

Consequently, all relevant documents filed under the matter in ISDS proceedings (including notice of arbitration, expert reports, tribunal orders, final awards, etc.) are often not public. In addition, a rule of thumb in investment arbitration is the "*in camera*" procedure, whereby hearings before ISDS tribunals shall be limited to the parties of the case. The tone on the matter was set in the *SD Myers* case in 1999.[12] In *SD Myers*, the NAFTA tribunal concluded that the *in camera* requirements under UNCITRAL rules grounded an implied right of confidentiality, which the tribunal could not overturn without the agreement of the parties to the arbitration. This assumption has not crystallised in ISDS arbitral practice. For instance, such implied duty of confidentiality was rejected by the *Methanex* tribunal, which moreover allowed for *amicus curiae* briefs to be submitted by civil society groups.[13] According to Article 37(2) of the ICSID Arbitration Rules, ISDS tribunals retain ultimate discretion on the admission of submission by non-disputing parties after consultation with the parties to the dispute. Controversial examples of ISDS tribunals inconsistency towards the matter can be drawn from two other landmark cases. In *Philip Morris,* which related to Uruguay's right to regulate plain packaging of cigarettes for public health, submission by third parties, namely the WHO and the Pan-American Health Organisation, has played an influential role in the tribunal's findings of fact in their final decision.[14] However, in the *Eco Oro* case related to the *Angostura* gold mining project in the *Santurbán Páramo*, the ICSID tribunal refused to accept submissions from the Committee for the Defense of Water and the *Páramo of Santurbán* and allied international organizations, thus sparking rebuffing from civil society.

Disclosure and transparency are now gaining ground in ISDS as a matter of existential need against the strong backlash from civil society, governments and the international community (Polanco 2018). Notwithstanding, confidentiality rules are still applied especially with regard to highly economically and politically sensitive cases. For example, in the context of energy and climate-change related ISDS disputes, both the *Rockhopper v. Italy* and *Vattenfall v. Germany* proceedings were put under full confidentiality—except for the final panels' decisions.

Already in 2014 UNCITRAL had adopted 'Rules on Transparency in Treaty-based Investor-State Arbitration' (the 'UNCITRAL Rules on Transparency'), the application of which is promoted by United Nations Convention on Transparency in Treaty-based Investor-State Arbitration, also known as the Mauritius Convention on Transparency. Such rules provide for meaningful improvements, thus including the

---

[12] *S.D. Myers Inc. v. Government of Canada,* NAFTA Arbitration under UNCITRAL Arbitration rules.

[13] *Methanex Corporation v. United States of America,* NAFTA Arbitration under UNCITRAL Arbitration rules.

[14] *Philip Morris Brand Sàrl, Philip Morris Products S.A. and Abal Hermanos S.A. v. Oriental Republic of Uruguay*, ICSID Case No. ARB/10/7.

publication of arbitral documents, *amicus curiae* submissions by third parties, and even the opening of arbitral hearings to the public. However, the UNCITRAL Rules do not apply automatically to previously existing IIAs before 2014, nor can trump conflicting provisions provided therein (Article 3a).

Progress has been made also under the newly adopted 2022 ICSID Arbitration Rules. Accordingly, parties are deemed to have consented to the publication of arbitration awards unless a party objects in writing within 60 days after the award has been dispatched. This rule also applies to the publication of orders and decisions. In addition, the 2022 ICSID Arbitration Rules provide that ICSID tribunals must allow persons in addition to the parties to observe hearings unless one of the parties to the dispute objects. Finally, the new ICSID Arbitration Rules have expanded the list and included an obligation for the tribunal to provide relevant documents to the non-disputing party unless a party objects.

Similar efforts are being put forward under recent treaty-making, including NAFTA (Art. 2017(4)) and the following USMCA Agreement, CETA, the latest Dutch Model BIT and, last in time but not least, the proposed modernized version of the ECT.

3.3 *Justiciability in ISDS*

As rightly pointed out by investment legal scholarship, if arbitration simply bow to formal enactments without examining the lawfulness of their awards under more fundamental principles, they may just end up giving sectarian conflicts (Paulsson 2013). Accountability more broadly encompasses several aspects of decision-making, including all the above issues of transparency, impartiality and justiciability. However, in addition to those two elements, the lack of judicial review stands out as a peculiar feature of ISDS. Once again, the underlying rational for a limited judicial review of ISDS decisions lie in core objectives of the investment regime, namely cost-effectiveness and, more importantly, time-efficiency. One of the asserted objectives of the ICSID Convention was precisely to provide speedy and prompt solutions to investment disputes as opposed to yield international proceedings with lengthy introductory and preliminary claims. Therefore, the opportunity of creating an appellate mechanism under the ICSID Convention was put on the agenda, yet neatly rejected precisely against concerns over excessive length of arbitration proceedings and increases of costs.

The length of proceedings and the resulting impact on costs has long been a topic of considerable importance for arbitration. According to a recent account by ICSID, carried out on cases filed between 2015 and 2017, the average duration of proceedings equaled 1336 days, where annulment proceedings take, on average, 1.91 years—approx. 697 days (ICSID Secretariat 2018; Zárate et al. 2020). Where investment tribunals' awards are enforceable before host States' national courts or foreign courts according to the 1958 New York Convention on the Recognition and Enforcement of Arbitral Awards, awards rendered by tribunals under the ICSID

Convention are binding upon host States signatories and therefore shall be immediately complied with regardless of any procedural issue arising in the course of enforcement (Article 53 ICSID Convention).

ISDS awards can only be challenged by either party to the disputes through annulment procedures. According to the ICSID Convention, under the annulment procedures a new arbitral panel shall be established after a preliminary screening by the ICSID Secretariat. However, such procedures are far from sharing the features of a proper judicial review of ISDS panels' decisions. More specifically, Article 52 of the ICSID Convention allows the parties to request the annulment of the award based on five exhaustive grounds, namely:

1. That the tribunal was not properly constituted.
2. That the tribunal manifestly exceeded its powers.
3. That there was corruption on the part of a member of the tribunal.
4. That there was a serious departure from a fundamental rule of procedure.
5. That the award failed to state the reasons on which it was based.

The above elements may also be extended to encompass issues of independence and impartiality of sitting arbitrators. For instance, an ICSID Annulment Committee quashed a 128 Mln € Award in one of the cases related to the renewable energy support schemes, *Eiser v. Spain*.[15] In this case, the Annulment Committee stressed that the annulment ground of the proper constitution of the tribunal includes the lack of independence and impartiality of arbitrators throughout the proceedings with particular regard to an undisclosed long-standing relationship between one arbitrator and an employee of a firm providing expert testimonies on damages assessment in the case.

The departure from any substantial merits of the dispute in the scope of review of annulment committees is however apparent. Moreover, it must be noted that in the spirit of the ICSID Convention the annulment procedure shall be deemed exceptional, and in principle should not affect the enforcement of the award and the overall duty on the parties of compliance therewith.

For decisions rendered outside the application of ICSID Convention (e.g., under UNCITRAL *ad hoc* tribunals), review by domestic courts applies. Yet the scope of review by domestic courts shall also be generally limited to jurisdictional issues and issues related to the overall integrity of the arbitration, thus bypassing any assessment on the merits of the dispute. The reference set of rules as adopted in several jurisdictions are the 2006 UNCITRAL Model Law. Under Article 34(1) of the UNCITRAL Model Law, as also further explained by UNCITRAL, the grounds for review shall mirror the provisions on enforcement under the New York Convention (Article V). Hence the grounds for annulment are the following:

- That the party to the arbitration agreement was under some form of incapacity or that said agreement is not valid under the law to which the parties subjected it.

[15] *Eiser Infrastructure Limited and Energía Solar Luxembourg S.à r.l. v. Kingdom of Spain*, ICSID Case No. ARB/13/36, Decision on Annulment, 11 June 2020.

- That the party making the application was not given proper notice of the appointment of an arbitration or of the arbitral proceedings or was otherwise unable to present its case.
- That the award deals with a dispute not contemplated by or not falling within the terms of the submission of the arbitration or contains decisions on matters beyond the scope of submission to the arbitration.
- That the composition of the arbitral tribunal or the arbitral procedure was not in accordance with the agreement of the parties.

Accordingly, the scope of appraisal that domestic courts shall carry out on the grounds of annulment is very similar to that laid down under the ICSID Convention.

Yet in addition, and importantly, while reviewing ISDS awards domestic courts may find that either the subject matter of the dispute is not capable of settlement by arbitration under the law of this state or the award conflicts with the public policy of said state. Contrariety to host States' public policy is the only material element that can provide a room for a proper review on the merits of the dispute and therefore ensure a proper review of any proportionality analysis taking into account, for example, the overarching obligations of host States to protect the environment and biodiversity under the UNCBD and tackle climate change under the UNFCCC. Unsurprisingly, this is being advocated and upheld within the European Union, with regard to the asserted conflict between the ISDS adjudicatory system and the supremacy and autonomy of EU law (López-Rodriguez 2019).

As notably pointed out by critical investment legal scholarship, however, even awards that have been successfully challenged before annulment committees are still being cited and referred to by other ISDS panels, as well as other means of review—including parties' interpretation and review by subsequent tribunals—are mostly disregarded (Aschner 2020).

Substantial reforms are currently in place to address the issue of justiciability of ISDS awards. The most far-reaching attempt in this respect refers to the establishment of multilateral investment court as an overhaul of the current ISDS system currently being advocated by the European Commission in its latest Free Trade Agreements (such as CETA and the EU-Vietnam FTA). Such multilateral investment Court would operate as a permanent, two-tiers panel with a first instance and an appellate tribunal—akin to the WTO Dispute Settlement Body. In addition, it would draw on a permanent pool of arbitrators (with fixed-term appointment) for both tribunals, which would operate under stringent transparency rules and provide openness to investment proceedings. In 2021, Canada and the EU agreed on general rules for the establishment, among others, of the Appellate Court under CETA. Accordingly, the Appellate Court shall be able to "uphold, modify or reverse" the legal findings and conclusions of the first instance tribunals in whole or in part. In so doing, the Appellate Court must apply "its own legal findings and conclusions" to the facts of the disputes. The Appellate Court may also refer the case to the first instance Tribunal for a reconsideration of the matter in case it is not able to deliver a full judgment thereupon. Awards by the Appellate Court must be rendered within 180 days from the lodging of the appeal by either disputing party in shall not exceed 270 days at any rate. This speaks for due consideration of the time-efficiency rationale behind settlement of investment disputes.

The above efforts by the European Union to establish a Multilateral Investment Court are being pursued also in the context of UNCITRAL Working Group III reform of ISDS. In the latest round of negotiations held in January 2023, draft provisions on the functioning of an Appellate Mechanism for ISDS disputes have been released. Accordingly, the scope of review of the Appellate Mechanism should include, alongside the above issues of lack of jurisdiction, unlawful constitution of the tribunal and excess of arbitral powers, the following elements:

- An *error in the application or interpretation of the law*; or
- A *manifest error in the appreciation of the facts, including the appreciation of relevant domestic legislation* and the assessment of damages.
- The decision by the first-tier tribunal is in *conflict with international public policy* (emphasis added).

The decision of the Appellate Tribunal shall be final and binding upon the parties to the dispute.

However, and notwithstanding the above developments, several key formal and substantial questions remain open, starting from the number of sitting arbitrators to their qualifications and appointment.

Pending the ongoing reform of ISDS by UNCITRAL Working Group III, therefore, any ongoing arbitration as well as any arbitration filed before the final adoption of the reform will see a very limited scope of review as to merits of the award—and in particular on the interpretation and application of standards of treatment or other doctrines, such as the police powers doctrine with regard to host States' right to regulate on matters of public policy—is left untouched and set in stone. Nor, in theory, this can be reviewed down the line of the enforcement procedure—where the only real backstop to the final recognition of arbitral awards relates to the contrariety of arbitral decisions to public order.

4 Conclusion

An honest appraisal of investment arbitration panels as adjudicatory bodies for environmental matters requires a proper understanding of the same FDIs protection system's objectives. Such objectives include the protection and promotion of foreign investment, as well as market liberalization and the building of closer economic and political relations among contracting states. These objectives are arguably geared towards enhancing contracting states economic welfare in the long-term. In this respect, as noted by investment legal scholarship, the international investment regime embraces investment protection as an application of the rule of law against arbitrary, unlawful States conducts (Schill 2017). At the same time, while pursuing this objective, ISDS tribunals also encroach upon domestic rulemaking by sanctioning States' regulatory choices and therefore curtailing their regulatory space.

Yet apparently "a shy, yet sufficiently discernible, change of mindset regarding the place granted to environmental considerations in investment litigation" can be noticed, which is further unfolding an "upgraded" traditional approach to the relations between international investment law and environmental protection (Viñuales 2016). However, as this Chapter has tried to point out, major setbacks remain within the core elements of the ISDS system to align it with a full-fledged environmental adjudicatory body, if not to comply with the rule of law altogether.

It is an established fact that as the fight against climate change ramps up, the biodiversity crisis unfolds and the pollution crisis remains largely unaddressed, both developed and developing countries will resort to more stringent and radical measures potentially affecting investments in a large range of economic sectors. Thus, sovereign States will hardly eschew the net of IIAs and their application through ISDS. Unprecedented efforts are being put in motion to make the system up to par with the global environmental and climate challenge. However, where such patterns of reform will yield results in the coming years and hopefully correct some of the flaws of the international investment regime, other inherent features of the arbitral system will likely remain untouched. Arbitrators hence have a clear, yet far-reaching task ahead in upholding environmental and climate law standards also to vouch for the whole legitimacy of the ISDS system while not hampering legitimate domestic action to protect our common home.

References

Arato J (2015) Corporations as Lawmakers. Harv J Int Law 56(2):229

Aschner W (2020) Correctness of investment awards: why wrong decisions do not die. Law Pract Int Court Trib 18(3):345–368

Behn D, Malcolm Langford M, Litourneau-Tremblay L (2020) Empirical perspectives on investment arbitration: what do we Know? Does it matter? J World Invest Trade 21(2–3):240–224

Born G (2014) International commercial arbitration, 3rd edn. Kluwer Law International

Brada JC et al (2020) Does investor protection increase foreign direct investments? A meta-analysis. J Econ Surv 35(1):34–70

Cordonier-Segger MC, Gehring MW, Newcombe A (eds) (2010) Sustainable development in world investment law. Kluwer Law International

Dezalay Y, Garth BG (1996) Dealing in virtue: international commercial arbitration and the construction of a transnational legal order. The University of Chicago Press, Chicago

Di Benedetto S (2013) International investment law and the environment. Edward Elgar, Cheltenham

Eberhardt P, Olivet C (2012) Profiting from injustice: how law firms, arbitrators and financiers are fuelling an investment arbitration boom. Brussels-Amsterdam, Transnational Institute

ICSID Secretariat (2018) Backgrounder on proposals for amendment of the ICSID rules. Available at https://icsid.worldbank.org/en/Documents/Amendment_Backgrounder.pdf

Intergovernmental Panel on Climate Change (2022) Mitigation of climate change. Contribution of Working Group III to the sixth assessment report of the intergovernmental panel on climate change. Cambridge University Press, Cambridges

Janeba E (2019) Regulatory chill and the effect of investor state dispute settlement. Rev Int Econ 17(4):1172–1198

Langford M, Behn D, Runar Hilleren Lie R (2017) The ethics and empirics of double hatting. ESIL Reflection 6(7):1–13

Levashova Y, Lambooy T, Dekker I (eds) (2015) Bridging the gap between international investment law and the environment. Eleven International Publishing

López-Rodriguez AM (2019) The sun behind the clouds? Enforcement of renewable energy awards in the EU. Transnatl Environ Law 8(2):279–302

Mehranvar L, Sasmal S (2022) The role of investment treaties and investor-state dispute settlement (ISDS) in renewable energy investments. Columbia Center on Sustainable Investment, New York

Miles K (2013) The origins of international investment law: empire, environment and the safeguarding of capital. Cambridge University Press, Cambridge

Park WW, Aguilar G (2003) The new face of investment arbitration. Yale J Int Law 28:365

Paulsson J (2013) Arbitral authority to reject unlawful laws. In: Paulsson J (ed) The idea of arbitration. Oxford University Press, Oxford, pp 231–255

Perrone N (2022) Local communities, extractivism and international investment law: the case of five Colombian communities. Globalizations 19(6):837–853

Polanco R (2018) The rise of and backlash against investor-state arbitration. In: Polanco R (ed) The return of the home state to investor-state disputes: bringing back diplomatic protection? Cambridge University Press, Cambridge, pp 29–52

Salacuse JW (2015) The law of investment treaties. Oxford University Press, Oxford

Schill SW (2017) International investment law and the rule of law. Amsterdam Law School Research Paper no. 2017-8

Schultz T, Dupont C (2015) Investment arbitration: promoting the rule of law or over-empowering investors? A quantitative empirical study. Eur J Int Law 25(4):1149

Stoll PT (2018) International investment law and the rule of law. Goett J Int Law 9(1):267–292

Stone Sweet A, Gisel F (2017) The evolution of international arbitration: judicialization, governance, legitimacy. Oxford University Press, New York

Tienhaara K (2013) The expropriation of environmental governance: protecting foreign investors at the expense of public policy. Cambridge University Press, Cambridge

Tienhaara K (2018) Regulatory chill in a warming world: the threat to global warming posed by investor-state dispute settlement. Transnatl Environ Law 7(2):229–250

UNCITRAL (2018) Report of Working Group III - investor-state dispute settlement reform on the work of its thirty-sixth session. UN Doc A/CN.9/964

UNCTAD (2021) Investor-State Dispute Settlement cases: facts and figures 2020. IIA Issues Note 4/2021

UNCTAD (2022) International investment in climate change mitigation and adaptation: trends and policy developments. UNCTAD

Van Harten G (2010) Investment treaty arbitration, procedural fairness, and the rule of law. In: Schill SW (ed) International investment law and comparative public law. Oxford University Press, Oxford, pp 627–658

Van Harten G (2018) Leaders in the expansive and restrictive interpretation of investment treaties: a descriptive study of ISDS awards to 2010. Eur J Int Law 29(2):507

Viñuales JE (2012) Foreign investment and the environment in international law. Cambridge University Press, Cambridge

Viñuales JE (2016) Foreign investment and the environment in international law: the current state of play. Cambridge Centre for Environment, Energy and Natural Resource Governance, University of Cambridge Working Papers 1:1–42

Viñuales JE (2019) Foreign investment and the environment in international law: current trends. In: Miles K (ed) Research handbook on environment and investment law. Edward Elgar, Cheltenham, pp 12–38

Zárate JMA et al (2020) Duration of investor-state dispute settlement proceedings. J World Invest Trade 21(2–3):300–335

Matteo Fermeglia is Assistant Professor in International and European Environmental Law and Post-Doctoral Assistant at Hasselt University, Faculty of law, where he teaches European Environmental Law and European Climate & Energy Law. His main research interest lies in the interplay between international investment law and the climate change legal regime. In 2017, he was visiting scholar at Columbia Law School, where he collaborated with the Sabin Center for Climate Change Law and the Columbia Centre for Sustainable Investments. He regularly (co-) authors internationally peer-reviewed journal articles in the field of environmental and climate law, and was awarded the Raúl Estrada-Oyuela Award for Emerging Scholars in Climate Law in 2017.

GPSR Compliance
The European Union's (EU) General Product Safety Regulation (GPSR) is a set of rules that requires consumer products to be safe and our obligations to ensure this.

If you have any concerns about our products, you can contact us on

ProductSafety@springernature.com

In case Publisher is established outside the EU, the EU authorized representative is:

Springer Nature Customer Service Center GmbH
Europaplatz 3
69115 Heidelberg, Germany

www.ingramcontent.com/pod-product-compliance
Lightning Source LLC
Chambersburg PA
CBHW071555220925
32987CB00004B/358
9783031415265